INDUSTRIAL ACCELERATORS

AND THEIR APPLICATIONS

INDUSTRIAL ACCELERATORS

AND THEIR APPLICATIONS

Edited by

Robert W. Hamm

Marianne E. Hamm

R & M Technical Enterprises, California, USA

World Scientific

NEW JERSEY • LONDON • SINGAPORE • BEIJING • SHANGHAI • HONG KONG • TAIPEI • CHENNAI

Published by

World Scientific Publishing Co. Pte. Ltd.
5 Toh Tuck Link, Singapore 596224
USA office: 27 Warren Street, Suite 401-402, Hackensack, NJ 07601
UK office: 57 Shelton Street, Covent Garden, London WC2H 9HE

British Library Cataloguing-in-Publication Data
A catalogue record for this book is available from the British Library.

INDUSTRIAL ACCELERATORS AND THEIR APPLICATIONS

ISBN 978-981-4307-04-8

DEDICATION

The editors wish to dedicate this book to our longtime colleague and dear friend Dr. Jerome L. Duggan, Regents Professor of Physics at the University of North Texas, whose urgings and suggestions ultimately led to the idea of producing it. Jerry is widely recognized for his tireless efforts to promote and support the use of accelerators through his work at UNT, where he has mentored and inspired countless students throughout his tenure, and through 40+ years of leadership of his highly successful conference series on the Applications of Accelerators in Research and Industry (CAARI). In fact, it was our attendance at CAARI conferences during our graduate studies at Texas A&M University that piqued our interest in the practical applications of accelerators and led to our successful careers in industry.

CONTENTS

Dedication v

Introduction to the Beam Business 1

Robert W. Hamm and Marianne E. Hamm

Chapter 1. Ion Implantation for Fabrication of Semiconductor Devices and Materials 9

Michael I. Current

1. Introduction 9
2. Applications of Ion Implantation: Devices and Materials 10
   - 2.1 Pre-amorphization 12
   - 2.2 Cocktail implants 13
   - 2.3 Carbon implants for tensile strained nMOS 14
   - 2.4 Oxygen implants for direct formation of Silicon-on-Insulator (SOI) wafers 15
   - 2.5 Hydrogen implants for formation of SOI wafers by layer transfer 15
3. Accelerator Designs 16
   - 3.1 Beamline system types 17
   - 3.2 Accel-decel beamlines 18
   - 3.3 MeV beamlines 22
   - 3.4 Plasma immersion and ion shower implanters 24
   - 3.5 SIMOX high-current, high-temperature implanters 25
4. Ion Source Designs 27
   - 4.1 Special ion sources: SIMOX, molecular ions, non-volatile elements, and large-area beams 31
5. Scanning Methods 33
   - 5.1 Beam deflection and wafer motion in orthogonal directions 34
   - 5.2 Spinning wheel and pendulum wafer scanning 36

6. New Directions: Gas Cluster Ions, Photovoltaic Cell Doping, and MeV Protons for Si Membrane Cutting 39
6.1 Gas cluster ions 40
6.2 Doping of Si-based photovoltaic cells 42
6.2.1 Alternatives to implant doping for PV cells 45
6.2.2 Advanced PV cells 46
6.3 High-current, multi-MeV proton beams for fabrication of thin Si PV membranes 47
7. Implantation into Metals and Biomaterials 49
7.1 Metals: hardness, friction, and corrosion 50
7.2 Biomaterials treated by plasma immersion implantation and deposition 51
8. Summary 53
References 54

Chapter 2. Electron Beam Materials Processing **57**

Donald E. Powers

1. Introduction 57
2. Electron Beam Equipment 59
3. Electron Beam Welding 61
3.1 Large steam turbines 66
3.2 High efficiency impellers 68
3.3 Speed gears 68
3.4 Drive rings 70
4. EB Cutting and Drilling 74
5. EB Heat Treating 76
6. EB Melting and Casting 78
7. Summary and Future Trends 80
References 83

Chapter 3. Electron Beam Materials Irradiators **87**

Marshall R. Cleland

1. Introduction 87
2. Physical Properties of High-Energy Electrons and X-Rays 89
2.1 High-energy electrons 89
2.2 High-energy X-rays 93
2.3 Radiation dosimetry 95

2.4 Dose versus electron beam power 97
2.5 Dose versus electron beam current 97
3. Industrial Electron Accelerators 98
3.1 Low-energy accelerators 98
3.2 Medium-energy accelerators 104
3.3 High-energy accelerators 108
4. Major Applications of Industrial EB Irradiators 114
4.1 Cross-linking of materials 114
4.1.1 Wire and cable insulation 115
4.1.2 Heat-shrinkable plastic tubing and film 116
4.1.3 Curing of inks, coatings, and adhesives 117
4.1.4 Automobile tires 118
4.1.5 Polyethylene foam 118
4.2 Radiation sterilization of medical devices 119
4.3 Irradiation of foods 124
5. Other EB Irradiation Applications 126
5.1 Treatment of waste materials 127
5.2 Cleaning of stack gases 128
5.3 Curing of composite materials 128
5.4 Silicon-carbide fiber manufacturing 129
5.5 Production of fuel cells 129
5.6 Cross-linking of PTFE and rubber sheeting 130
5.7 Seed and soil disinfestation 130
5.8 Human tissue sterilization 130
5.9 Direct food contact coatings 131
6. Summary 131
References 132

Chapter 4. Accelerator Production of Radionuclides 139

David J. Schlyer and Thomas J. Ruth

1. Introduction 139
2. Applications of Radionuclides 143
2.1 Radiotracers 143
2.2 Nuclear medicine imaging 145
2.3 Therapeutic and other medical applications 148
2.3.1 Brachytherapy 148
2.3.2 Targeted radiotherapy 150
2.3.3 Other medical applications 151
2.4 Industrial 151

3. Accelerators for Radionuclide Production . . . 152
3.1 Cyclotrons . . . 156
3.1.1 Ion source . . . 158
3.1.2 Ion injection . . . 159
3.1.3 Beam extraction . . . 159
3.1.4 Beam transport . . . 160
3.1.5 Targets . . . 161
3.1.6 Radiation shielding and facilities requirements . . . 161
3.2 Linear accelerators . . . 162
3.3 Choice of accelerator . . . 163
4. General Principles of Radionuclide Production . . . 164
4.1 Nuclear reactions . . . 164
4.1.1 Coulomb barrier . . . 165
4.1.2 Reaction models . . . 165
4.1.3 Q value and threshold energy . . . 167
4.1.4 Cross section . . . 168
4.2 Optimizing production . . . 169
4.2.1 Production rate . . . 170
4.2.2 Saturation factors . . . 171
4.2.3 Specific activity . . . 172
5. Accelerator Targetry . . . 172
5.1 Stopping power and energy loss . . . 173
5.2 Energy straggling . . . 173
5.3 Small angle multiple scattering . . . 175
5.4 Beam heating and density reduction . . . 175
5.5 Ionization of target materials . . . 176
5.6 Radiation damage and activation . . . 176
5.7 Chemical reactions . . . 176
5.8 Pressure increases . . . 177
5.9 Beam focusing . . . 178
6. Conclusions and Future Directions . . . 178
References . . . 179

Chapter 5. Industrial Aspects of Ion Beam Analysis 183

Ragnar Hellborg and Harry J. Whitlow

1. Ion Beam Analysis in Industry . . . 183
2. Accelerators for IBA . . . 188
3. Quantity Analysis . . . 193
3.1 Introductory remarks . . . 193

3.2 PIXE technique 194
3.3 PIGE technique 199
3.4 Accelerator Mass Spectrometry (AMS) 201
4. Depth Profiling Methods 208
4.1 Introductory remarks 208
4.2 Fundamentals of RBS and ERDA measurements 209
4.3 Basis of the quantitative analysis in ERDA and RBS 214
4.4 Rutherford Backscattering Spectrometry (RBS) 215
4.4.1 Determination of the thickness of the TiO_2 film 219
4.4.2 Determination of the composition of the titanium oxide film 222
4.4.3 Analysis of complex RBS spectra 223
4.5 Elastic Recoil Detection Analysis (ERDA) 224
4.6 Nuclear reaction analysis (NRA) 230
4.7 Charged particle activation analysis (CPAA) 231
5. Industrial Facilities for Ion Beam Analysis 233
References 239

Chapter 6. Production and Applications of Neutrons Using Particle Accelerators **243**

David L. Chichester

1. Introduction 243
2. Neutron Production 247
2.1 Ion reactions 247
2.1.1 $^2H + ^2H \rightarrow ^3He + n$ 251
2.1.2 $^2H + ^3H \rightarrow ^4He + n$ 253
2.1.3 $^1H + ^7Li \rightarrow ^7Be + n$ 254
2.1.4 $^2H + ^7Li \rightarrow ^8Be + n$ 255
2.1.5 $^1H + ^9Be \rightarrow ^9B + n$ 255
2.1.6 $^2H + ^9Be \rightarrow ^{10}B + n$ 257
2.1.7 Other two-body ion reactions 257
2.2 Reactions using photons 258
3. Industrial Neutron Production Accelerators 261
3.1 Open-vacuum systems 261
3.2 Sealed-tube systems 264
3.2.1 History 265
3.2.2 Industrial sealed-tube ENGs 267
3.2.3 Long-lived DD neutron generator design considerations 270

3.3 Photoneutron systems 271
3.4 Other electronic neutron sources 272
4. Industrial Applications 273
4.1 Neutron interactions 274
4.2 Geophysical exploration 274
4.3 Gauging and radiography 279
4.4 Laboratory activation analysis 281
4.5 Biomedical applications 282
4.6 Bulk material analysis 282
4.7 Radiation effects testing 284
4.8 Detection of contraband, high explosives, and chemical weapon agents 284
4.9 Fissionable material analysis for safeguards 287
4.10 Fissionable material detection for screening and security 288
4.11 Other applications 291
5. Summary and Future Trends 292
References 295

Chapter 7. Nondestructive Testing and Inspection Using Electron Linacs **307**

William A. Reed

1. Introduction 307
1.1 History of X-rays for NDT 308
1.2 X-ray tubes vs. electron linacs 310
2. Market Overview 311
2.1 Computed tomography 313
2.2 Metrology and reverse engineering 314
2.3 Hazardous waste identification 314
2.4 Portable inspection applications 315
2.5 Isotope replacement 315
2.6 Homeland security 316
2.7 Air cargo 318
3. NDT Electron Linac Technology 319
3.1 NDT accelerator basics 320
3.2 Major subsystems 322
3.3 Beam properties 325
3.3.1 Radiation leakage 326
3.3.2 Energy spectrum 326

3.3.3 Beam flatness 327
3.3.4 Beam symmetry 328
3.3.5 Focal spot size 328
3.3.6 Collimation 329
4. Digital Detectors 331
4.1 High energy digital detectors 333
4.2 Detector performance 334
5. Traditional Radiographic Testing and Inspection Applications 336
5.1 The radiographic inspection facility 337
5.2 Radiographic principles 339
5.3 Radiographic procedures 341
5.3.1 Radiography of castings 342
5.3.2 Radiography of welds 342
5.3.3 Radiography of assemblies 343
5.3.4 Radiography of rocket motors 343
6. Security Inspection Applications 344
6.1 X-ray cargo inspection 344
6.2 Recent innovations 347
6.2.1 Interlaced dual-energy accelerators 348
6.2.2 Advanced detector technology 350
6.2.3 More powerful software 351
6.3 X-ray cargo screening trends 352
7. High Energy Industrial CT Applications 354
7.1 Engineering and design applications 356
7.2 Production environment 356
8. Quality Standards and Measurements 356
8.1 Half-value layer energy measurements 357
8.2 Cargo and vehicle screening standards 359
8.3 Digital data standardization 364
9. Summary 365
References 367

Chapter 8. Industrial Use of Synchrotron Radiation: Love at Second Sight **371**

Josef Hormes and Jeffrey Warner

1. Introduction 371
2. Synchrotron Radiation: History and Properties 374
3. SR-based Techniques for Industrial Applications 377

4. Synchrotron Radiation for Quality Control and the Control of Regulatory Requirements . . . 380
4.1 Quality control . . . 380
4.2 Control of regulatory requirements . . . 383
4.2.1 Workplace aerosols . . . 384
4.2.2 Tailings management . . . 385
5. Synchrotron Radiation for Production . . . 387
5.1 X-ray lithography for the fabrication of microelectronic devices . . . 387
5.2 Deep etch X-ray lithography and microfabrication . . . 388
6. Synchrotron Radiation for Research and Development . . . 390
6.1 Biotechnology, pharmaceuticals, and cosmetics . . . 391
6.1.1 Cosmetics and the human hair . . . 392
6.1.2 Metals containing drugs . . . 393
6.1.3 Protein crystallography . . . 395
6.2 Automotive . . . 399
6.2.1 Automotive catalysts for emission control . . . 400
6.2.2 Rubber research . . . 402
6.3 Mining . . . 404
7. Concluding Remarks . . . 407
References . . . 408

Index **413**

INTRODUCTION TO THE BEAM BUSINESS

Robert W. Hamm and Marianne E. Hamm
R&M Technical Enterprises, Inc.
Pleasanton, CA 94566, USA
rmtech@comcast.net

The goal of this book is to present for the first time in more than 25 years a review of the broad range of technologies and applications of "Industrial Accelerators". For purposes of this work, we consider any charged particle accelerator that generates an external electron or ion beam for any beam process other than direct medical treatment or basic research to be an industrial accelerator. We do not include self-contained low-energy devices such as cathode ray tubes, X-ray tubes, radio frequency and microwave power tubes, and electron microscopes in this category even though they are used mostly for industrial purposes. However, we do consider the accelerator production of therapeutic or diagnostic radionuclides for nuclear medicine an industrial application because these are either stand-alone ingredients or end products that are for the most part produced by for-profit businesses using commercially-built accelerators.

This review covers the current status and future prospects of a large number of the diverse but well established industrial accelerator applications, not only from the perspective of the industrial processes and technologies used but also from the commercial and economic impact of what we refer to as the "beam business" — the business of both producing and employing accelerators for commercial purposes. As the reader will observe, the list of materials and parts that have been touched in some way by an accelerator-generated particle beam is extensive, ranging from shrink-wrap materials and wire coatings to automobile/aircraft tires and components to virtually all modern consumer electronics. Industrial accelerator beams are also employed to sterilize disposable medical products, food, and wastewater; search for oil, gas, and minerals; inspect critical assemblies for cracks and corrosion; identify explosives and contraband in cargo and luggage; monitor pollution; and much, much more. Our estimate is that industrial accelerators are used to produce, sterilize, or examine more than US$500B of end products annually worldwide.

We estimate that more than 24 000 particle accelerators have been built worldwide over the past 60 years to produce charged particle beams for use in the industrial processes described in this book. It should be noted that this number does not include the more than 11 000 particle accelerators that have been produced exclusively for medical therapy with electrons, ions, neutrons, or X-rays. The bar graph below shows the breakdown of the cumulative number of industrial accelerators according to the application categories covered here. These numbers are an update to those published a few years ago by one of us[a] and are based on the most current statistics available from various sources, including (1) the authors of the book chapters, (2) production and sales data published in presentations or market surveys, and (3) data provided to us by a number of manufacturers in response to our inquiries. Because many industrial accelerator vendors have exited the business over the years, some of the historical data are only rough estimates; therefore, these totals should be viewed as approximations that only indicate the relative size of each category.

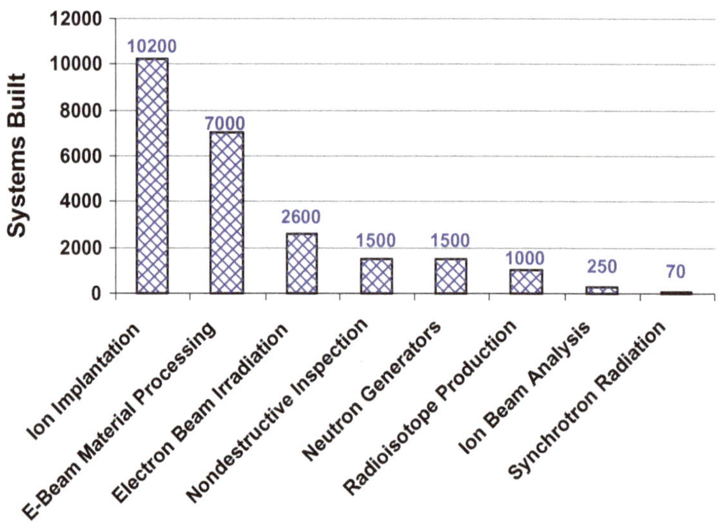

Most of these industrial accelerators have a useful lifespan of 20 to 40 years, and it is estimated that more than 75% of those built are still in operation today. Even though the technology has changed slowly, the rate of adoption of accelerators as industrial processing tools has steadily increased over the years. This has resulted in an escalating growth in the

[a]Robert W. Hamm, *Reviews of Accelerator Science and Technology* **1**, 163 (2008).

accelerator manufacturing business, which is now carried out by at least 70 companies and institutes around the world. The number of new vendors is increasing as the use of accelerators is adopted by growing industries in emerging economies, but the list is in a continuous state of flux as new vendors appear and others are purchased by competitors or other businesses in their application areas.

Currently, the majority of industrial accelerators are produced by only a few large vendors in North America, Europe, and Japan, but the number of vendors in Russia, India, Korea, and China is growing rapidly. These new vendors serve specific geographic regions or niche markets, and their numbers will continue to grow as the technology is more widely accepted by industries in their market areas. However, as a particular application matures, the size of some vendors will increase while the number will dwindle due to the competitive nature of the business. Some companies are reluctant to publicize their sales data, but by combining published statistics with responses from some major vendors, we estimate that collectively these accelerator manufacturers ship more than 1100 industrial systems per year — almost twice the number produced for research or medical therapy — at a market value of ~US$2.2B.

Each of the eight chapters in this book covers one of the broad categories of industrial accelerator applications given in the bar graph above. Although each of these areas may have been the subject of several technical books and reports meant for experts in a particular discipline, our main goal was to combine them into a single volume targeted to a more general audience. To that end, the authors have given an overview of the history, the technology, and the status of the specific applications in each category in a way that should be useful to both non-experts and experts alike. The authors have also endeavored to include, where reliable data are available, the market size by application segment and, in some cases, the market value of the end products. The intended audience includes students, educators, and accelerator scientists, as well as professionals in industry, research, and government agencies.

However, as is the case with most review books dealing with a very broad topic, it is not possible to include everything — some emerging or very specialized applications and technologies may not have been covered. Two specific industrial accelerator applications that we have not included are focused ion beams (FIB), a key technology used in the semiconductor industry for the inspection and ablation of materials, and ion beam figuring (IBF), a relatively new technique for preparing optical and nano-material surfaces. These two application areas are very specialized uses of a specific accelerator technology, not broad application categories as are the others described in this book. But it should be noted that FIB is a relatively large application dominated by only a few vendors who have produced in excess of 3000 systems over the past 20 years, and the uses of the technology are expanding. In contrast, IBF is a relatively new application in the booming nanotechnology business. Both of these areas may need to be included in future reviews of industrial accelerators.

Many of the accelerators described here are modern industrialized versions of the "atom smashers" invented in the early 1900s to study the laws of physics and the basic properties of matter, while others were invented or developed in the later decades of the 20th century, sometimes with practical applications in mind. They cover essentially all types of accelerating methods and structures, including electrostatic systems, RF linacs, betatrons, cyclotrons, Rhodotrons, and synchrotrons. The electron and ion beams produced by these devices span more than nine orders of magnitude in both particle energy and current — from eV to GeV and from nano-amperes to amperes. Correspondingly, beam powers range from microwatts to megawatts.

Each industrial application presented here relies on one or more of the basic interactions of electrons, ions, neutrons, or photons with matter, with the specific types of accelerators and beams being chosen to utilize the interactions that give the desired end results. Some electron beam applications use energetic electrons directly to modify the chemical or physical properties of materials through heat deposition, ionization, and/or atomic interactions, while others rely on the secondary production of bremsstrahlung X-rays or synchrotron radiation that is then used to process, modify, or examine materials. Many applications of ion beams rely on specific direct nuclear interactions with nuclei in the target materials. Others rely on the stopping power of ions in materials for ion deposition or on the production of neutron beams that are then used for examination and identification of materials.

As can be seen from the bar graph, the use of ion beams for "ion implantation" into materials (primarily semiconductors), as described in Chapter 1, is by far the largest of all the categories. In fact, with more than 10,200 systems having been built for this purpose over the past 30 years, it surpasses all of the other categories both in terms of the number of systems in use and the economic value of the end products. Essentially all modern-day electronic devices contain semiconductor components and display screens made possible with modern ion implantation systems. In addition, this process is also being used in emerging applications such as the fabrication of silicon wafers for photovoltaic solar cells and in the production of biocompatible devices for implants. It is estimated that the annual sales of ion implanter accelerators alone is greater than US$1B and the market value of semiconductor devices produced is approaching US$300B worldwide.

While ion implantation is usually performed at energies below the Coulomb barrier for most nuclear reactions, with ion deposition in the material being the primary goal, most of the other industrial ion beam applications rely on nuclear reactions actually taking place. These include the production of radionuclides for tracers, diagnostic imaging, and cancer therapy discussed in Chapter 4, the production of neutrons for many analytical applications described in Chapter 6, and many of the techniques used for industrial ion beam analysis described in Chapter 5. More than 1000 cyclotrons have been built for radionuclide production, with almost 800 in current use. At least 1500 neutron production accelerators are currently in use, mostly in the petroleum industry. Smaller but growing applications of neutrons include radiography and explosive detection. About 250 electrostatic accelerators, mostly Van de Graaff systems, are now used for ion beam analysis by industry. Some of these are still at the universities where they were originally used for nuclear physics research but now are used by industry through contracts and collaborations.

The largest industrial use of electron accelerators is for "materials processing" applications. The oldest and most widespread of these is reviewed in Chapter 2. This category, referred to as EB processing in the chapter, utilizes well-defined beams of relatively energetic electrons to very precisely transmit thermal energy into a material or part for precision welding, cutting, drilling, brazing, annealing, glazing, and surface hardening. EB processing is also used for the precise melting of refractory metals (e.g. tungsten or molybdenum) and the production of pure superconducting metals such as niobium in industrial furnaces. At least 4000 of the 7000 EB processing systems built over the last 50 years are in use today in a large number of industries. EB processors are particularly important in the automotive industry where, among other things, they are used to make gears and to weld and harden camshafts and tie-rod ends.

Another type of material processing using electron beams is the irradiation of materials and products with ionizing energy in the form of high energy beams of electrons or X-rays to enhance or modify their physical, chemical, or biological properties as described in Chapter 3. This form of radiation processing is by far the most diverse of all the categories in the book, both in terms of the technology used and the items produced or processed. The accelerators in this category are referred to as electron beam "irradiators". These systems cover a very wide range of accelerator technologies, beam currents, and energies to perform such processes as polymer grafting and cross-linking, and the curing of monomers, oligomers, and epoxy-based composites. Applications include the production of heat-resistant wire insulation, heat-shrinkable tubing and food-packaging films, polyethylene foams, tire rubber, and hydrogels for wound dressings, as well as the curing of inks, coatings, and adhesives for paper, wood, metals, and plastics. High energy beams of electrons and X-rays are also used for the sterilization of single-use medical products and waste water, disinfection and preservation of food stuffs, decontamination of chimney and flue gases, and degradation of plastics for use in coatings and inks. It is estimated that more than 2600 industrial electron beam irradiators have been built worldwide, and at least 1800 of these are currently in use providing an added value to numerous products in excess of US$80B annually.

In addition to the electron beam material processing applications described in Chapters 2 and 3, high-energy electron beams are also used by industry to generate secondary radiation for materials inspection and processing. This can be in the form of bremsstrahlung X-rays generated by bombarding metal targets with electrons from high-energy linacs, or it can be synchrotron radiation generated by relativistic electrons circulating in an electron synchrotron. In fact, as described in Chapter 7, the use of radiographic inspection techniques for airline and border security has caused an "explosion" in the use of electron linacs for nondestructive testing compared to the traditional radiography applications of inspecting thick castings and complex parts. More than 1500 of these systems have been produced to date, and the production rate is rapidly increasing. Synchrotron radiation (SR) is used by industry in an ever-increasing number of processing and inspection techniques as described in Chapter 8. This work is performed at more than 70 light sources (both synchrotron and free electron laser facilities) around the world, with a few of these facilities actually built exclusively for industrial work. Industrial applications include semiconductor-device lithography and the studies of material interfaces. The chemical industry uses SR to study stress and texture patterns in materials produced as well as chemical reactions. Biomedical firms use SR for protein crystallography, molecular imaging, and the study of molecular dynamics in tissue cells. Drug development using protein crystallography is by far the largest use of synchrotron radiation by industry.

There is no doubt that as advances are made in existing accelerator technology and as new technologies mature into commercial products, the applications of industrial accelerators will continue to expand. Most accelerator manufacturers and industrial users are busy even now working on the development of systems for new applications and markets in the beam business.

Chapter 1

ION IMPLANTATION FOR FABRICATION OF SEMICONDUCTOR DEVICES AND MATERIALS

Michael I. Current
1729 Comstock Way, San Jose, CA 95124, USA
currentsci@aol.com

Ion acceleration of dopants and other ions is a critical and nearly universally employed tool for the fabrication of transistors in IC devices and the various forms of electronic, photovoltaic, and photonic materials. This chapter reviews the major types of accelerators, ions sources, and scanning methods used in contemporary practice for ion implantation for electronic, photovoltaic, and biomedical applications.

1. Introduction

The use of accelerated ions to dope and modify semiconductor materials, generally called ion implantation, is a core technology for fabrication of integrated circuits (IC) that form the backbone of the global communications and advanced computation capabilities that have transformed modern life over the last half-century. The use of ion implanters in the IC industry, starting in the 1970s for precision doping of channel structures at doping densities of $\sim 10^{-4}$ monolayers to set the threshold conditions for transistor switching, enabled the practical fabrication of complementary metal-oxide semiconductor (CMOS) transistors that are now the dominant form of IC devices. Development of accelerator technologies capable of delivering highly stable and collimated ion beam currents ranging from a few μA to 100 mA and incident ion energies ranging from 100 eV to $\sim$10 MeV have provided a broad and capable technology for fabrication of ICs for logic, memory and analog operations as well as an increasingly varied array of optical sensors and imaging devices. Ion implantation is now used for the overwhelming majority of applications for doping of semiconductor materials and in an increasing array of applications for creation and modification of electronic and photonic materials.

The yearly revenues from the sale of industrial ion implantation and materials modification tools constitute a US\$1.5B market (in "normal"

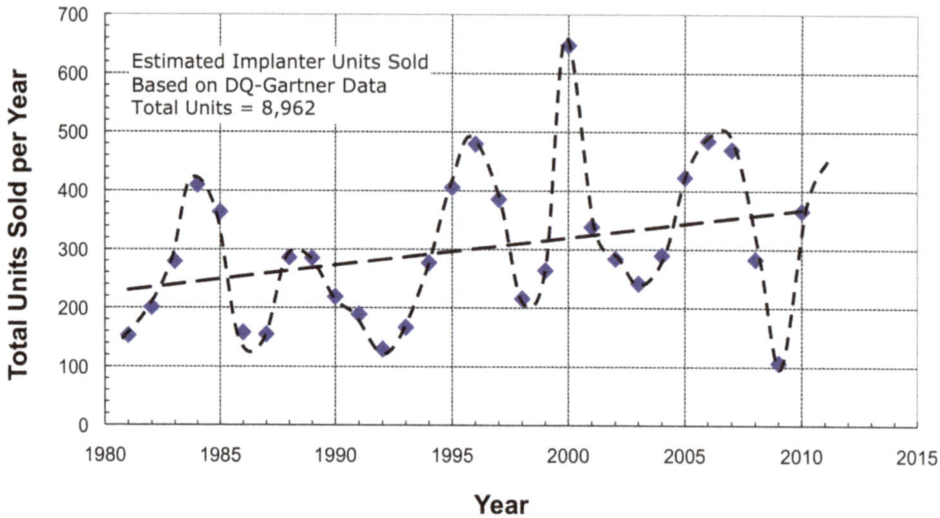

Fig. 1. Estimates of the number of commercial ion implanters sold per year, mainly for IC fabrication. The data shown here are an underestimate, particularly for the earlier years.

economic environments). In addition, there are substantial markets for suppliers of dopant species materials (~US$140 M/year) and a diverse array of suppliers of spare and upgrade parts and services. Additional parts of the ion implantation commercial infrastructure include the suppliers of system components, magnets, power supplies, and vacuum pumps, and a varied array of metrology tools focused on process characterization and control applications.

The average number of commercial ion implantation systems sold per year, predominantly for fabrication of Si-based IC devices, has increased from ~250 to ~350 per year since 1980 as seen in Fig. 1. The year-to-year sales show large fluctuations due to the highly cyclic nature of IC fab construction trends especially when driven by additional factors, such as shifts in dominant wafer sizes, introduction of new implanter machine types and IC devices, expansion into new global regions, and variations driven by general economic cycles. The combination of these technology and market factors has resulted in a rather steady "5 year" cycle in units sold per year over the last 3 decades.

2. Applications of Ion Implantation: Devices and Materials

The planar CMOS transistor, illustrated in Fig. 2, has been the workhorse device structure for IC devices for logic and memory applications since the mid-1980s.

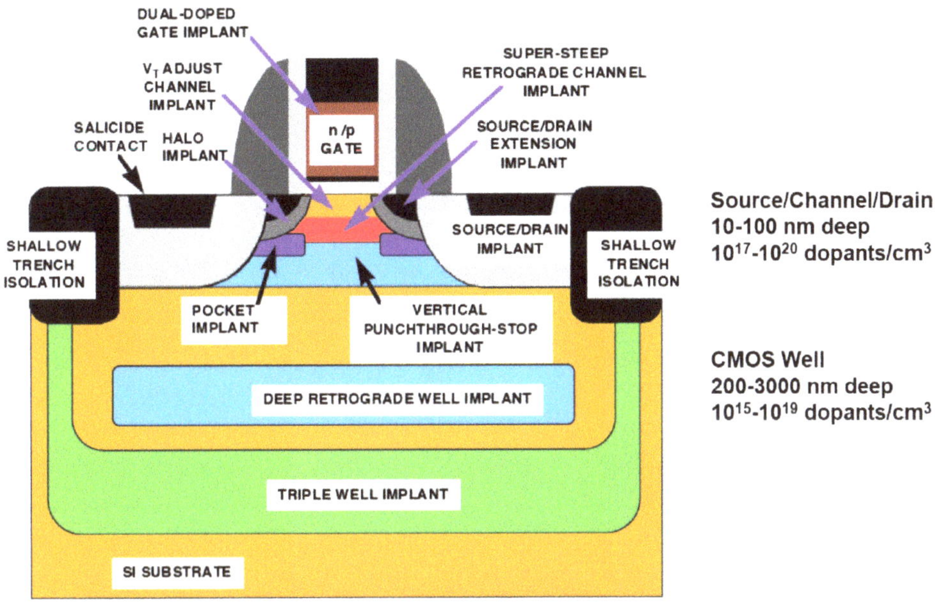

Fig. 2. Sketch of major doped regions for a planar CMOS transistor.

For doping of near-surface (<100 nm) layers on Si with the common dopants B, As, and P, the beam energy requirements for atomic species is a fraction of a keV to a few tens of keV. Increasingly the ion species of choice for low energy (sub-keV) implants is some form of molecular ion containing up to 18 dopant atoms (for $B_{18}H_{22}$), allowing for beamline operation at proportionally higher energies and significantly higher total beam flux for higher production throughputs. For deeper profiles to form CMOS wells for transistors and optical imagers, ion energies range up to several MeV.

The initial applications for ion implantation were for low dose (10^{11} to 10^{12} ions/cm^2) doping of CMOS channels to set the transistor switching, or "threshold", conditions. These implants required relatively modest beam currents with 100 to 500 μA being common for "medium-current" implanters. By 1980, higher current sources and beam lines were developed for "high-current" machines to produce beams of 1 to 30 mA of dopant ions. These machines were capable of efficient implants with doses of $\sim 5 \times 10^{15}$ ions/cm^2 used for CMOS source/drain junctions as well as emitter, buried layer, and collector implants for bipolar devices.

By the late 1980s, implantation machines capable of ion energies of $\sim$1–2 MeV were being used for doping deep "retrograde" wells in CMOS devices to suppress latch-up instabilities between neighboring well structures and to reduce radiation-generated "soft failures" in DRAM arrays. With the evolution of complex doping schemes to suppress lateral punch-through between source and drain junctions in short channel (<100 nm) transistors, ion implanters were developed with capabilities for mid-10^{14} ions/cm^2 doping at beam incidence angles up to 45° to provide additional well doping in a "halo" surrounding shallow source/drain extension (SDE) junctions.

The dose range for implant doping expanded above 10^{16} ions/cm^2 with the use of very high dose implants to counter-dope poly-Si gates in pMOS transistors in DRAM circuits. To reduce the overall processing cost, an n-doped poly-Si thin film was deposited over the entire wafer and used for the nMOS transistor gates. For the pMOS transistors, a very high dose ($3-5 \times 10^{16}$ B/cm^2) implant was needed to counter-dope the n-type dopants and transform the pMOS gates into a highly-doped, p-type material. This led to the use of "plasma immersion" and "ion shower" types of doping tools, which were especially suited for very high dose implants at low (several keV and less) energies.

With the wide-spread use of Si-based imagers for cell-phone and other cameras, fabrication of charge-coupled devices (CCDs) and CMOS imagers called for the use of implanters with energy ranges of 2–8 MeV for formation of doped regions up to 4 μm deep. These dopant implants for CMOS devices, with many additional specialized applications for analog and bipolar components, fill in the space occupied by the left side and lower segments on the dose–energy chart shown in Fig. 3.

2.1. *Pre-amorphization*

In recent years the implantation of non-dopant ions (Ge, C, N, F, H, etc.) has become a standard part of the range of ion species in use for CMOS device fabrication. The most direct form of these applications is the implantation of Si or Ge ions at doses high enough ($\sim 10^{15}$ ions/cm^2) to transform a surface layer of Si into an amorphous form through accumulation of lattice recoil damage during the stopping of the Si or Ge ions. The initial purpose of these "pre-doping amorphization implants" (PAI) was to eliminate the channeling of light ions, mainly B, in crystalline Si and to enable the formation of shallow dopant profiles with low-energy implants. While suppression of ion channeling effects is still important, the uses of PAI implants have broadened to include activation of dopants during Si re-crystallization growth in the low temperature (<700°C) furnace and microwave anneals.[1] In addition, PAI implants are used to take advantage of the increased absorption of light in

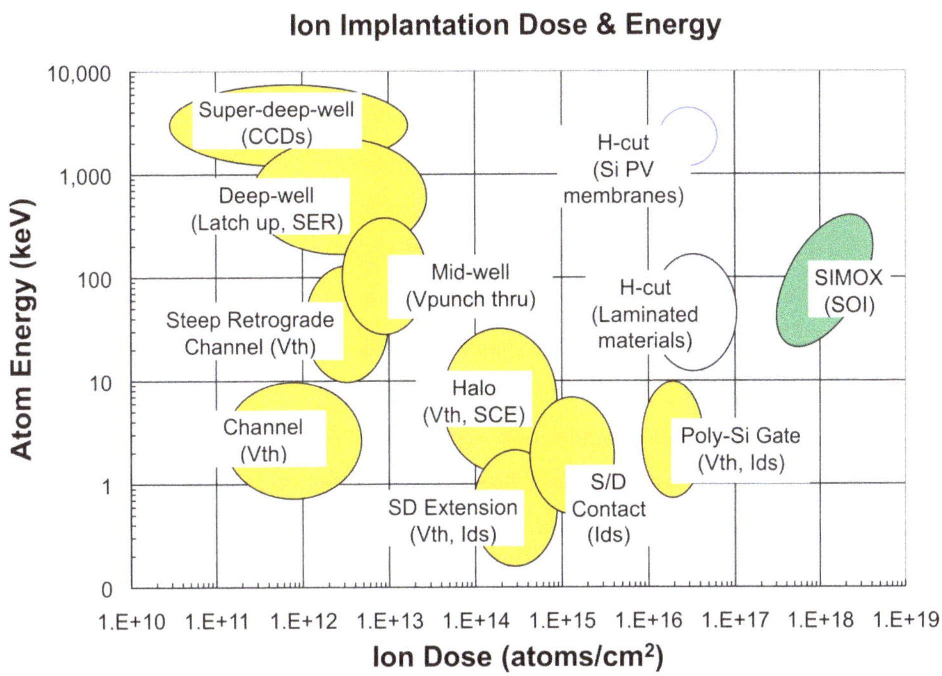

Fig. 3. Dose and atom energy regions for CMOS transistor doping, high dose hydrogen implants for Si layer splitting (H-cut), and direct implantation of oxygen to form SOI wafers (SIMOX).

amorphous-Si layers during ms-timescale annealing with laser or flash lamp sources.

PAI implants are also used for materials modification of poly-Si gates to boost the net strain in CMOS channel regions in order to improve carrier mobility and overall transistor drive currents. There are many variants of "stress memorization" techniques using PAI steps to amorphize poly-Si gate electrodes, resulting in an overall tensile strain (good for nMOS) in the transistor channel. The "remembered" channel strain levels are held in place in the final CMOS device through the replacement of the poly-Si with a metal gate electrode and addition of nitride overlayers.[2]

2.2. *Cocktail implants*

Non-dopant atoms, mainly C, F, and N (commonly referred to as "cocktail atoms"), are implanted as part of an implant-and-anneal sequence to control the diffusion of shallow dopant profiles through trapping of Si interstitials at the cocktail atom profile location.[3] The process starts with a deep PAI implant (with Ge or Si) followed by the implantation of cocktail atoms at

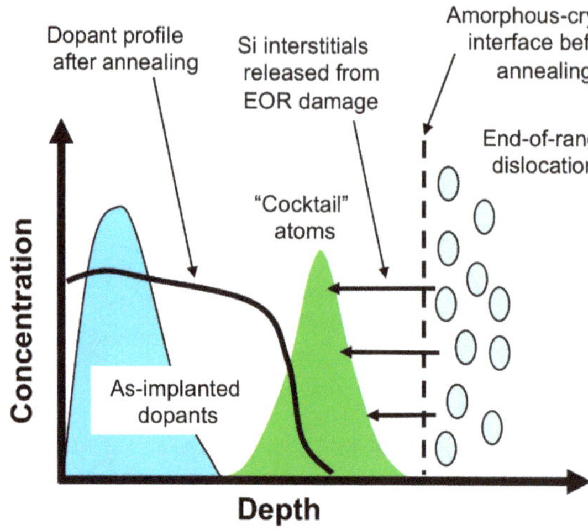

Fig. 4. Schematic diagram of the implant and damage profiles for the use of "cocktail" atoms (C, F, N) to reduce the diffusion of dopants (B, P) that diffuse rapidly when combined with Si interstitials.

a mid-depth location and dopants near the surface (see Fig. 4). When this combination of implants is followed by a rapid thermal process (RTP) anneal at ~1050°C for ~1 s, the amorphous layer re-grows early in the anneal cycle into near-perfect crystalline Si, leaving a deep layer of defects at the end-of-range (EOR). As the high-temperature cycle continues, the EOR damage anneals and releases Si interstitial atoms. If the shallow dopants are B or P, which diffuse rapidly when combined with enhanced concentrations of Si interstitials, then the presence of a Si interstitial trapping layer "starves" the dopant diffusion process resulting in shallower dopant profiles at the end of the RTP anneal cycle. These sorts of combination implants with dopant and non-dopant atoms can be used as a method to control dopant diffusion and fabricate activated shallow junctions with the use of RTP anneal equipment, forestalling the need to shift to more costly and complex ms-timescale anneals with lasers or flash lamps.

2.3. *Carbon implants for tensile strained nMOS*

Another use for C implants is to create local strain regions to establish tensile strain conditions (which increase electron mobilities in Si) in the channels of nMOS transistors. The conditions to establish a stable C-substitutional population of ~2% in the source/drain (SD) contact region, needed to obtain adequate tensile strain in sub-30 nm nMOS channels, include the formation of a deep uniform amorphous layer, a C implant into the amorphous layer, and then a rapid re-crystallization anneal to limit C diffusion. Early versions of this process combined a PAI step with Ge or Si atoms with C ion implants and an RTP type anneal.[4] Advanced versions of the use of C implants to generate tensile strain in nMOS channels employ molecular ions, such as $C_7H_7^+$ and $C_{14}H_{14}^+$, that are massive enough to create dense and uniform amorphous layers directly without the need for a separate PAI step.[5] Tensile strain levels can be increased by the use of ms-timescale anneals with scanned laser beams in combination with molecular C ion implantation.

2.4. *Oxygen implants for direct formation of Silicon-on-Insulator (SOI) wafers*

The early production of thin "Si on Insulator" (SOI) wafers for CMOS applications was made possible by direct implantation of oxygen at very high doses ($\sim 5 \times 10^{18}$ O/cm^2 — see the upper right region of Fig. 3), sufficient to form a buried SiO_2 layer during a very high temperature anneal (>1350°C for ~1 h).[6] The high doses required for this "Separation by Implantation of Oxygen" (SIMOX) process spurred the development of specialized implanters capable of sustaining beam currents of ~100 mA. These systems in turn required the use of RF-powered filamentless ion sources as well as efficiently cooled beamline walls, apertures, and beam stops to deal with ion beam powers in the 10 kW range.[7] In addition, SIMOX implanters need to heat the target wafers to ~600°C at the start of the beam exposure and then dynamically control the wafer temperature at that level throughout the high-power implantation. The purpose of the high implantation temperature is to provide a sufficient rate of dynamic annealing so that the surface Si layer can remain crystalline throughout the implant cycle and a high-quality Si(100) device layer can be formed during the anneal. Beginning in the late 1980s, the commercial use of SOI for fabrication of high-speed logic and microprocessors was made possible by the development of sophisticated multi-implant SIMOX cycles at both high and room temperatures followed by anneals at temperatures approaching the melting temperature of Si.[8]

2.5. *Hydrogen implants for formation of SOI wafers by layer transfer*

By the mid-1990s, the need for a more efficient and versatile method for fabrication of SOI wafers led to the use of high dose hydrogen ($\sim 5 \times 10^{16}$ H/cm^2) implants to form a buried hydrogen-saturated layer in semiconductor materials. When the H-implanted wafer is bonded to an oxidized Si wafer and then subjected to either thermal or mechanical stress, the wafer pair can be separated along the peak of the H profile leaving a thin Si (or Ge) layer bonded to the oxidized "handle" wafer.[9,10] After removal of the heavily-damaged H-rich separation region layer, by either Chemical-Mechanical-Polishing (CMP) or a high-temperature hydrogen etch process, a high-quality SOI wafer with typical Si device layer thicknesses of 10 to 100 nm is produced.[11] Variations on these "H-cut" processes combined with innovative wafer bonding techniques can be used to fabricate many types of laminated structures, integrating diverse types of heterogeneous materials that would be difficult or impossible to combine in high-quality forms with standard deposition methods.[12] An example is the bonding of H-implanted Ge wafers to thick (Si thickness of 2–5 μm) SOI wafers to form a Ge/SOI multilayer "photonic master template". The bonded Ge layer can be doped to form a detector structure or can be used to provide a compatible substrate layer for deposition of III–V layers to be later formed into vertical cavity semiconductor laser emitters (VCSLs). The thick SOI layer can be etched to form waveguide multiplexer structures suitable for creation of photonic elements for processing 1.5 μm wavelength light signals that can travel across the global optical network of the Internet.

3. Accelerator Designs

Given the broad range of energies and doses in use for CMOS transistor doping and other materials processing, a variety of accelerators with quite different designs and beam scanning methods are used for each of the specific application segments shown in Fig. 3. The typical classifications for these general machine types are:

- Medium current — ion beam currents between 10 μA and ~2 mA.
- High current — ion beam currents up to ~30 mA; ~100 mA for SIMOX beams.
- High energy — ion energies above 200 keV and up to 10 MeV.
- Very high dose — efficient implant of doses greater than 10^{16} ions/cm^2.

Accelerator systems for ion implantation are provided commercially by a number of vendors, with the business being dominated by a few long-standing companies: Varian Semiconductor Equipment Associates (Glouchester, MA), the market leader; Nissin Ion Equipment (Kyoto, Japan), with a strong Asian market position; Axcelis Technologies (Danvers, MA), leading the high-energy market; and SEN Corporation (Shikoku, Japan), former Axcelis partner also with a strong Asian market presence. Other smaller vendors include Advanced Ion Beam Technology

(Hsinchu, Taiwan), a new high-current vendor; Ion Beam Services (Rousset, France), a new PIII and SiC implanter company; Ulvac Technologies (Japan), long-time vendor for R&D tools; High Voltage Engineering (Netherlands) and National Electrostatics Corporation (Madison, WI), both supplying specialty R&D tools; and several RF linac vendors known to be supplying R&D units for MeV-proton beams for deep H-cut.

Beam current requirements for each machine type have increased as the wafer sizes and implanted areas increased from the early days of 75 mm diameter wafers to the present standard of 300 mm, with the contemplated increase to 450 mm after 2012. In addition, the increased requirements for dose accuracy (1% or better) and control of ion incidence angle ($<0.5°$ for critical implants) place tight constraints on ion beam stability, scan uniformity, and beam collimation in the area near the wafer surface.

3.1. *Beamline system types*

The wide variety of ion implantation beamline designs all contain certain general groups of functional components (see Fig. 5). The first major segment of an ion beamline includes a device known as an ion source to generate the ion species. The source is closely coupled to biased electrodes for extraction of the ions into the beamline and most often to some means of selecting a particular ion species for transport into the main accelerator section. The ion source usually is an arc-discharge plasma type fed by a vapor of the atoms to be ionized, accelerated, and implanted. The "mass" selection is often accomplished by passage of the extracted ion beam through a magnetic field region with an exit path restricted by blocking apertures, or "slits", that allow only ions with a specific value of the product of mass and velocity/charge to continue down the beamline.

If, as is often the case, the desired ion energy is different than the potential used to extract ions from the plasma source, the "selected" ion beam is accelerated (or decelerated) by additional sets of electrodes or RF fields. If the target surface is larger than the ion beam diameter and a uniform

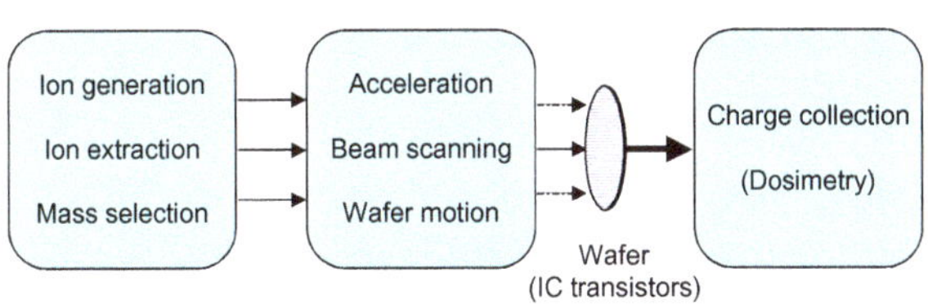

Fig. 5. Major ion implantation beamline components.

distribution of implanted dose is desired over the target surface, then some combination of beam scanning and wafer motion is used.

Finally, the implanted surface is coupled with some method for collecting the accumulated charge of the implanted ions so that the delivered dose can be measured in a continuous fashion and the implant process stopped at the desired dose level. Sometimes the wafer is encased within a Faraday cup assembly with appropriate methods of exclusion of extraneous charges from electrons and other ions within the beam plasma and charge losses due to secondary electrons and sputtered ions.

The design of an ion implantation beamline often needs additional components to satisfy the target requirements. These include components to make the scanned beam parallel at the surface of the target wafer and sources of electrons and ions to be injected into the ion beam plasma in the region close to the final target surface. These electrons and ions are needed to improve the space-charge balance in the ions beam, thus stabilizing the beam collimation and controlling the local potentials and current flows on the target surface.

An example of a beamline where the additional ion acceleration is obtained by floating the ion source and selection magnets to a high dc potential (often called a "Cockcroft–Walton" accelerator from the initial design developed at the Cambridge Cavendish lab in 1930), followed by horizontal beam scanning and vertical wafer motion in the beam path, is shown in Fig. 6.

The Varian VIISta900XP,[13] shown in Fig. 7, is an example of a medium-current beamline with many of the components shown in Fig. 6. This beamline is capable of overall acceleration of single-charged ions to 300 kV (and P^{+3} ions to 900 keV) and contains a number of additional components, including a small deflection magnet and slit assembly placed directly after the source/extraction region and an electrode array for either accelerating or decelerating the selected ion beam prior to entering the main mass-selection (analyzer) magnet. The main function of the source filter magnet and slit is to prevent the P_2^+ and P_3^+ multimer ions (which can lead to energy and dose errors in P^{+2} and P^{+3} beams) from entering the main selection magnet and slit assembly and travelling to the target wafer.

3.2. *Accel-decel beamlines*

High-current beamlines, which are designed to maximize the ion flux delivered to the target wafer surface and often operate at low energies

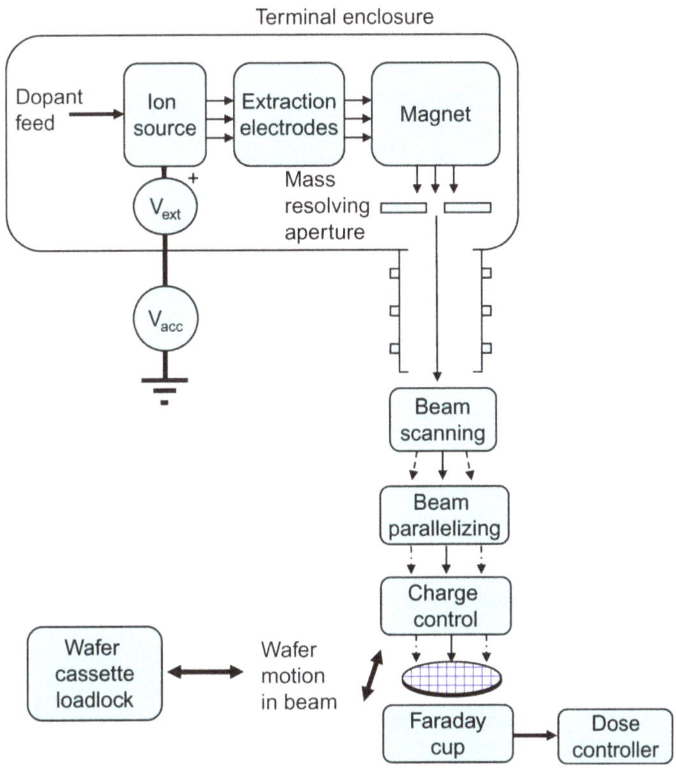

Fig. 6. Schematic of a 2-stage dc accelerator beamline with beam and wafer scanning.

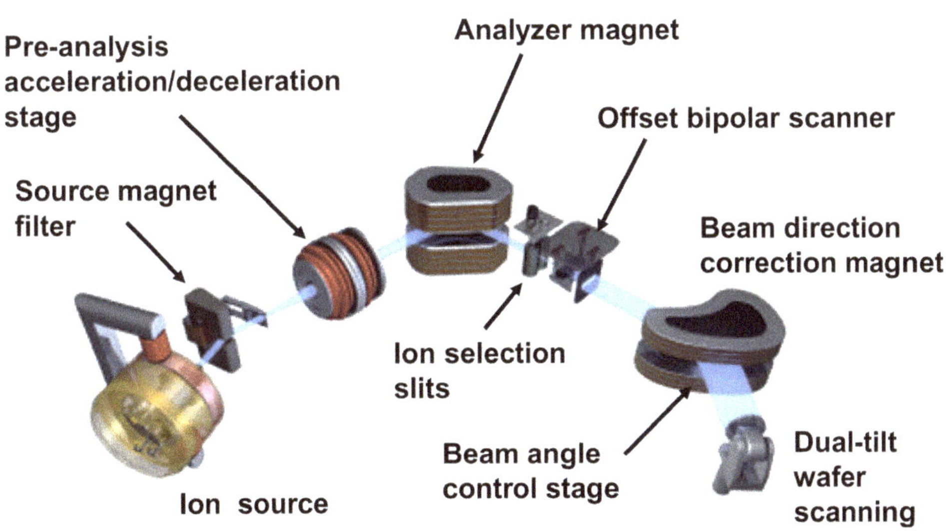

Fig. 7. Beamline components of a Varian VIISta900XP medium-current ion implanter.[13]

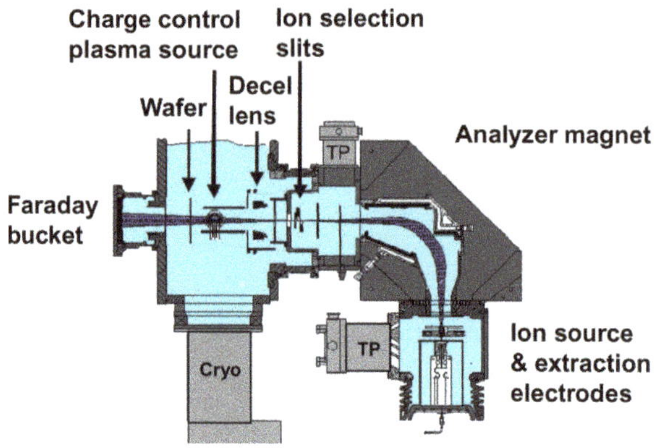

Fig. 8. Beamline for the Applied Materials xR-80 high-current ion implanter with ion acceleration up to 80 keV at the source extraction stage and deceleration to as low as 0.2 keV after the mass resolving slits.[14]

(1 keV or less), are usually shorter and contain fewer beamline components than medium-current designs. A classic example of a high-current beamline is the Applied Materials xR-80 shown in Fig. 8, which was introduced in 1996. The short beamline path (~1.6 m from source to wafer) and open geometries in the magnet analysis section resulted in improved ion transmission in "drift" mode with the only acceleration stage being at the source extraction. To further increase the available ion flux for low energy (<1 keV) implants, an electrostatic deceleration lens was added just after the mass resolving slits. With the decel lens powered up, boron ion beams at energies as low as 0.2 keV could be delivered to the wafer surface with beam currents comparable to drift mode currents at 10x higher energies.

A limitation of this "accel-decel" mode of beam transport is that if dopant atom neutrals are formed by collisions with background gas molecules in the region following the mass analysis magnet and before the decel lens, neutral dopant atoms would be implanted at the extraction voltage energy along with the decelerated ions. This type of "energy contamination" is minimized by the operation of a turbopump in the region around the mass resolving slits (to reduce the ion neutralization rate from collisions with background gases) and by restricting the difference between the drift and decelerated ion energies.

Modulation of the ion energy along the beamline path continues to be a useful strategy to maximize beam transmission and facilitate beam shaping and focus. In the example of the Varian HCP high-current implanter beamline[15] shown in Fig. 9, a ribbon-shaped beam is formed by the extraction electrodes and focused at a mass resolving slit located downstream from

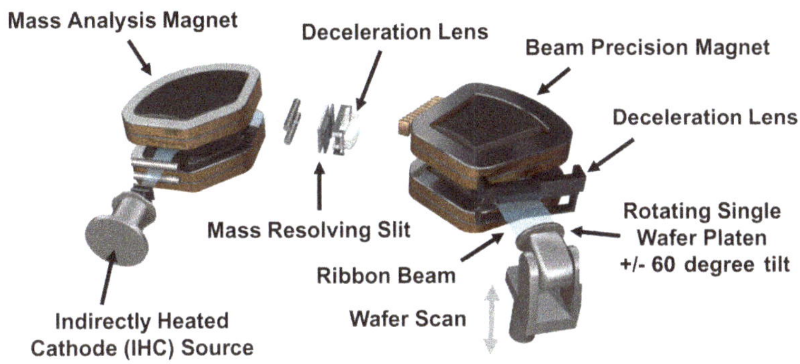

Fig. 9. Schematic view of the ribbon-shaped beam in the Varian HCP high-current implanter with operation range from 0.2 to 60 keV.[15]

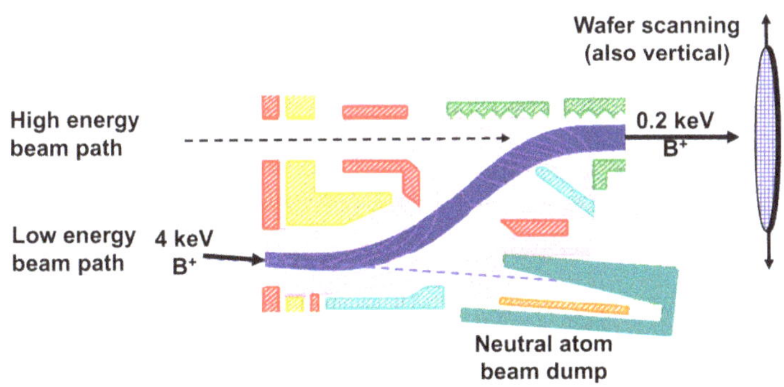

Fig. 10. Schematic top view of the "chicane" beam path for deceleration of a vertical ribbon beam in the iPulsar AIBT high current implanter with a beam dump for neutral atom contaminants.[16]

the 90° mass analysis magnet. The beam is then directed and shaped by another downstream magnet prior to being implanted into a wafer scanned up and down in the ribbon beam. The ion beam can be decelerated at two stages along the way: directly after the mass resolving slits and after the final beam shaping magnet.

In a sophisticated combination of beam optics and system design, the AIBT high-current implanter[16] shown in Fig. 10 uses a number of electrodes to decelerate and deflect low-energy ions in a vertical ribbon beam along a "chicane" path (similar to turns in motorcycle racing tracks) before emerging in front of the wafer scanning position. Neutral dopant atoms that enter the decel stage continue undeflected into a beam trap. When higher energy ions are desired, the ribbon beam is switched back to a straight beam bath, thus bypassing the chicane decel electrode assembly.

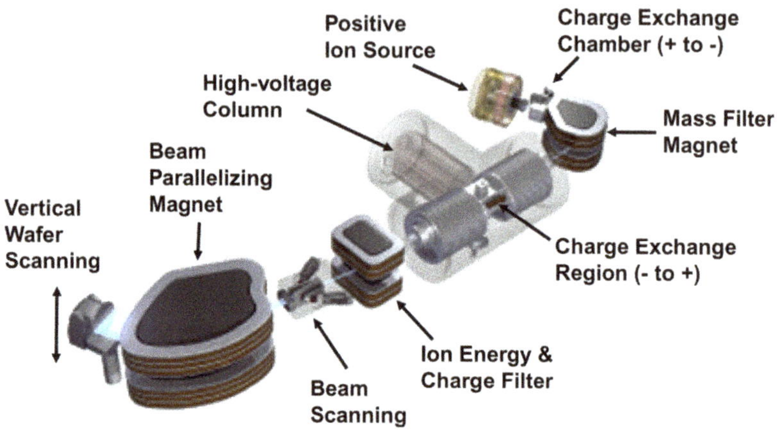

Fig. 11. Schematic of the high-energy tandem accelerator beamline for the Varian 3000 XP implanter[17] with an operational range from 10 keV to 3 MeV.

3.3. *MeV beamlines*

When high-energy ions (>200 keV to several MeV) are needed, variations on either tandem or linear accelerators developed in the 1930s for nuclear physics research are used. In the tandem approach, multiple changes in the ion charge state along the beam path are induced to obtain a multiplication of the acceleration by a dc high-voltage potential as seen in Fig. 11. First, positive ions are extracted from a plasma source and passed though a metallic vapor stream in order to form a beam of negative ions via collisional electron pickup. After undergoing mass selection in a magnet and slit assembly, the desired negative ions are accelerated toward a positive potential region, usually encased within a pressurized SF_6 atmosphere to allow for isolation of high potentials (up to several MV) in a compact chamber. At the center of this tandem beamline, the fast negative ions pass through a high level of Ar gas atoms resulting in collisional charge exchange that transforms the negative ion beam to positive charged ions with various charge states. As the positive ions leave this shielded charge exchange region, they are accelerated away from the high positive potential in the center of the tandem. Single-charged positive ions leave the tandem beamline with a kinetic energy equal to twice the center positive potential. Double-charged positive ions have a final energy of three times the center potential. Useful, although low level, dopant beam currents with charge states up to +4 (with energy equal to 5 times the center potential) can be obtained for some ions with commercial tandem implanters. A magnetic selector segment placed at the exit of the tandem accelerator section allows only the ions with a desired charge state (and energy) to be passed to the target wafer.

In the Varian 3000 XP tandem high-energy implanter[17] shown in Fig. 11, positive ions are filtered by a magnetic field after the high-voltage acceleration stage to reject unwanted ions with undesirable energies and charge states. The selected ions are then scanned along a horizontal beam path and directed through a magnetic beam shaping element to produce a series of parallel beam paths at the wafer, which is moved up and down to produce vertical wafer scanning similar to the design shown in Fig. 7.

Commercial tandem accelerators are in operation with 2 MV potentials providing 8 MeV B^{+3} ions for doping profiles extending $\sim 10\,\mu m$ into Si. At these boron ion energies, control of gamma and neutron radiation calls for careful shielding and selection of beamline component materials, active radiation monitors, and limits to operating beam currents.[18]

Phase-synchronized linear accelerators have also been used for high-energy ion implanters for over 20 years. In the Axcelis Optima XE high-energy implanter[19] shown in Fig. 12, twelve RF cavities resonating at 13.57 MHz with maximum potentials up to 80 kV each combine to provide a beamline with mA-level beam currents for single-charged dopant ions at energies of ~1.5 MeV. Double-charged P^{+2} ions can be accelerated to 2.9 MeV with 0.5 particle-mA beam currents for deep n-well applications. The positive ion beam from a plasma ion source enters the linac section after

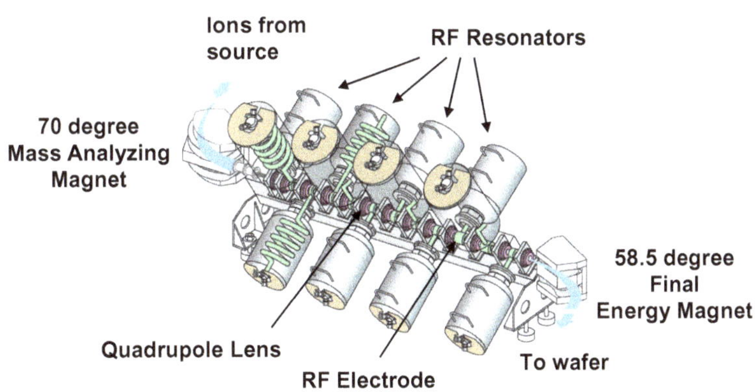

Fig. 12. RF linac section of the Axcelis Optima XE high-energy implanter[19] with an operation range from 10 keV to 1.5 MeV for B^+ (1.2 MeV for P^+ and 2.9 MeV for P^{+2}).

passing through a 70° mass-selection magnet (on the left in Fig. 12). After the beam is bunched and accelerated through the linac assembly, it is filtered with a 58.5° magnet (on the right in Fig. 12) that is tuned to select ions with the desired energy for transmission to the beam and wafer scanning stages.

3.4. *Plasma immersion and ion shower implanters*

The systems used for very high dose (>10^{16} ions/cm^2) and low-energy (a few keV and less) implantation employ a radically simplified design. In "plasma immersion ion implanters" (PIII), the usual beamline components are eliminated, leaving only a means for generating a plasma containing dopant ions and a wafer mounted on a biasable platen. The ions in the plasma formed near the wafer are accelerated into the wafer surface by short (a few μs) negative potential pulses applied to the wafer platen at energies up to the maximum desired value.

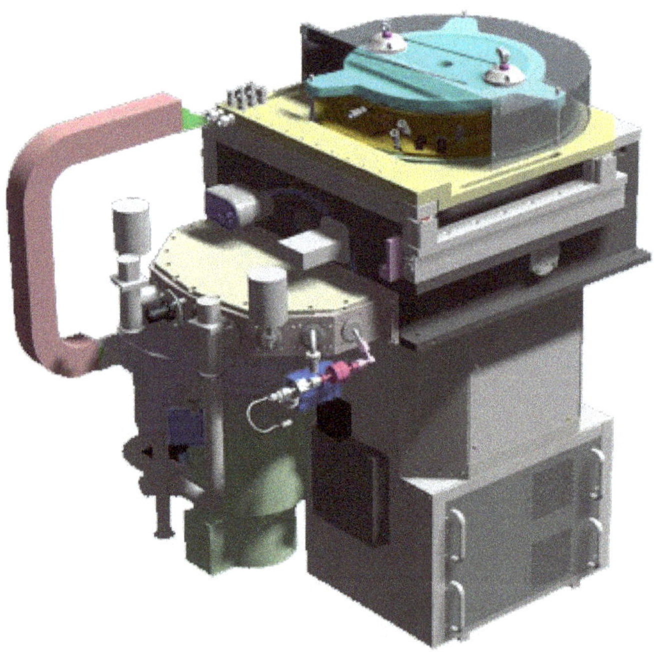

Fig. 13. The Varian VISta PLAD is an RF plasma doping tool that operates in a pulsed-plasma mode with the plasma formed only when the target wafer is biased for implantation. The wafer bias voltages range from 0.1 to 10 kV.[20]

PIII systems are remarkably compact in comparison to beamline accelerators, resembling the look of CVD and plasma etch systems. An example is shown in Fig. 13. The applications for these sorts of implanters include:

- High dose ($\sim 5 \times 10^{16}\,B/cm^2$) boron implants at energies of a few keV to counter-dope n^+ poly-Si to form p^+-doped gate electrodes in DRAM devices.
- High dose ($\sim 5 \times 10^{16}\,H/cm^2$) hydrogen implants using mono-ionic "protonic" plasmas and wafer biases in the 20–50 kV range to form hydrogen-saturated layers for thermal or mechanical separation as part of a multi-layer lamination process for fabrication of SOI and photonic wafer structures.
- High dose doping of 3-D CMOS device structures such as source/drain regions of finFETs.
- High dose implantation of non-dopant ions for sealing of low-k dielectric surfaces and other materials modification applications.

Another form of simplified ion implanters are the "ion shower" systems designed for high dose implantation of large target areas, such as flat panel arrays containing poly-Si thin-film transistors (TFT) for televisions and other displays, at energies up to ~100 keV. To provide a closely controllable doping process, the large area ion beam is filtered through a tall beamline analysis magnet that allows only specific ion species to proceed to the target plates (see Fig. 14). The beam strip at the target surface is 80 cm tall with a width of 3–5 cm. The target materials, 73×94 cm glass plates containing TFT or OLED devices, are scanned in a horizontal direction across the beam.

3.5. *SIMOX high-current, high-temperature implanters*

The direct implantation of oxygen ions at doses sufficient to form buried SiO_2 layers, ranging from 5×10^{17} to $3 \times 10^{18}\,O/cm^2$, provided the commercial source of sufficient Silicon-on-Insulator (SOI) wafers for companies such as IBM, AMD, and others to convert the core of their high-performance logic IC devices to SOI wafers in the late 1990s. The evolving class of SIMOX implanters involves operational conditions of truly "heroic" proportions. To operate at commercially viable throughputs with the high implanted oxygen dose needed for SIMOX processes, systems with ion beam currents up to 100 mA and energies of up to 240 keV have been developed for continuous production operations.[22,23] The high beam currents needed and the chemical reactivity of oxygen ions require the use of filamentless

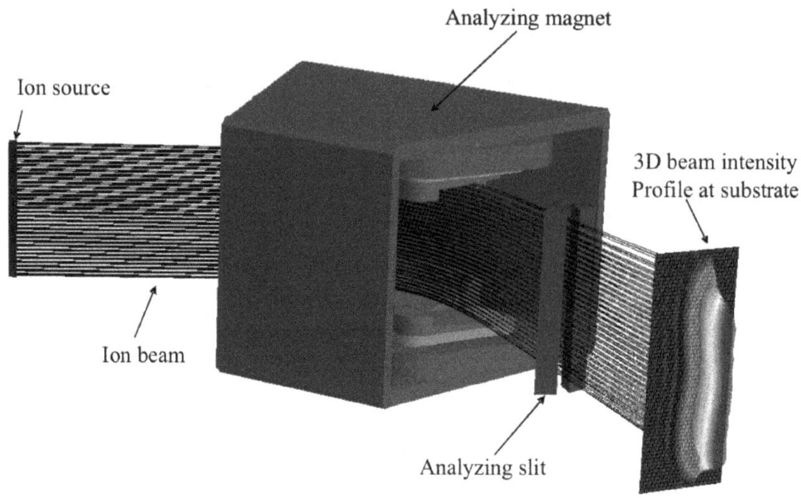

Fig. 14. Tall magnet analysis section of a Nissin Ion iG4 system[21] for implantation of devices in poly-Si films or OLEDs on large (73 × 94 cm) glass plates. The ion beam at the target plane is 80 cm tall and 3–5 cm wide with a B^+ beam current of over 40 mA. The mass resolution of the magnet section is adjustable over a range of $M/\Delta M = 4$ to 10.

ion source technologies using microwave and RF excitation to produce the source plasma. In order to maintain a crystalline, although highly damaged, Si surface layer, the wafer temperature is held between 500 and 650 °C during implantation to encourage sufficient "dynamic" annealing of the accumulating ion damage. Following the implant step, the buried SiO_2 layer is formed during annealing at temperatures between 1350 and 1400 °C, within a few degrees of the melting point of Si.

The structural integrity of the buried oxide (BOX) layer can be significantly improved by a number of modifications of the implant and anneal conditions. For example, if a "hot" (>500 °C) implant cycle with a relatively low ($\sim 2 \times 10^{17}$ O/cm^2) dose is followed by a second implant at close to room temperature with a lower dose ($\sim 10^{16}$ O/cm^2) and an ion energy designed to intentionally amorphize the region of the oxygen profile closest to the wafer surface, and then by an oxidizing anneal designed to further develop the buried SiO_2 layer, a continuous pin-hole-free BOX layer can be formed with lower total ion dose than could be achieved with a standard SIMOX process. In addition to a higher quality BOX layer, the thickness of the layer can be adjusted by direct modification of the ion dose and oxidizing anneal with the use of these "modified low dose" SIMOX processes.[24]

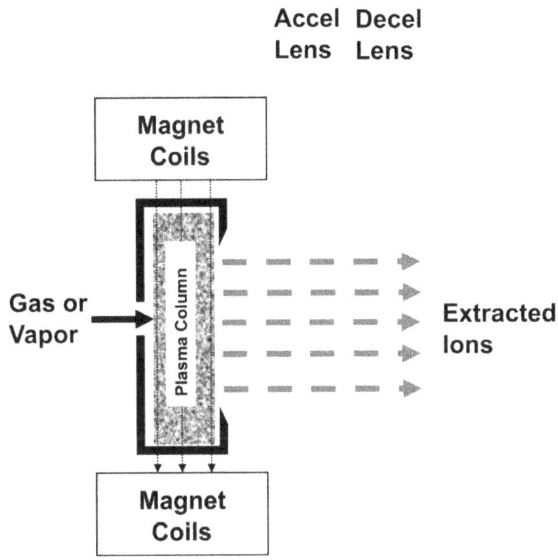

Fig. 15. Schematic of a sustained plasma column and extraction electrodes.

4. Ion Source Designs

Designs for ion sources are as diverse as the many beamlines used to accelerate and transport ions to the wafer and glass plate target.[25,26] The majority of ion sources that provide ionized dopant atoms for ion implantation use a form of sustained plasma column into which a vapor stream of dopant-containing molecules are fed and ionized by electron bombardment in the plasma. An example of such a source is shown in Fig. 15.

The typical form of the plasma is a column contained within a rectangular volume with a shaped opening, or slit, aligned in close proximity to a negatively biased extraction electrode to pull positive ions from the plasma and inject them into the beamline. Ionization within the plasma column is accomplished by bombardment of the incoming gas or vapor stream atoms by electrons accelerated within the plasma confinement volume, or arc chamber. The path length of the energetic electrons is often increased by imposing an external magnetic field along a major axis of the arc chamber, encouraging electrons to travel in spiral trajectories around the magnetic field lines and thereby increasing the probability that the electrons will encounter and ionize a dopant atom.

There have been many forms of arc discharges designed for use as dopant ion sources. Four of these are depicted in Fig. 16. The "Calutron" ion source, developed for use in the large-scale uranium isotope separators at Oak Ridge in the US during World War II, had a heated filament that was

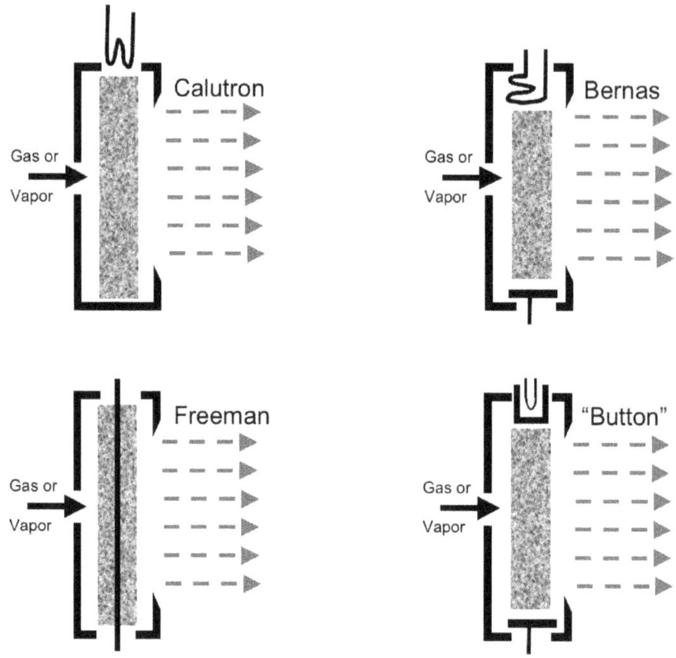

Fig. 16. Schematic of the arc column, extraction slit and heated electron sources for Calutron, Freeman, Bernas, and Button ion sources. The external axial magnetic fields are not shown.

external to the arc chamber that served as a source of electrons to ionize the incoming UF_6 vapor stream. In the Oak Ridge isotope separators, the ion source was situated within the same magnetic field that provided the ion bending along an 8 foot long beam path from the ion source to the collection grids. The use of multiple source chambers provided ion currents of more than 300 mA of U ions in continuous operation in over 1100 separators by mid-1945.

As the interests in isotope separation shifted to the full diversity of elements, smaller and more stable ion sources were developed for high-current operations. One of the most versatile and successful designs was developed by Harry Freeman at Harwell, England. The Freeman ion source featured a heated filament rod, usually made of tungsten (W), placed along the axis of the arc chamber box. The additional circular magnetic field provided by the ohmic heating current in the filament rod, combined with an axial magnetic field from an external set of electromagnet coils (see Fig. 15), increased the path length in the arc chamber volume of electrons emitted from the heated filament and thereby increased the ionization rate for the incoming vapor feed atoms.[27,28]

One of the limitations of the Freeman source was that since the filament rod was immersed in the arc discharge plasma and biased negatively to promote emission of electrons, the filament was continually sputtered by low-energy impacts by the source plasma ions. The sputter removal of filament atoms results in a gradual decrease of the filament cross section leading to over-heating and eventually a melt failure of the rod. The sputtering and chemical removal of W atoms from the heated filament leads to corresponding deposition of W-rich layers on the arc chamber walls, extraction slits, and various insulating components in the arc chamber, all leading to source failures related to serious distortion of the arc plasma conditions and shorting out of various electrical current flows. An opposite-tending effect is the buildup of deposits and growth of W whiskers on hot filament surfaces when fluorine-containing molecules were used as dopant vapor sources.[28]

The effect of ion sputtering on the source filament can be significantly reduced with the use of a coiled filament placed at one end of the arc chamber, as in the Bernas source design shown in Fig. 16. Since the heated filament is located at the edges of the arc discharge column, the ion bombardment rate is much lower than in the Freeman filament case, resulting in increased filament and ion source lifetimes. To increase the overall electron path length in the arc chamber volume, the Bernas source also incorporates an anti-cathode electrode opposite to the heated filament that is intended to reflect electrons back into the arc plasma column, resulting in increased ionization rates of the incoming vapor atoms.

Another advantage of the Bernas ion source is the increased extracted beam currents for multiple-charged ions. In the Freeman design, the negative bias on the filament rod placed close to the extraction slit opening results in a significant capture of multi-charged positive ions as they pass into the region near the extraction slit by the "gate keeping" filament rod. In the Bernas design, the biased filament is located far from the extraction slit region, and the full range of highly-ionized ions can be extracted from the source plasma.

Bernas source filaments are still vulnerable to sputter erosion and chemical layer build up, both of which lead to source lifetime limitations although at a lower rate than with the Freeman design. In addition, the forming of the coiled shape of the Bernas filament design leads to a high level of strain, particularly for a brittle material such as tungsten. These limitations of both Freeman and Bernas filaments are avoided in large part by using a heated plate or "button" as the electron source cathode (Fig. 16, lower right). These cathodes are made of a W cap roughly the diameter of the arc chamber plasma column. This large area cap provides a low resistance electrical source but is not efficiently heated by ohmic current flows as were the filament sources. The W cap, or "indirectly heated cathode" (IHC), is heated to sufficient temperatures for adequate thermionic emission of electrons through backside bombardment by electrons emitted from a small filament source. Because the hot filament in a button source is completely isolated from the arc chamber plasma, this filament can be long lived without limitations from ion bombardment and chemical reactions with the dopant source materials. Because the IHC is located at the edge of the arc chamber and far from the extraction slits, this source also shares the high production efficiency for extraction of high charge-state ions with Bernas designs. With its long source life and other advantages, the W button or IHC source is now the favored design for an arc plasma source for high-current implantation production machines.

Among the many conditions that must be satisfied for successful operation of a continuous operation arc plasma ion source, one of the most basic is the production of sufficient numbers of electrons with enough energy to strip electrons from (i.e., ionize) source atom molecules by collisions. The threshold energies for ionization vary systematically with atomic number and ionized charge state as seen in Fig. 17. For the common dopant elements for Si, the first ionization potentials (needed to produce +1 charge-state ions)

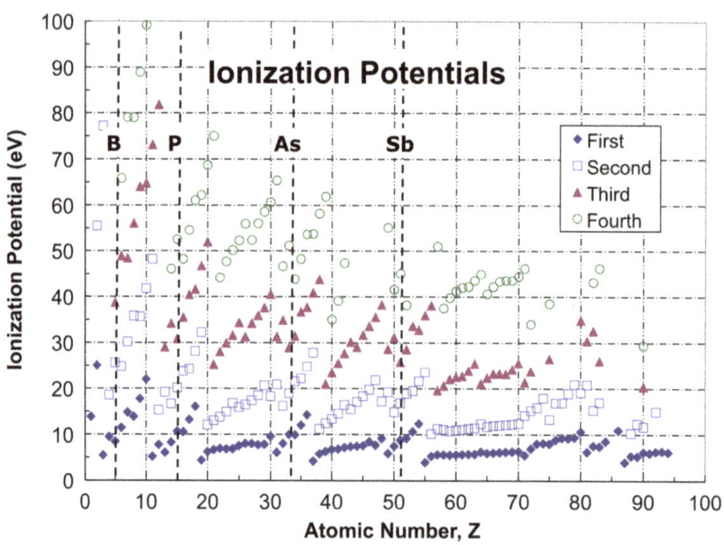

Fig. 17. Ionization potentials for all elements for charge states up to +4, highlighting the cases for four common dopant elements (B, P, As, and Sb) in Si.

are about 10 eV. Increasingly higher electron energies are needed to form the higher charge-state ions often used to obtain energetic ions for deep implant profiles using Cockcroft–Walton style accelerating columns.

The usual way to obtain higher electron energies for increasing the number of high charge-state ions is to increase the negative bias on the heated button or filament electron source, or to increase the electron emission rate by operating the heated source at higher temperatures. Both of these choices increase the rates of damage and wear on the heated electron sources by increasing the ion sputter bombardment energy and chemical reaction rates at the electron source surfaces. In addition, one also needs to limit the operating pressure of gases in the arc chamber volume in order to control the loss of high charge-state ions through transfer of electrons during collisions with background gas atoms. Another way to boost the collisional ionization rate is to increase the strength of the axial external magnetic field (see Figs. 15 and 16), thus increasing the path length for higher energy electrons within the arc chamber volume and the probability for ionizing collisions with dopant-containing vapor molecules. In addition, in order to avoid the "gate keeping" actions of a Freeman source filament if high numbers of high charge-state extracted ions are desired, one should use a "remote" electron source such as in a Bernas or IHC design.

4.1. *Special ion sources: SIMOX, molecular ions, non-volatile elements, and large-area beams*

Not all implantation requirements are well served by the arc discharge ion sources described so far. As mentioned earlier, for high-current (~100 mA) oxygen beams used in SIMOX implanters to form buried SiO_2 layers, the high chemical reactivity of oxygen ions with heated metal surfaces prevents the use of heated filaments or W caps as electron sources. For these types of implantation systems, various forms of microwave and RF excitation are used to ionize the O_2 or H_2O feed gas molecules without the use of a filament.[29]

When large-scale molecular ions such as $B_{10}H_{14}^+$ and $B_{18}H_{22}^+$ are desired for low-energy high-dopant flux implants to form shallow p-n junctions, the high thermal environments of operating arc discharge sources result in undesired "cracking" of the dopant molecule. For example, $B_{10}H_{14}$ decomposes into H and small B_xH_y components at temperatures above 350°C, well below the 700 to 1100 °C operating range of arc discharge ion sources.[30] To produce sufficient beam currents of $B_{10}H_{14}^+$ and similar high-mass molecular ions, a vapor stream of source molecules is ionized by collisions with an energetic

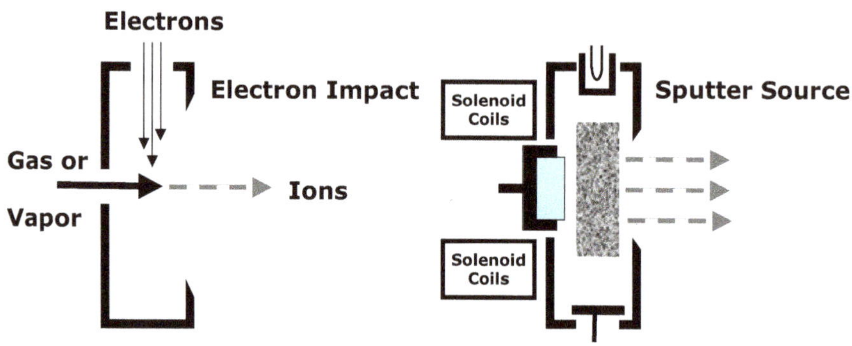

Fig. 18. Schematic of the ionization chambers and extraction slits for an electron impact source (left) used for generation of high-mass molecular ions and a Hill-Nelson source (right), where atoms and ions are supplied to an arc discharge plasma by sputter erosion of a solid target by plasma ions guided to a solid target by a local solenoid magnetic field.

electron beam directed across the vapor stream path as shown in Fig. 18 (left panel).[31]

When the desired elements are not easily available as an atom or molecular vapor, atoms of interest can be sputtered from a solid source placed at the edge of an arc discharge plasma (Fig. 18, right panel).[32] The sputter removal of the solid target atoms into the arc plasma is encouraged by applying a negative bias to the solid target with respect to the plasma sheath and by guiding plasma ions toward the target with a local solenoid magnetic field. These types of ion sources are particularly useful as sources of ions of metallic and high-melting point elements.

When the target area is considerably larger than a Si wafer, such as is the case for implantation of large Si-coated glass plates for fabrication of high performance flat panel computer displays and television screens that are currently fabricated on glass panels of up to $920 \times 730\,\text{mm}^2$ area, substantially different methods for ion beam formation and transport are needed. A widely used method is to form and transport a "broad brush" ribbon ion beam up to 90 cm long and move the large area panels back and forth across the beam in an "ion shower" tool (see Fig. 14). In the large area ion source shown in Fig. 19, a large volume ($95 \times 20 \times 20\,\text{cm}^3$) arc chamber contains an ionizing plasma within a multi-cusp magnet confinement with electrons provided by an array of filaments distributed along the back wall of the ion source.[21] Ions are extracted though a multi-grid array with small extraction openings, and the resulting beamlets are formed into an 80 cm tall by 3–5 cm wide ribbon beam for transport through a tall mass analysis magnet before entering the

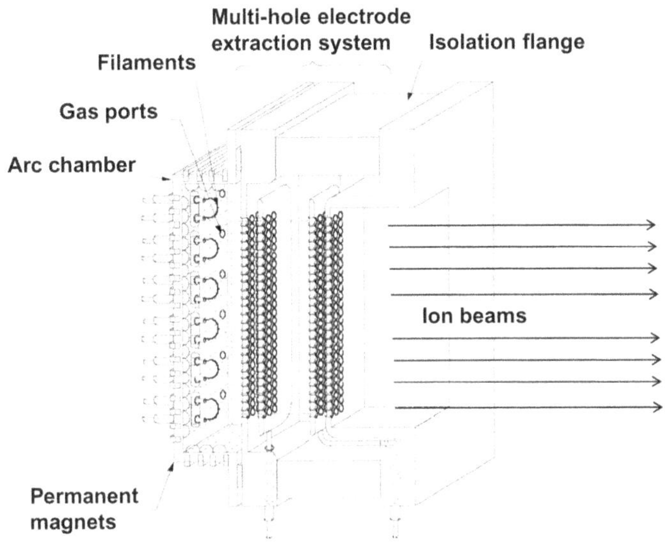

Fig. 19. Large-area ion source for a Nissin iG4 "ion shower" implanter with a multi-filament arc plasma chamber confined by a multi-cusp magnet array and producing an 80 cm tall by 3–5 cm wide extracted beam with up to 40 mA of B^+ ions.[21]

glass plate scanning chamber. By adjusting the bias on the potential on the grid plate closest to the arc plasma, the fraction of extracted (and desired) B^+ ions compared to heavier BF_2^+ ions can be increased, allowing the source power supplies to run at lower power and still obtain B^+ beam currents of 0.4 mA/cm, or more than 40 mA.

5. Scanning Methods

Since except for the case of the plasma immersion implant (PIII) systems, the ion beam is smaller than the target surface area, some combination of scanning of the ion beam and target surface is needed to distribute a uniform dose of ions across the surface. An extensive variety of methods have been used to achieve this goal. Many of the early ion implanters introduced in the 1970s held a wafer in a fixed position and used vertical and horizontal electrostatic deflection electrodes to scan the ion beam over the wafer surface. As the ion beam currents and wafer sizes increased, combinations of beam scanning in one direction and wafer motion in another led to more effective system designs. Another method widely used for high-power beams in high-current and high-energy implanters involved mounting many wafers on spinning wheels and scanning the axis of the wheels so that the wafers moved in front of a stationary ion beam. This resulted in the beam heating and charging effects being distributed across a large area, with the wafers being exposed only to short pulses of beam charge and energy.

Along with the requirement that the ion dose be distributed across the wafer surface in a controlled and highly uniform fashion, the angle of incidence of the ion beam must also be restricted to a distribution with less than 1° variation and often adjustable over a range from normal to up to 60° incidence. This means that the scanning methods used must result in a nearly parallel scanned beam path with minimal beam divergence at the wafer plane. Other requirements, such as the delivery of monoenergetic ions to the target wafer as well as control of current flows and charge build up in the device structures on the wafer surface, also demand innovative consideration in the beamline design.

5.1. *Beam deflection and wafer motion in orthogonal directions*

A widely used strategy in contemporary ion implanters is to scan the ion beam in a horizontal direction with either electrostatic or magnetic deflection and to move a single wafer in an approximately vertical direction. An example of this beam/wafer scanning combination is the SHX-III implanter[33] from SEN, which uses an electrostatic beam deflection and a curved dual-electrode electrostatic lens to scan the ion beam in a parallel set of paths along a horizontal plane and mechanical wafer motion across the scanned beam, as shown in Fig. 20. This design, which evolved from medium current systems developed in the early 1990s by Eaton (now Axcelis), provides beam currents comparable to high-current implanters for ion energies above 10 keV (e.g., 7 mA for B^+ and 10 mA for P^+). It can also use these high beam transport efficiencies for higher energy ions to deliver substantial beam currents to the wafer at low energies by decelerating the ion beam to 0.2 keV with a decel stage directly after the beam parallelizing lens. The SHX design can operate with much higher accel-decel energy ratios than in-line decel beamlines (see Figs. 8 and 9) because of the presence of a deflection electrode placed after the decel stage that directs low energy ions toward the wafer while allowing any higher-energy beam neutrals to pass undeflected to a beam dump. The deflection is done by electrostatic electrodes for high-energy, low-current ion beams and by a magnetic field for high-current, low-energy beams. This "energy filter" allows the SHX to deliver monoenergetic dopants to the wafer surface with accel-decel ratios as high as 40 with final ion energies at $\leq$1 keV. The orientation of the wafer in the wafer motion

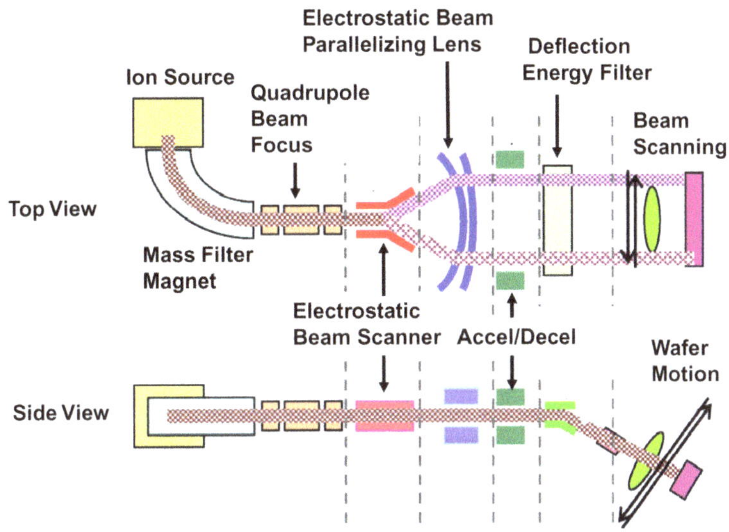

Fig. 20. Top and side-view schematics for the SHX high-current implanter[33] with electrostatic beam scanning and mechanical wafer motion.

plane is adjustable for tilt angles of up to 45° for high beam incidence angle implants.

A similar scanning concept is used in the high-energy beamline in the Optima-XE[19] from Axcelis shown in Fig. 21, where the electrostatic dual lens for directing the scanned beam along parallel paths is replaced by a 60° magnetic field. The ion beam scanning in a horizontal plane is done with electrostatic electrodes, and wafers are mechanically scanned across the scanned beam plane on a variable tilt handler similar to the SHX.

When high-current beams need to be scanned at low energies, electrostatic methods present many difficulties because they tend to separate the fast ion beams from the space-charge electrons in the beam path, leading to Coulomb beam expansion and other deleterious effects. Magnetic field deflection is much more effective in this situation since it deflects the energetic ion beam while allowing the slower beam electrons to follow the moving ion beam and maintain good local space-charge balance. An example of a magnetic-scan beamline is the Nissin Ion 9600A[34] shown in Fig. 22, which uses a series of magnetic field components to separate the extracted ions into paths of various curvatures in order to (1) select the desired mass and charge state with a variable aperture, (2) adjust the beam shape in a quadrupole triplet stage, (3) scan the ion beam in the horizontal plane, and (4) deflect and direct the ion beam into a set of parallel scanned paths at the wafer plane. This beamline when coupled with an electron beam ionization source

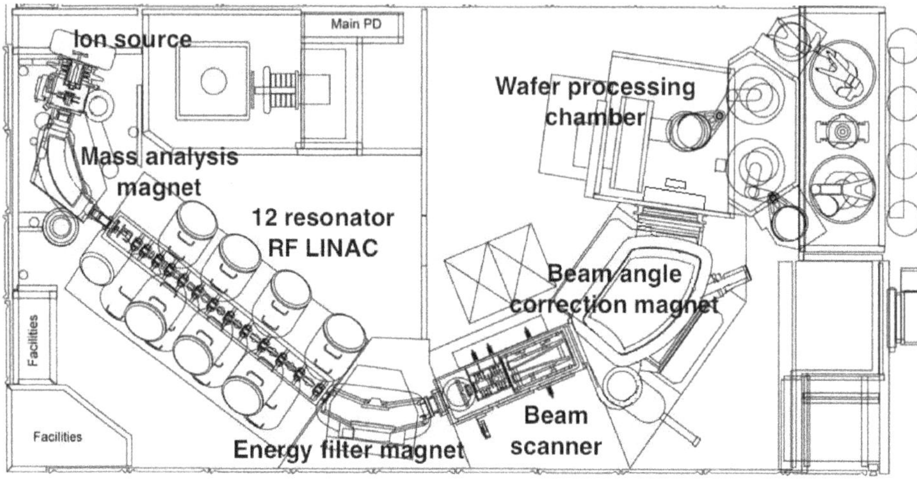

Fig. 21. Top view schematic of the high-energy Optima-XE beamline[19] with RF linac acceleration followed by electrostatic beam scanning in the horizontal plane, a 60° bending magnet to form parallel beam paths across the wafer, and vertical mechanical wafer scanning.

(such as shown in Fig. 18) developed by SemEquip is optimized for delivery of low-energy molecular ions, providing 20 particle-mA beams of 2 keV boron in the form of $B_{18}H_{22}$ ions for high dose, shallow junction doping.[35] Lower beam currents are delivered for decelerated beams down to dopant atom energies of 0.2 keV at dopant flux levels much higher than can be delivered with single-atom ion beams.

5.2. *Spinning wheel and pendulum wafer scanning*

For many years, the scanning method of choice for implantation with high-current and high-energy ion beams was to mount a number of wafers on a rapidly spinning wheel like the one shown in Fig. 23 (rotating at up to 1250 rpm) and to translate the axis of the wheel so that the wafers scanned across a stationary ion beam. This scanning method allowed the beam power and delivered charge to be distributed across the wafers and limited beam exposures to short pulses (<1 ms for some systems), thereby greatly reducing problems related to charge build up and beam heating of the wafers. The high centrifugal forces on the wafers also allowed for designs that avoided the need for front-side clamps to hold the wafers in place on the spinning wheel.

However, the spinning wheel design had limitations for control of beam incidence angles that led to problems with control of ion channeling effects.

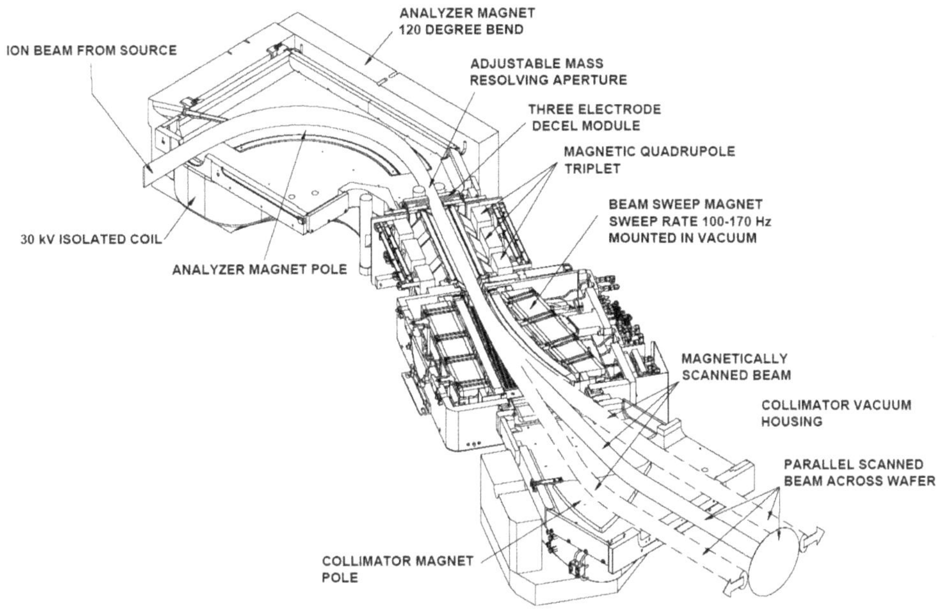

Fig. 22. Schematic of the magnetic beam elements for mass analysis, beam focusing, scanning, and formation into parallel, collimated beam paths at the wafer plane in the Nissin Ion 9600A implanter.[34] The wafer is scanned up and down across the beam path plane by a mechanical, adjustable wafer tilt mechanism. This beamline is optimized for low-energy beams using molecular ions such as $B_{18}H_{22}$. It also supports high-energy single-atom ions up to energies of 320 keV and higher for multiple-charged ions.[35]

Also, the high wafer speeds (up to 90 m/s) led to destructive effects on delicate device structures such as thin (<100 nm) poly-Si gate electrodes when collisions occur on the wafer surface with particles transported in the ion beam. With contemporary advanced gate widths now at 30 nm, this collisional damage from particles is an unacceptable yield loss mechanism.[36]

Reducing the wheel spinning rates and wafer velocities leads to numerous problems with dosimetry, wafer charging, and beam heating and so was an interim solution. Several contemporary implanters use "partial wheel" pendulum designs where the wafers are rocked back and forth across the beam at far slower speeds on a swing arm that is also translated vertically so that the beam "paints" across the full wafer, as depicted in Fig. 24. Since the wafer motion is now too slow and non-uniform, centrifugal forces are now replaced by electrostatic chucks to hold the wafer in place and provide for sufficient heat and charge flow paths out of the wafer.

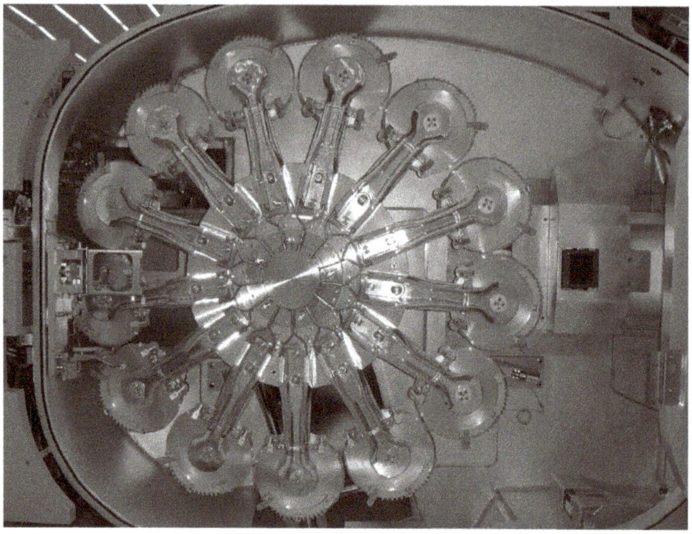

Fig. 23. A backside view of the spinning wheel for 300 mm wafers in the Applied Materials PI9500 implanter. When operating in a vacuum, the wheel was spun at 1250 rpm and the axis of the wheel was rocked left and right so that the wafers scanned in front of a stationary ion beam coming through the beam port, visible on the right.

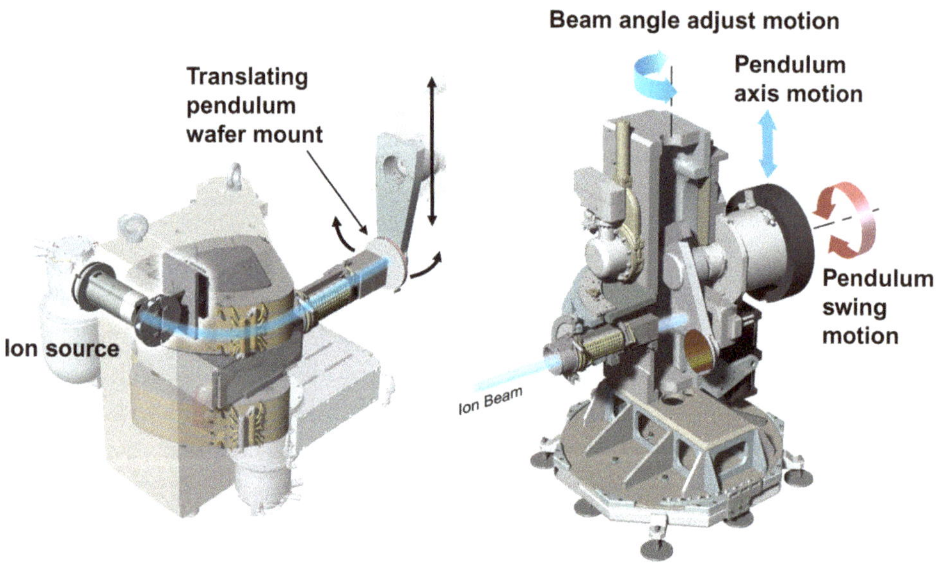

Fig. 24. Beamline (left) and wafer motion (right) on a translating pendulum in the Axcelis Optima-HD high-current implanter.[37] High tilt beam incidence implants are obtained by rotating the wafer motion assembly around the vertical axis during setup.

Some aspects of "R-theta" scanning extend into these modern machine designs. Pendulum scans share the same requirement as spinning wheel systems that the wafer rocking motion be linked to the axis translation speeds so that the delivered dose across the wafer is uniform, translating faster as the ion beam traverses the portions of the wafer closer to the rotation axis. In the example shown in Fig. 24, large tilt angle incidence implants are achieved by rotating the entire wafer motion assembly around the vertical axis prior to the implant.[37]

6. New Directions: Gas Cluster Ions, Photovoltaic Cell Doping, and MeV Protons for Si Membrane Cutting

Ion implantation tools have been widely used to investigate the formation methods and properties of advanced technological materials and devices. The key advantage of ion implantation here is that new chemical and structural combinations can be formed without undue restrictions from chemical processes and structural interfaces. For other materials deposition techniques such as diffusion from gas sources, chemical vapor deposition (CVD), or sputtering, the incorporation of atoms into a material is limited by the conditions at the material surface, such as the presence of oxides and other diffusion barriers, and by the rates of diffusion and chemical reactions within the material volume. With ion implantation techniques, known quantities of atoms can be placed into the volume of a material without concern for surface diffusion barriers. By implanting high dose distributions of atoms, meta-stable alloy combinations can often be achieved at much higher concentrations than can be achieved by near-equilibrium processes such as diffusion and chemical reactions.

An additional key factor for new materials applications is the action of the ion beam itself. In the process of slowing down the incident ions in the target material, a significant amount of energy can be deposited into the material surface, often achieving temperatures far above the melting point during the short time scale (less than 10 ps) of the ion stopping process. When large numbers of atoms arrive at the material surface at the same time, as with the use of large cluster ions containing tens to thousands of atoms, entirely different materials processes occur than in the case of single atom ion irradiation. Occasionally a favorable outcome of these studies results in the development of new types of accelerators and an expansion of the meaning of "ion implantation".

6.1. *Gas cluster ions*

The interactions of energetic gas cluster ion beams (GCIB) with surfaces have been studied in detail for over 30 years, most consistently by a series of groups at Kyoto and nearby universities in Japan.[38] Gas cluster ions are formed by ionization of large atomic clusters formed by adiabatic cooling of gases expanding through a nozzle between a pressurized (10 Torr or so) chamber and a vacuum. This effect, also used in the operation of refrigerators and air conditioners, results in the "freezing" of even inert gases such as Ar into clusters that can number several thousand atoms. When the forward-directed gas clusters are selected by a small opening in a skimmer aperture that rejects the off-axis gas flows containing smaller atomic clusters and monomers, a supersonic beam of large atomic clusters can be ionized by low-energy electron bombardment and accelerated toward a target surface, as depicted in Fig. 25.

The kinetic energy of the ions is determined by the acceleration potential, and the energy per atom depends on the cluster size. Therefore, the energy per atom of a 1000 atom cluster accelerated to 1 keV is only 1 eV per atom. This dependence of the energy per atom on cluster size can have dramatic consequences. In the molecular-dynamic modeling of 20 keV Ar clusters on Si shown in Fig. 26, small (n = 20 to 2000) cluster impacts result in a dense damage region and craters. However, for a large (n = 20 000) Ar cluster impact, where the energy per ion is far lower than the threshold for

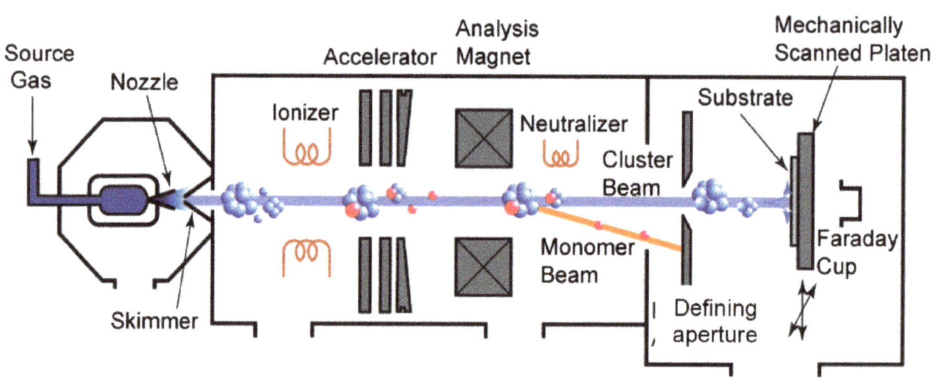

Fig. 25. Schematic of a gas cluster ion beam (GCIB) system[38] where (i) large atomic clusters are formed by expansion of gases through a nozzle into a vacuum region, (ii) smaller cluster sizes are rejected by a skimmer aperture, (iii) ions are created by low-energy electron bombardment and accelerated, (iv) residual monomers are deflected out of the beam path by a magnetic field, and (v) the ions are neutralized before impact on a target surface.

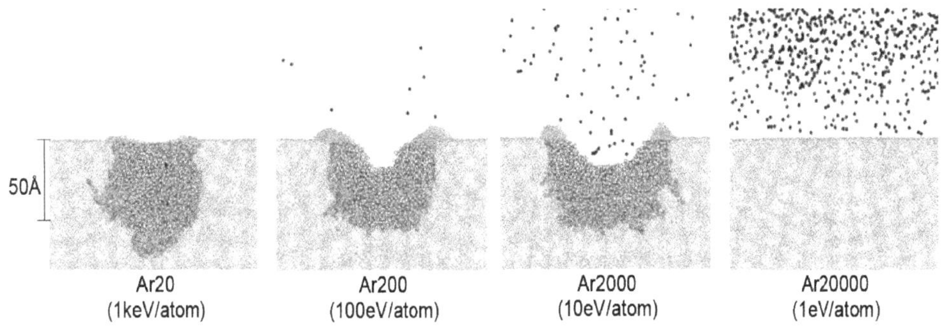

Fig. 26. Molecular dynamic modeling of the impact of 20 keV Ar_n clusters, shown 16 ps after initiation of the impact. For the largest (n = 20 000) cluster impact, the energy per Ar atom is far lower than the threshold for recoil displacement of Si atoms, resulting in reflection of the cluster atoms with high local heating but no recoils in the Si.

recoil displacement of Si atoms from their lattice (~12 to 15 eV), the entire cluster is reflected from the Si surface with no recoil damage. The momentum exchanged with the surface from the impact of the 20 keV ion cluster results in energy being transferred with intense local heating in a short (~10 ps) time. When this local heating exceeds the melting temperature, atoms from the cluster debris are incorporated into the melted region with a high efficiency, leaving a highly-doped amorphous layer on the surface. Most importantly, since in this case there are no recoil displacements of Si atoms, the doped amorphous layer is not surrounded by a dense distribution of lattice defects, and no "end of range" damage is present after re-growth of the amorphous layer during thermal annealing. Since junction leakage currents from carrier recombination and tunneling effects are strongly increased by residual damage near the junction location, GCIB doping offers the potential of shallow depth implantation with no defect-related leakage current.

The applications of GCIB impacts are not limited to shallow junction doping. When smaller clusters of inert atoms are accelerated into Si, the strong lateral sputtering and mass transfer associated with surface crater formation (see the middle examples in Fig. 26) result in a powerful method for micro-polishing surfaces to atomic-level smoothness for such applications as fabrication of X-ray mirrors and tuning of surface acoustic wave devices. When the gas cluster ions contain chemically reactive elements, such as SF_6 clusters on Si and metals, the local surface heating by the cluster ion impact results in etch rates that are orders of magnitude higher than for monomer beams or plasma exposures. Prolonged exposure of massive GCIBs can also result in deposition of films at room temperatures. The use

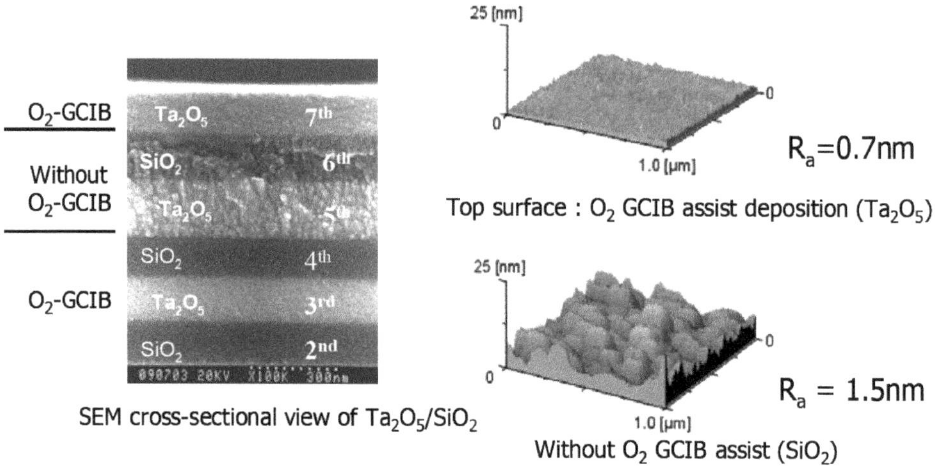

Fig. 27. SEM cross section (left) and AFM scans (right) of a Ta_2O_5/SiO_2 multi-layer structure formed with e-beam evaporated Ti and Si in an O_2 atmosphere with and without the operation of a 7 keV O_2 GCIB.

of GCIB exposures to circumvent the thermal cycles associated with CVD deposition allows for the formation of films with fragile components, such as local room temperature deposition of Ge on Si using polymer photo-resist patterning.

When the local ion-beam heating associated with GCIB impacts is combined with molecular beam deposition, unique forms of dense, high-quality multi-layer structures can be formed. In the example shown in Fig. 27, when e-beam evaporated beams of Si and Ta are deposited in the presence of an O_2 atmosphere, multi-layer films of SiO_2 and Ta_2O_5 can be formed. When the film growth is assisted by impacts of energetic GCIB of O_2 molecules, the grown films are denser with more planar interfaces, producing in this case higher-quality optical reflection components for photonic devices. Similar GCIB processes have been used to form diamond-like C films on polymers and other unique structures, as well as sealing of the surfaces of porous materials such as low-k dielectric films in IC metallization structures.[38]

6.2. *Doping of Si-based photovoltaic cells*

Doping of Si-based photovoltaic (PV) cells by ion implantation is a major new application for implanters. The principal driving forces for the use of implantation doping for Si-PVs are (1) increased solar power efficiency for moderately-doped junctions and (2) reduction of the number of steps in the PV cell production flow with corresponding reductions in cost. Additional

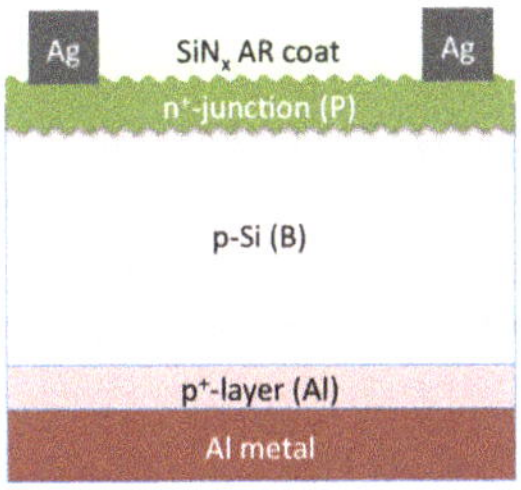

Diffusion Doping Process

1. Saw damage etch, front surface texturing
2. $POCL_3$ deposition & diffusion (all sides)
3. Removal of phosphosilicate glass
4. SiN_x deposition on front side (PECVD)
5. Print & dry Al contact metal (backside)
6. Print & dry Ag contact lines (frontside)
7. Metal sintering anneal
8. Edge isolation (laser removal)

Fig. 28. Schematic of a diffusion-doped PV cell (left) and process flow (right). The front-side n^+ junction is doped by diffusion from a doped glass layer deposited by a $POCL_3$ gas flow. The backside contact is enhanced by the formation of an extra p-doped region formed by out-diffusion of Al during the metal sintering anneal.

advantages can be seen for implant doping in more advanced PV cell designs utilizing backside p- and n-junctions and contact grids, leaving the front side 100% unblocked for full light incidence.

Standard PV cell doping for both c-Si (crystalline silicon) and poly-Si blocks begins with the formation of a front-side n^+ junction in p-type substrates by deposition and drive-out diffusion of P from a surface glass coating formed by reactions with $POCl_3$ vapor in a furnace tube (see Fig. 28). $POCl_3$ vapor doping is a well-developed method of n^+ junction formation resulting in surface dopant concentrations in the range of $\sim 3 \times 10^{20}\,\mathrm{P/cm^3}$. The desired junctions depths, 0.5 to 1 μm, are achieved by diffusion anneals in the $POCL_3$ furnace. After $POCL_3$ doping, the n^+ doped regions on the cell backside and edges are etched and/or ground off, exposing the p-type bulk for later processing.

A problem with the high-doping levels that result from $POCL_3$ diffusion doping is that the high P levels degrade the conductivity of the n^+ junctions through mobility decreases (Coulomb scattering from the dopant ion cores) and reduction of the free carrier concentration through recombination mechanisms. So as the P dopant concentration increases above $1 \times 10^{20}\,\mathrm{P/cm^3}$, the PV cell power efficiency decreases by 2 to 5%, depending on the dopant level and n^+ junction depth, as is seen in Fig. 29. This loss in PV cell efficiency is seen in the reduced sensitivity to shorter ("blue") light wavelengths absorbed in the high-dopant concentration surface regions. At dopant levels below $\sim 1 \times 10^{19}\,\mathrm{P/cm^3}$, efficiency also falls off due to insufficient carrier concentrations, increasing the junction resistivity and again limiting photocurrent efficiencies.

As the use of PV power cells moves beyond large-area installations to area-limited locations such as industrial and residential rooftops, the efficiency of the PV cell becomes a prime concern. Reducing P concentrations while still using $POCL_3$ doping is doable by running $POCl_3$ deposition with increasing O_2 flows and etching major portions of the P-containing oxide coating before the extended drive-in anneal. But both methods increase process time and complexity.

With implantation, the efficient and accurate delivery of P dopant to the front PV surface to provide dopant concentrations of $\sim 1 \times 10^{19}\,\text{P/cm}^3$ after drive-in anneal is easily achieved with modern high-current implant tools. The increase in PV efficiency with an implanted, moderately-doped junction can be substantial (as can be seen in Fig. 29) with the added benefits of (1) substantially improved doping level uniformity ($POCL_3$ doping varies with the vapor flow rates, furnace temperature, and cleanliness of the P-glass/Si interface) and (2) reduction of PV process steps, especially when combined with local implant doping of n^+ regions under the metal grid line emitters (see Fig. 30).

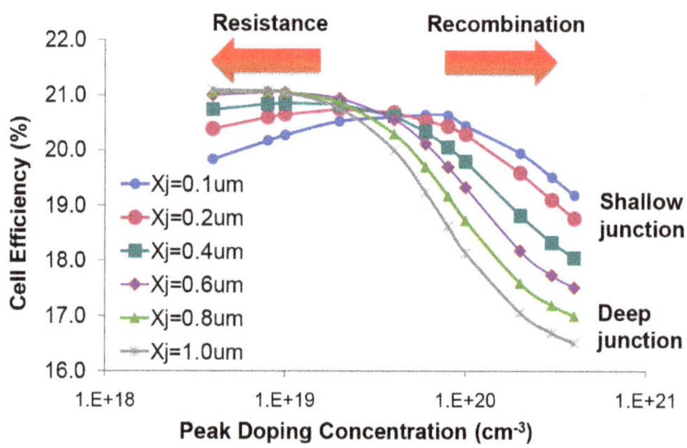

Fig. 29. PV cell power efficiency as a function of dopant concentrations in the topside n^+ junction for a variety of junction depths.[39]

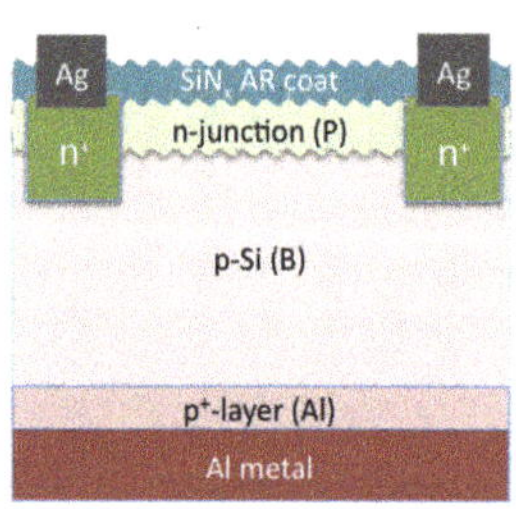

Implant Doping Process: Frontside contact

1. Saw damage etch, front surface texturing
2. Implant frontside n-junction
3. Align shadow mask & implant n^+ emitter lines
4. SiN_x deposition on front side (PECVD)
5. Anneal implanted dopants
6. Print & dry Ag grindline metal (frontside)
7. Print & dry Al contact metal (backside)
8. Metal sintering anneal

Fig. 30. Schematic of moderately-doped n-type front junction with local high dopant concentration under the metal contact lines (left) and process sequence (right).[40]

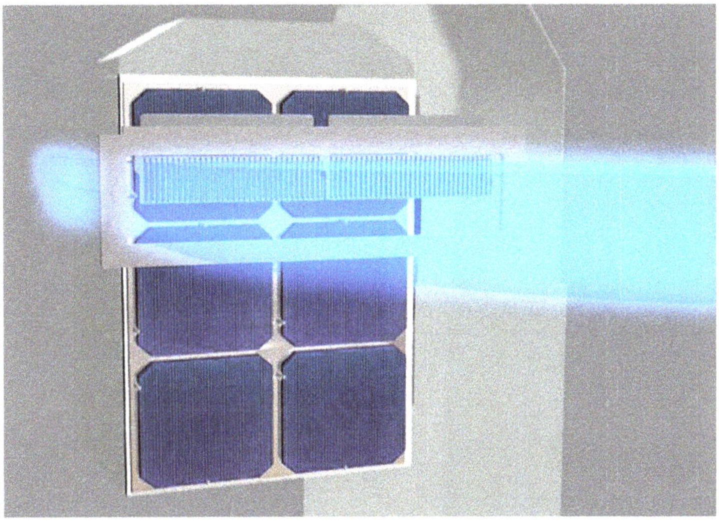

Fig. 31. Shadow mask used in the Varian Solion PV cell doping tool, based on the Varian high-current implanter, featuring a horizontal ribbon beam (Fig. 9) where a shadow mask with openings aligned to the position of metal grid line paths is placed in front of the PV cell during the doping step. The ribbon beam also passes through the open part of the mask, doping the junction and the higher-dose emitter regions at the same time the PV cells are scanned vertically under the mask.[41]

The low contact resistivity needed to obtain Ohmic contact between the topside n-junction and the metal contact line grid can be recovered to levels similar to the $POCL_3$ case by addition of extra dopants to "selective" or "double" emitter junctions along the paths of metal grid line depositions. In the case of the Varian Solion implanter, this local extra doping can be achieved with a shadow mask placed in front of the wafer, aligned to the location of the metal grids, thus avoiding the need (and costs) for on-cell patterning (see Fig. 31).

6.2.1. *Alternatives to implant doping for PV cells*

Implant methods can provide tight controls on junction doping levels, with direct controls of adjustments for optimization, and with the use of non-contact shadow masking, a cost efficient method for adding the extra "double emitter" doping under the metal grid lines on topside junctions. However, implant tools are considerably more complex than diffusion furnaces and require care for high-voltage and vacuum environment operation, toxic dopant gases, and the challenging mechanics of fragile PV cell handling.

One alternative being used for selective or "double" emitter doping is to drive in extra dopant atoms from the P-doped glass layer with a scanned laser along the path of the metal contact grid. This is a non-vacuum, in-line process involving no patterning steps for the extra emitter doping.

Another technique is to locally deposit doped "Si inks" using the same screen printing methods used for metal line deposition. The dopants are driven into the Si body during the metal sintering anneal after the screen printing of the metal (Ag) contact grid. Versions of these doped "inks" are being developed for non-contact spraying with ink-jet printers onto more fragile PV materials such as poly-crystalline and ultra-thin Si cells.

6.2.2. *Advanced PV cells*

While implant doping faces several competing alternatives for front junction doping, including the incumbent $POCL_3$ diffusion doping, next generation Si-based PV cells with efficiencies greater than 20% require the avoidance of front-side metal contacts and the placing of both p- and n-type junctions on the cell backside. In this case, implant doping has a significant advantage over all forms of diffusion doping because it can use room temperature dopant delivery with easily-removed PR masking. CVD-based doping requires deposition, patterning, and removal of high temperature compatible "hard" masks of oxide and nitride layers that add significant cost and complexity to the PV cell fabrication process.

Next generation PV cells will incorporate (1) thinner and higher carrier lifetime and mobility Si, such as exploiting the higher electron mobility of n-Si bulk cells, (2) backside optical coating to reflect light back into the bulk volume of the cell, and (3) better passivation interfaces at the Si bulk surfaces and cell edge to reduce carrier loss from recombination (see Fig. 32). Some of these features are already incorporated in high-efficiency c-Si cells from PV vendors such as SunPower.

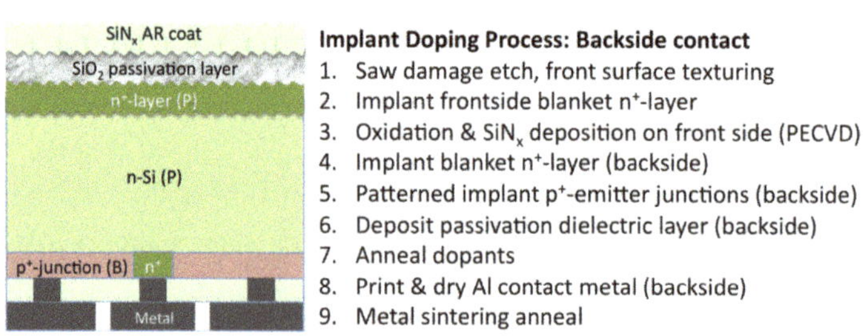

Fig. 32. Advanced PV cell design with backside p^+ and n^+ junctions and contacts, and no front-side metal contacts (left), and implant doping process flow (right).

In Fig. 32, the body of the PV cell is n-type, taking advantage of the higher carrier mobility for electrons. All of the junctions on the front and backsides are doped by ion implantation. Note the presence of passivating dielectric films adjacent to the front and back junctions with thickness and optical properties designed to be an anti-reflective coating on the front side and reflective (along with the metal layer) on the backside.

Doped backside contacts for next generation PV cells could be manufactured by extensions of the screen printing methods in the incumbent processes using local laser heating for diffusion from doped glass films or ink-jet printed doped "inks". Among the challenges for these approaches will be the narrower contact lines — present day front-side metal lines are $>100\,\mu$m wide — desired for backside contacts (to allow for maximum area available for backside optical reflecting coatings and passivation interfaces). If the lateral dimensions of the doped contacts, likely to be in the form of interdigitated p- and n-type junctions interspersed with reflecting and passivating surfaces, are too difficult for screen printing patterning, the need for optical litho and PR films would give strong incentive to use implantation doping for the junctions.

Although the active use of ion implantation for PV doping is relatively new, at least three vendors have designed tools for this market. Varian (VSEA) has modified its widely used high current tool by incorporating an adjustable shadow mask (see Fig. 31) for front-side junctions and emitter doping. Varian is also actively developing methods for backside PV doping using both beamline (Fig. 9) and plasma immersion (Fig. 13) technologies.

New implantation tool designs specific to the needs of PV doping are under development by Intevac and Amtech/Kingstone. As the PV doping market develops, other implant tool vendors may choose to modify the various plasma immersion tools (from Varian, Applied Materials, and Ion Beam Services) used for CMOS doping as well as "ion shower" (Fig. 14) tools (from Nissin Ion and SEN) presently used for doping flat panel displays and electronics.

6.3. *High-current, multi-MeV proton beams for fabrication of thin Si PV membranes*

The usual energy range for H implants for formation of SOI and photonic structures, typically 20–100 keV, has recently been expanded to 1–4 MeV. With high dose implants at these higher proton energies, deeply buried H-saturated splitting layers can be formed capable of separating Si membranes 20–150 μm thick. For H-rich layers at these depths, self-supporting Si

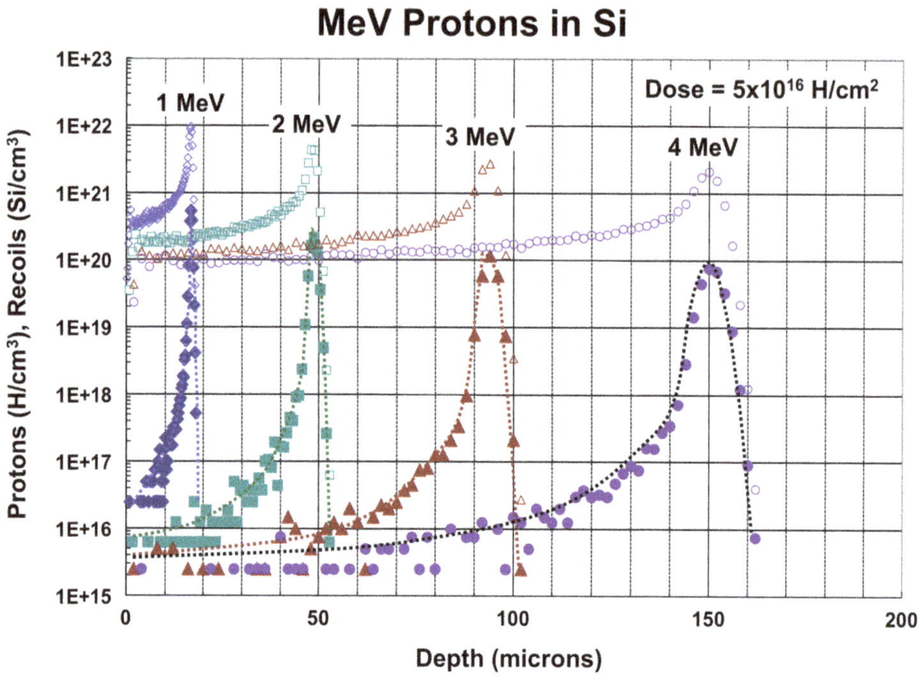

Fig. 33. Monte Carlo (SRIM08) calculations of hydrogen (solid symbols) and Si recoil profiles (open symbols) for 1 to 4 MeV proton implants into Si at a dose of $5 \times 10^{16}\,\mathrm{H/cm^2}$.

membranes can be separated without the need of a bonded handle wafer to suppress formation of surface blisters. The intent of these multi-MeV proton implants is to form low-mass photovoltaic Si solar cells with minimal loss of Si in the cutting process.[42]

The energy range for 20–150 μm deep proton implantation can be estimated with the Stopping and Ranges of Ions in Matter (SRIM) Monte Carlo simulation code,[43] as seen in Fig. 33. Beam energies of ~4 MeV are required for proton profiles with 150 μm depths. Hydrogen ions lose most of their energy in Si by collisions with target electrons, with a small fraction of the energy deposition going into creation of Si recoils at the end of the ion path. The location of the ion damage peak is important for H in Si and Ge because the final proton distribution is determined by the depth of the implant damage layer that traps the otherwise highly-mobile H atoms and holds them in place. Given the importance of the location and density of the implant damage layer, factors that control the damage accumulation rate such as the ion beam current density, scan rate, and Si temperature also need to be closely controlled for success in this application.

Fig. 34. SEM image of a 50 μm thick Si membrane separated after high dose H implantation at ~2 MeV.[42]

The strongly peaked proton distributions result in highly-localized splitting interfaces with little or no loss of Si in the separation process, leaving the bulk of the membrane with minimal ion damage. This results in high-quality, long carrier lifetime, ~50 μm photovoltaic materials, a sample of which is shown in Fig. 34. Given the high proton dose ($\sim 5 \times 10^{16}$ H/cm^2) needed for H-cut splitting, implanters based on new RFQ (Radio Frequency Quadrupole) linear accelerators and high beam transmission dc accelerators have been developed to serve this application to provide the mA-range beam currents required for cost effective fabrication of these thin Si membranes for photovoltaic applications. There are at present two companies, Silicon Genesis and Twin Creeks Technologies, who are operating high-current, high-energy proton accelerators for this application.

7. Implantation into Metals and Biomaterials

Although this chapter has focused on the mainstream applications of ion implantation, i.e., the implantation of dopants and reactive elements (O, H, C, F, etc.) into semiconductor materials, industrial accelerators and plasma immersion systems are also used for a variety of implants into metals, ceramics, and a rapidly diversifying range of biomaterials.

7.1. *Metals: hardness, friction, and corrosion*

Strengthening of the wear and corrosion resistance of metals, such as Fe, Cr, and Ti, through formation of hard nitride precipitates after high dose mid-10^{17} ions/cm^2 implantation of nitrogen has long been used to improve the performance and longevity of metal tooling for drill bits and other materials-cutting and die-stamping applications. The use of this technique is limited by the high doses required for significant improvements in materials properties and the difficulty of treating the surfaces of complex shapes with a strongly directional ion beam. Many of these problems can be mitigated by using variations of PIII.[44]

In the PIII applications for doping of IC devices on semiconductor wafers (discussed in Sec. 3.4), the wafer target is mainly a planar surface placed at the boundary of a dopant-containing plasma region and implanted by drawing ions out of the plasma by short (μs) negative voltage pulses applied to the wafer. For metallurgical applications with large complex shapes, the target workpiece is fully immersed in the central volume of the plasma and plasma ions are drawn to the target from all directions, at roughly normal incidence to the shape of the local surface, when the target is pulsed to a negative bias. This original form of PIII was developed by John Conrad at the University of Wisconsin in the mid-1980s, roughly at the same time as the initial PIII work on semiconductor applications was started by Bunji Mizuno at Matsushita.

A large scale PIII demonstration unit was developed at the US Los Alamos National Laboratory to explore the use of PIII for surface modification of large-scale structures (such as shown in Fig. 35) that could process parts in a large volume (1.5 m diameter $\times$ 4.5 m deep) plasma chamber capable of pulsing currents of 55 A at 100 kV.[45]

The extraction of metal ions from plasmas formed by vacuum arc melting of solid targets, developed extensively at the US Lawrence Berkeley Laboratory (LBL) since the mid-1980s, has been used to create intense metal ion beams, often with high charge state for high energy implantation.[46,47] In metal vapor vacuum arc (MEVVA) ion sources, a plume of mixed ions and liquid metal particles are evaporated from small (several microns) intensely heated regions on a metal cathode surface. The resulting metal ion plasma can be guided by gas flows and external magnetic fields toward an extraction grid assembly where the ions can be accelerated to the desired energies. Metal particles can be filtered from the ions by guiding the metal plasma along a curved magnetically confined path before the ion beam is exposed to the target surface. MEVVA-type ion sources can be used to form metal-rich

Fig. 35. Large volume (1.5 m diameter, 4.5 m deep) PIII chamber with racks of Al engine piston parts being implanted at energies up to 100 keV.[45]

films and deposited layers on diverse surfaces with excellent adhesion and film coverage. Metal ion beams have been formed with over 50 metallic elements and a wide range of alloys.[47]

Metal ion sources involving multiple cathode targets have been developed with large area extraction grids. In the example shown in Fig. 36, a 50 cm diameter grid has been demonstrated to support 100 keV pure Ti ion beams at currents of 7 A with an arc current of 300 A.[47]

7.2. *Biomaterials treated by plasma immersion implantation and deposition*

With changes in the gas flow rates and composition, vacuum pressure, plasma density, and target pulsing conditions, PIII chambers can be used to deposit and etch films as well as implant ions. This operational complexity, which often presents numerous challenges to the use of PIII for controlled implantation of dopants into IC devices, can be a considerable advantage when preparing diverse combinations of materials on complex surfaces for fabrication of biomaterials.

Ion implantation of N^+ into Ti and Ti alloys has long been used to improve the wear and corrosion resistance of medical implants for hip and knee replacements. The present day production rate of implanted medical-implant parts is in excess of 10^5 per year. In the last few years the range of applications for formation of biomaterials has expanded into many new areas.[48]

Fig. 36. A 50 cm diameter metal ion source developed at LBL that is capable of providing beams of pure Ti ions at currents of 7 A and mean energies of 100 keV.[47]

Ti materials are also well suited for use in dental and orthopedic replacements where formation of strong bone re-growth around the implanted part is desired. When PIII is used to implant Na or Ca ions formed into a plasma plume from a local evaporated source in the chamber, the wear resistance and rate of bone growth when soaked in simulated body fluids is strongly increased. The ability of PIII to implant complex surfaces, which are difficult to expose to a directed ion beam from a conventional implanter, is a strong advantage in these applications.

For implantation of IC devices used to sense body conditions and transmit data to external receptors, it is often desirable to encourage bone growth around the Si material to stabilize the chip in the body. Implantation of high doses of H^+ into Si, forming a H-rich amorphous-Si layer, strongly increases the rate of growth of bonelike apatite when soaked in simulated body fluids.[49] PIII is particularly well suited for this application since the high proton dose can be implanted with high chip part throughput without subjecting the bio-sensor device and circuitry to high temperatures or contaminating environments.

PIII chambers with carbon-rich plasmas formed with acetylene and other organics can also be used to form surface coatings of diamond-like carbon (DLC) with high film adhesion and coverage. Additions to the C-rich plasma, including metal ions formed with cathodic arc sources discussed earlier, can be used to dope the DLC film for improved bioactive performance. Doping of DLC films with elements such as Ca and P can further improve the wettability of the surfaces and inhibit the formation of platelet adhesion and damage, which is a key requirement for bio-surfaces that are in contact with blood flows.[50]

The number of PIII-type systems used to form useful biomaterials can clearly be expected to increase over the next several years.

8. Summary

The four-decade history of the use of ion accelerators for fabrication of IC devices and materials contains a continuous cycle of innovation and sophisticated engineering to provide tools for the worldwide industry of modern electronics. Beginning with roots in accelerators designed for nuclear physics experimentation and isotope separation reaching back to the 1930s, the sale of commercial accelerators beginning in the late 1960s for use in manufacturing of IC devices and materials is now a greater than US$1B-a-year enterprise with follow-on links to many support and process metrology businesses. There are approximately 14 vendors worldwide that currently produce and sell these commercial systems, with 4 companies dominating the market for implanters used for the fabrication of semiconductor devices and materials. The range of available beam currents extends from μA levels to $\sim$100 mA and ion energies from a few eV (for GCIBs) to $\sim$10 MeV (with linear and tandem accelerators).

The driving force for this diversity is the ability of ion implantation to deposit dopant and other useful atoms at precise depths and locations at concentrations spanning six orders of magnitude with dose controls of less than 1% in well-managed, continuous production environments. This ability has resulted in the nearly universal use of ion implantation for doping applications in IC devices.

The present trends seem to indicate a continued expansion of the range and capabilities of accelerators and associated components used for doping of ICs and fabrication of semiconductor and optical materials. The continued development of photonic applications, which calls for increasingly sophisticated materials and structures, seems to offer many new applications of the skills learned in the doping of transistors with ion beams. The continued exploration of ion implantation applications for treatment of metal surfaces, selective doping of high-efficiency photovoltaic cells, and the rapidly expanding area of fabrication of specialized biomaterials promises to extend the use of accelerators in many forms in the future.

Acknowledgements

Preparation of this chapter was assisted by numerous beamline and source diagrams and illustrations provided by Varian Semiconductor Equipment Associates (VSEA), Nissin Ion Equipment, Axcelis, Advanced Ion Beam Technology (AIBT), Applied Materials, and the SEN division of Sumitomo Heavy Industries. Their support is gratefully acknowledged.

References

1. E. J. H. Collart, *et al.*, in *Proc. 16th Int. Conf. on Ion Implantation Technology (IIT 06)*, AIP Conf. Proc. Vol. 866 (American Institute of Physics, Melville, 2006), p. 37.
2. T. Ghani *et al.*, in *IEDM Technical Digest* (2003), pp. 978–980.
3. A. Vanderpool and M. Taylor, *Nucl. Instr. Meth. Phys. Res.* **B237**, 142 (2005).
4. Y. Liu, *et al.*, in *Symp. on VLSI Circuits Technical Digest, 2007* (IEEE, New York, 2007), pp. 44–45, doi:10.1109/VLSIT.2007.4339720.
5. A. Li-Fatou *et al.*, *ECS Trans.* **11**(6), 125 (2007).
6. K. Robertson *et al.*, *Nucl. Instr. Meth. Phys. Res.* **B55**, 555 (1991).
7. J. Blake and S. Richards, *14th Int. Conf. on Ion Implantation Technology Proc. (IIT 02)*, IEEE Cat. No. 02EX505 (IEEE, New York, 2002), p. 391.
8. D. K. Sadana and M. I. Current, in *Ion Implantation Science and Technology, 2000 Edition*, Ed. J. F. Ziegler (Ion Implantation Technology, Chester, 2000), p. 341.
9. M. Bruel, *Electron. Lett.* **31**, 1201 (1995).
10. M. I. Current *et al.*, *Electrochemical Soc. Proc. SOI Technology & Devices X*, **PV-2001-3**, 75 (2001).
11. http://www.soitec.com.
12. *Integration of Heterogeneous Thin-film Materials and Devices*, MRS Symp. Proc. Vol. 768 (Materials Research Society, Pittsburgh, 2003).
13. A. Renau, in *Proc. 16th Int. Conf. on Ion Implantation Technology (IIT 06)*, AIP Conf. Proc. Vol. 866 (American Institute of Physics, Melville, 2006), p. 345.
14. C. Lowrie *et al.*, in *Proc. Int. Conf. on Ion Implantation Technology (IIT 96)*, IEEE Cat. No. 96TH8182 (IEEE, New York, 1996), p. 447.
15. G. Redinbo *et al.*, in *Proc. 16th Int. Conf. on Ion Implantation Technology (IIT 06)*, AIP Conf. Proc. Vol. 866 (American Institute of Physics, Melville, 2006), p. 614.

16. N. White *et al.*, in *Proc. 16th Int. Conf. on Ion Implantation Technology (IIT 06)*, AIP Conf. Proc. Vol. 866 (American Institute of Physics, Melville, 2006), p. 335.
17. http://www.vsea.com/products.nsf/docs/viista3000xp.
18. N. R. White *et al.*, in *Proc. 17th Int. Conf. on Ion Implantation Technology (IIT 08)*, AIP Conf. Proc. Vol. 1066 (American Institute of Physics, Melville, 2008), p. 277.
19. S. Satoh *et al.*, in *Proc. 17th Int. Conf. on Ion Implantation Technology (IIT 08)*, AIP Conf. Proc. Vol. 1066 (American Institute of Physics, Melville, 2008), p. 273.
20. E. Winder *et al.*, in *Proc. 16th Int. Conf. on Ion Implantation Technology (IIT 06)*, AIP Conf. Proc. Vol. 866 (American Institute of Physics, Melville, 2006), p. 511.
21. Y. Inouchi *et al.*, in *Proc. 17th Int. Conf. on Ion Implantation Technology (IIT 08)*, AIP Conf. Proc. Vol. 1066 (American Institute of Physics, Melville, 2008), p. 316.
22. C. McKenna *et al.*, in *Proc. 16th Int. Conf. on Ion Implantation Technology (IIT 06)*, AIP Conf. Proc. Vol. 866 (American Institute of Physics, Melville, 2006), p. 622.
23. K. Tokiguchi *et al.*, in *Proc. 14th Int. Conf. on Ion Implantation Technology (IIT 02)*, IEEE Cat. No. 02EX505 (IEEE, New York, 2002), p. 629.
24. D. K. Sadana and M. I. Current, in *Ion Implantation Science and Technology, 2006 Edition*, Ed. J. F. Ziegler, (Ion Implantation Technology, Chester, 2006), 8-1 to 8-28.
25. I. G. Brown, Ed., *The Physics and Technology of Ion Sources* (Wiley–VCH, Weinheim, 1989).
26. I. G. Brown, Ed., *The Physics and Technology of Ion Sources, 2nd Edition* (Wiley–VCH, Weinheim, 2004).
27. J. H. Freeman, *Radiation Effects 100*, 161 (1986).
28. M. Farley *et al.*, in *The Physics and Technology of Ion Sources, 2nd Edition*, Ed. I. G. Brown (Wiley–VCH, Weinheim, 2004), pp. 133–161.
29. N. Sakudo, in *The Physics and Technology of Ion Sources, 2nd Edition*, Ed. I. G. Brown (Wiley–VCH, Weinheim, 2004), pp. 177–201.
30. D. Adams *et al.*, in *Proc. 16th Int. Conf. on Ion Implantation Technology (IIT 06)*, AIP Conf. Proc. Vol. 866 (American Institute of Physics, Melville, 2006), p. 178.
31. T. N. Horsky, in *Proc. 16th Int. Conf. on Ion Implantation Technology (IIT 06)*, AIP Conf. Proc. Vol. 866 (American Institute of Physics, Melville, 2006), p. 159.
32. K. J. Hill and R. S. Nelson, *Nucl. Instr. Meth.* **38**, 15 (1965).
33. M. Sugitani *et al.*, in *Proc. 17th Int. Conf. on Ion Implantation Technology (IIT 08)*, AIP Conf. Proc. Vol. 1066 (American Institute of Physics, Melville, 2008), p. 292.

34. H. F. Glavish *et al.*, in *Proc. 16th Int. Conf. on Ion Implantation Technology (IIT 06)*, AIP Conf. Proc. Vol. 866 (American Institute of Physics, Melville, 2006), p. 167.
35. S. Umisedo *et al.*, in *Proc. 17th Int. Conf. on Ion Implantation Technology (IIT 08)*, AIP Conf. Proc. Vol. 1066 (American Institute of Physics, Melville, 2008), p. 296.
36. L. Pipes *et al., Nucl. Instr. Meth. Phys. Res.* **B237**, 330 (2005).
37. R. Tieger *et al.*, in *Proc. 17th Int. Conf. on Ion Implantation Technology (IIT 08)*, AIP Conf. Proc. Vol. 1066 (American Institute of Physics, Melville, 2008), p. 336.
38. I. Yamada, in *Ion Implantation Science and Technology, 2008 Edition*, Ed. J. F. Ziegler (Ion Implantation Technology, Chester, 2008), 14–1 to 14–62.
39. V. Moroz, http://www.avsusergroups.org/joint_pdfs/2011-2moroz.pdf.
40. A. Rohatgi and D. Meier, *Photovoltaic International* **10** (2010).
41. Varian/Solion patent: US 2010/0041176A1 (Feb. 2010).
42. F. Henley *et al.*, in *Proc. 23rd European Photovoltaic Conf.*, 2BO.2.3 (2008).
43. J. F. Ziegler *et al.*, *SRIM, the stopping and range of ions in matter*, ISBN-13: 978-0-965407-1-6 (2008). SRIM program is downloadable from: http://www.srim.org.
44. A. Anders, Ed., *Handbook of Plasma Immersion Ion Implantation and Deposition* (Wiley & Sons, New York, 2000).
45. Available at http://microserf.lanl.gov/bpw/psii.html. ©Copyright 2011 Los Alamos National Security, LLC. All rights reserved.
46. E. Oks and I. G. Brown, in *Physics and Technology of Ion Sources, 2nd Edition*, Ed. I. G. Brown (Wiley–VCH, Weinheim, 2004), pp. 257–284.
47. I. G. Brown, *J. Vac. Sci. Technol.* **A11**(4), 1480 (1993).
48. X. Liu, K. Y. Fu and P. K. Chu, in *Biomaterials Fabrication and Processing Handbook*, Eds. P. K. Chu and X. Liu (CRC Press, Boca Raton, 2008), pp. 573–631.
49. P. K. Chu, in *Trends in Biomaterials Research*, Ed. P. L. Pannone (Nova Science Publishers, 2007), pp. 81–108.
50. S. C. H. Kwok and P. K. Chu, in *Biomaterials and Surface Modification-2007*, Eds. P. K. Chu and X. Liu (Research Signpost, 2007), pp. 99–136.

Chapter 2

ELECTRON BEAM MATERIALS PROCESSING

Donald E. Powers
PTR-Precision Technologies, Inc.
120 Post Road, Enfield, CT 06082-5625, USA
dpowers@ptreb.com

In electron beam processing, a well-defined beam of relatively energetic electrons produced by a high voltage acceleration gap is used to transmit thermal energy into a material in a precise manner. This controlled deposition of heat is employed in a wide variety of industrial applications for precision cutting, drilling, and welding of materials as well as annealing, glazing, and surface hardening. This chapter will describe the equipment used and the most prominent industrial applications for this process.

1. Introduction

The use of an electron beam (EB) to perform material processing tasks began in 1907 when an electron beam was initially used and patented by Pirani as a heat source for melting refractory metals under vacuum.[1] While development of EB applications continued, the general employment of electron beams for commercial material processing tasks was not really demonstrated as a truly viable industrial tool until some 50 years later. By the 1950s, the two main supporting technologies of the process (vacuum engineering and electron optics) had reached a high level of technical maturity and provided the basis for construction of viable EB processing equipment. Work on the development of equipment by Steigerwald in Germany[2,3] and Stohr in France,[4] first published in 1957, led to the birth of this industrial technique, with work then taking place in several other countries and a number of commercial companies being formed to provide EB systems.

Prior to the end of the 1950s, use of the process as an industrial tool was introduced into the US by the purchase of an EB unit from Germany by Westinghouse's Bettis Plant. This was in response to articles published by the French Atomic Energy Commission regarding their success in utilizing the EB process to join components used in the nuclear energy industry.[5] Interest grew quickly in the US, and by the end of the decade, numerous EB processing symposia, seminars, and conferences were being held regularly.[6] These meetings provided attendees with a continuous source of information from both US and foreign users and manufacturers of early EB processing equipment. Concurrently, a number of suppliers of this equipment began exhibiting at various Welding and Vacuum Technology Shows held throughout the US. The combination of these meetings and shows led to the growing interest in the US in EB processing and its rapidly expanding use in industry.[7]

The continuing evolution of supporting technologies, coupled with a continuing enhancement of the electron beam's application and beam power delivery capabilities, provides today's users of this technique in industries such as aerospace, automotive, nuclear, machinery fabrication, medical equipment, and many others with a powerful tool for performing a wide range of material processing tasks. These thermal processing applications span the full spectrum of an electron beam's output characteristics, from very shallow (surface-type) tasks attainable with low power density beams to very deep (volumetric-type) tasks attainable with high power density beams as detailed in Table 1.[8] Although a wide range of material processing capabilities using EB technology currently exists,[9] this chapter provides the reader with an overview of its most prevalent industrial uses in welding, drilling, and heat treating tasks. EB melting and casting will also be briefly

Table 1. Outline of the wide range of EB capabilities in thermal processing applications.

Application	Power density (W/cm^2)	Beam power (kW)	Spot size (mm)	Beam voltage (kV)
Surface Processing				
Heat Treat Processes: Hardening, glazing, etc.	10^2–10^3	0.1–1	5×10^{-4}–30	20–150
Volumetric Processing				
Welding: High vacuum, partial vacuum and non-vacuum	10^5–10^7	0.1–100	0.3–3	30–300
Machining Processes: Cutting and drilling	10^7–10^9	0.1–10	0.03–1	100–200
Melting and Evaporation: Vacuum refinement and material coating	10^3–10^5	1–2000	3–50	10–50

described. The use of electron beam irradiators for non-thermal materials processing applications is covered in Chapter 3 of this book.

2. Electron Beam Equipment

The basic components of a typical EB processing device are illustrated in Fig. 1. An electron gun (consisting of a cathode, grid, and anode) is used to generate and accelerate the primary beam. A magnetic optical (focusing and deflection) system is used for controlling the way in which the electron beam impinges on the material being processed (the "workpiece"). In operation, the gun cathode (an emitter that can be either directly or indirectly heated) is the source of thermally-emitted electrons that are both accelerated and

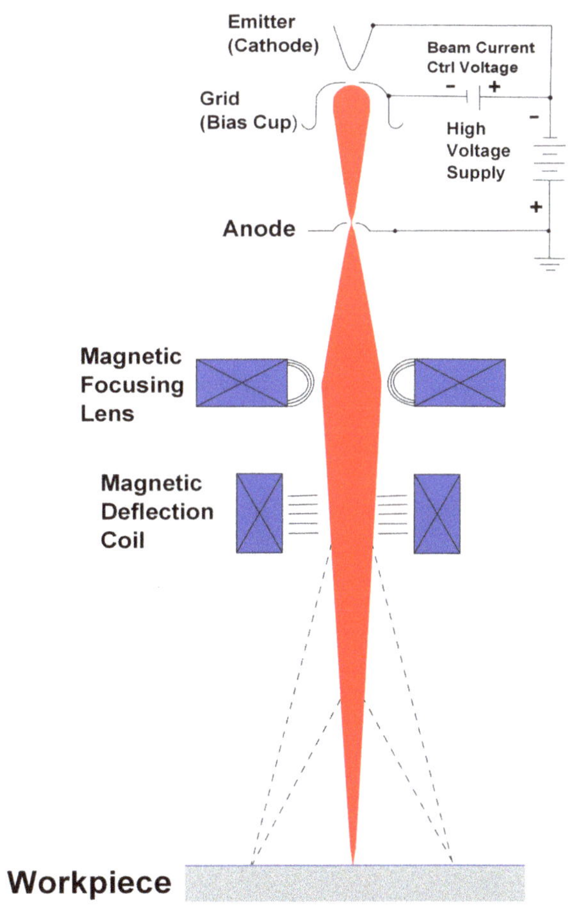

Fig. 1. Schematic showing basic components and operation of EB materials processing systems.

shaped into a collimated beam by the electrostatic field geometry established by the gun electrode (grid and anode) configuration used. The electron beam then emerges from the gun assembly through an exit hole in the ground-plane anode with an energy equal to the value of the negative high voltage (gun operating voltage) being applied to the cathode. This use of a direct high voltage to produce a high energy electron beam allows the conversion of input ac power to beam power at greater than 95% efficiency, making EB material processing a highly energy-efficient technique. After exiting the gun, the beam passes through an electromagnetic lens and deflection coil system. The lens is used for producing either a focused or defocused beam spot on the workpiece, while the deflection coil is used to either position the beam spot on a stationary location or provide some form of oscillatory motion.

As indicated in Table 1, the electron gun operating voltage depends on the processing method being performed and can vary over a broad range. With the triode gun arrangement shown in Fig. 1 (the form most commonly used in EB welding, drilling, and heat treating), the magnitude of electron flow, or operating beam current, can be easily adjusted (i.e., turned ON/OFF or raised/lowered) by simply varying the negative potential applied as a differential between the cathode and grid. This voltage differential can thus be used to pulse the output current of the gun instead of operating at a continuous output.

In addition to the triode gun described above, diode style guns are often employed. The type of gun chosen generally depends on the particular application, i.e., welding, drilling, melting, etc. In contrast to the triode gun, a diode gun employs a dual electrode (cathode & anode only) arrangement. This is accomplished by tying the cathode (heated filament emitting the electrons) and the field-shaping electrode surrounding it to a single voltage source. Thus with a diode gun, the beam output current has to be ramped up to and down from the desired operating level by varying the heating power supplied to the filament. For either style gun, the emitter (cathode/filament) may be of the directly heated wire or ribbon variety, or it may be of the disk/rod variety that requires some form of indirect heating, e.g., by electron bombardment or other means of inducing heat.

For EB welding, drilling, or heat treating tasks, gun operating voltages in the range of 30 to 300 kV are generally employed with beam currents ranging from less than 1 mA to ~1000 mA. Consequently, present day users of these processes can obtain EB systems capable of producing beam output power levels (the product of gun operating voltage V and beam current I) that span a wide range from less than 1 kW to ~100 kW.

The effect of changing the gun's electrode shapes and spacing on the output beam can be precisely predicted using modern digital computer simulations and plotting. Today's advanced computer technology and modern advances in beam deflection technology also provide present day EB equipment users and suppliers with the ability to both quantify and define the geometric characteristics of the final beam being delivered to the workpiece.[10] For example, there are devices now available that rapidly scan an electron beam across some form of aperture shielding an electrically isolated current collector, and then analyze and plot (by means of specially developed software programs) the collected data to provide a definitive power density distribution profile of the beam being delivered.[11]

Major present-day manufacturers of electron beam equipment include Steigerwald Strahltechnik and PTR–Praezisionstechnik (Germany), Sciaky and PTR-Precision Technologies (USA), Cambridge Vacuum Engineering (CVE, UK), Techmeta (France), Nippon Electric (NEC) and Mitsubishi Electric (MELCO) (both in Japan), and Paton Welding Institute (PWI, Russia). PTR–US, PTR–Germany, and Steigerwald are all members of the GBT group (Global Beam Technologies, formerly All Welding Technologies). It is estimated that roughly 7000 EB welding, drilling, and heat treating units have been manufactured and delivered worldwide over the last 50 years, with some 4000 still in operation today. Over 1000 of these are located in the US.

3. Electron Beam Welding

First used as a research tool in the early 20th century, the joining of materials with an electron beam was accepted by the nuclear industry in the 1950s and soon afterward by the aircraft and aerospace industry as a viable manufacturing process.[12] These industries adopted electron beam welding (EBW) primarily for the improved quality and reliability it provided for manufactured parts. In comparison to other more conventional methods of welding, a primary advantage of EBW is its ability to perform what is referred to as the "keyhole" method of joining.[13] When an energetic electron beam strikes a metal surface, the energy is converted into heat that instantly vaporizes the metal (most refractory metals melt and boil at temperatures up to 5000 °C), forming a vapor channel as the beam penetrates into the workpiece. As the beam advances, molten material flows around this "keyhole" and solidifies behind it to form the weld joint, as illustrated in Fig. 2. This process is capable of producing an extremely deep weld with a very high depth-to-width aspect ratio up to 25:1 (as will be seen in Fig. 7).

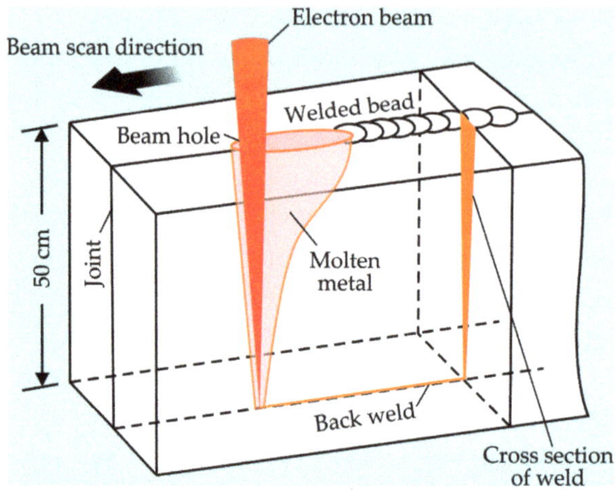

Fig. 2. A schematic illustration of the keyhole welding process. Welds depths of 30 cm have actually been achieved, but 50 cm depths as depicted in the drawing are thought to be possible. (Drawing courtesy of Elliott Plotkin, Physics Today, based on original drawing from PTR.)

Another key advantage of the highly-controlled deposition of energy in EBW is the ability to weld dissimilar materials and parts of two very different masses. In addition, its employment in a vacuum environment makes the process particularly useful for many reactive and refractory metals where contamination during the welding process causes damage in the material. Since EBW is a high power density process that produces extremely localized heating, it can be used to precisely join materials of quite different melting temperatures as well as minimize the extent of distortion and shrinkage that might be introduced by the welding process. EBW also provides users with the versatility of being able to easily adjust the heat input delivered to the workpiece, either by lowering the beam current or by slightly enlarging (defocusing) the beam spot on the workpiece. Thus by utilizing readily available process variables, either keyhole (deep welding) style or conductive (shallow arc variety) type welds can be produced.

A primary disadvantage of EBW is the capital equipment cost, which has significantly increased since the process was introduced some 50 years ago. This has been primarily due to the numerous advancements made in the technology, with such features as solid-state and computer-style controls, high beam power capacity, enhanced beam deflection, and diagnostic capability being fairly standard on present units. Consequently, whereas the relatively small size EB welders of the early 1960s (less than 3 kW and total manual operation) might have been purchased for ~US\$100 k or less, by the late

1990s a relatively large size unit (power output in the range of 40–60 kW with PLC/CNC controls and numerous optional capabilities) could cost ~US$1 M or more.

As with all present-day sophisticated machine tool purchases, acquisition of an EB welder today requires justification on the basis of economic viability. Thus, EBW equipment suppliers now offer their customers a large choice of systems, both in terms of price and capability, ranging from less sophisticated (moderately priced) units to highly specialized (more expensive) models.[12]

Initially the use of electron beams for welding required a high vacuum environment (pressure of $\leq 10^{-4}$ Torr) in both the beam generation region and the area where the beam is applied. Although high vacuum conditions are still required for generating the beam, the EBW process can now be performed in either a high vacuum (HVEBW), partial vacuum (PVEBW), or non-vacuum (NVEBW) mode of operation as illustrated in Fig. 3.[14]

The main markets for EBW when it first became an industrial tool, namely the aeronautical and energy industries, were primarily interested in the high degree of weld quality attainable from the process when applied under high vacuum conditions rather than the rate at which it could produce parts. However, as more production-orientated industries (e.g., automotive

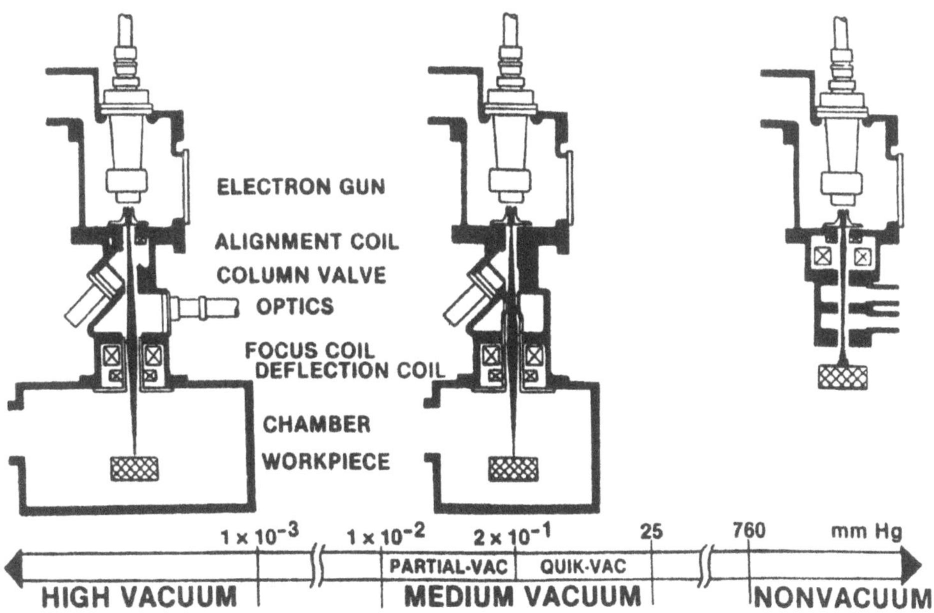

Fig. 3. A schematic illustration of the different modes of electron beam welding usage.

parts and tool manufacturers) became aware of the advantages EBW might offer them, methods to increase part throughput (i.e., production rate) became more important. Initially, "batch loading" techniques (methods for putting more than one part at a time in a chamber per pump-down cycle) were tried, but the "dead times" (i.e., non-productive times) associated with chamber pump-down times of 3–15 min needed to achieve high vacuum (depending on chamber size) were still unacceptable in extremely high-volume production industries such as those involved with manufacturing automotive parts.

Consequently, the partial vacuum (so-called "soft" or "medium" vacuum) and non-vacuum (i.e., atmospheric) EBW methods were developed. In PVEBW, the vacuum chamber is sized to fit the part to be welded (along with any required part tooling), and then it is pumped down to a vacuum level of only slightly below 10^{-1} Torr as opposed to the 10^{-4} Torr (or less) level required for HVEBW.[15] Pump-down times for PVEBW are on the order of 5 to 15 s — a substantial fraction of the times associated with HVEBW. Subsequently NVEBW was developed, whereby a dynamic differential pumping scheme allowed the electron beam (generated under high vacuum conditions) to be transported through a series of intermediate vacuum stages and applied directly in atmosphere thereby eliminating work chamber pump-down times entirely.[16]

EBW has its highest degree of flexibility in the relatively good vacuum environment ($\leq 10^{-1}$ Torr) utilized for HVEBW and PVEBW. In such a vacuum, the beam can easily traverse over long distances (in excess of 0.5 m) and still be focused to a highly intense beam spot on the workpiece. The focal spot formed on the workpiece can then be easily electromagnetically deflected in either a static or dynamic pattern at very high speeds, as illustrated in Fig. 4.[17] Figure 5 illustrates the types of weld modifications that can be produced using a dynamic beam deflection pattern during welding.[18] For this figure, the beam current values were varied to produce welds of similar depth, and the top views of the weld patterns have been enlarged for clarity and are not exactly to scale with the weld profiles shown on the bottom.

When used directly in atmosphere or not in a reasonably good vacuum, the electron beam's overall versatility is more limited. This is due to a beam-broadening effect resulting from collisions between the electrons in the beam and the gas molecules present in the higher pressure regions in the beam's path. The primary consequence of this effect is that it appreciably limits the standoff distance that can be utilized and the extent of penetration that can be achieved when employing the NVEBW mode.[19,20] However, there are a

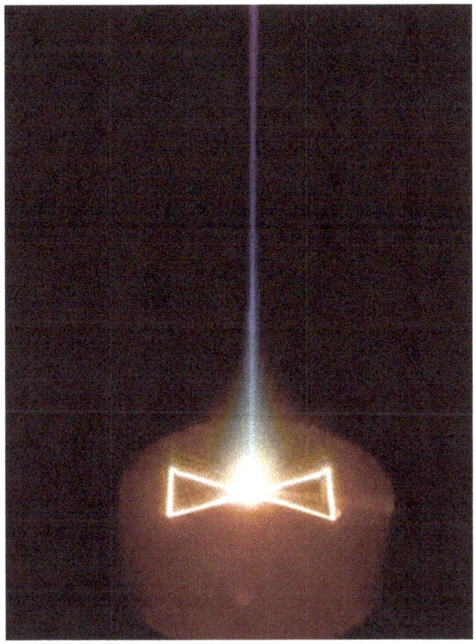

Fig. 4. Illustration of an electron beam's ability to be dynamically deflected.

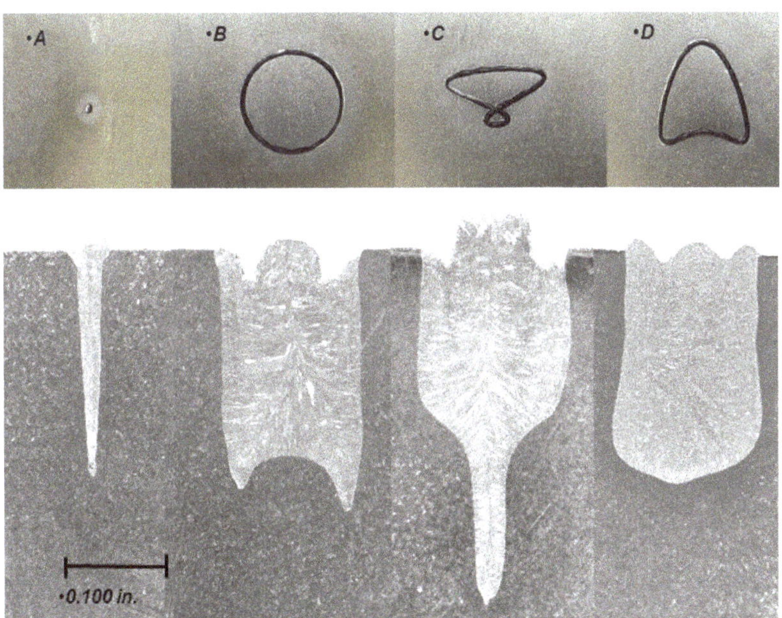

Fig. 5. Electron beam oscillation patterns (top) and their associated weld bead profiles (bottom). A: no oscillation; B: circular; C: "lazy-8"; D: parabolic.

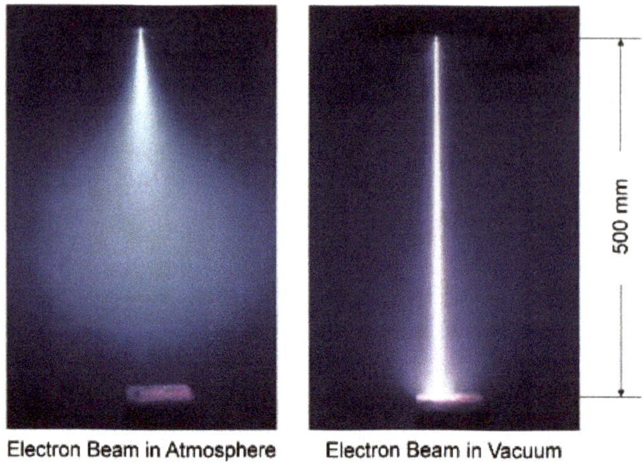

Fig. 6. Beam glow profiles produced by an electron beam in atmosphere and in vacuum.

number of applications for EBW in industry that can be efficiently carried out in a rapidly-achieved partial vacuum or in atmosphere.

Electron beams in vacuum are not visible to the human eye. However, as the ambient pressure increases, a "beam glow" is produced from the collisions between electrons and gas molecules, resulting in the subsequent excitation of the ambient gas molecules as shown in Fig. 6.[21] This figure illustrates the effect of beam broadening with increasing pressure, an effect that limits the distance over which the beam can effectively be focused into a high density beam spot in poor or no vacuum conditions. Several examples that illustrate the present day usage of EBW under both vacuum and non-vacuum conditions are described below.

3.1. *Large steam turbines*

A high vacuum, high power EBW system built specifically for processing very large workpieces was installed over two decades ago at GELSTG (GE Large Steam Turbine Generators, now GE Power Systems), a major manufacturer of large steam turbine generators, and is still welding those workpieces today. This unit has a weld chamber with a total volume of approximately 36 000 liters, an electron beam generation system capable of producing 60 kW of beam output power, and a work motion system that can handle workpieces up to 3 m in diameter and 10 000 kg in weight.

Figure 7 shows the profile of a portion of a weld made with this system.[22] Thick-section carbon and low-alloy steel material segments were joined with a roughly 130 mm deep weld and a depth-to-width aspect ratio of ~20.

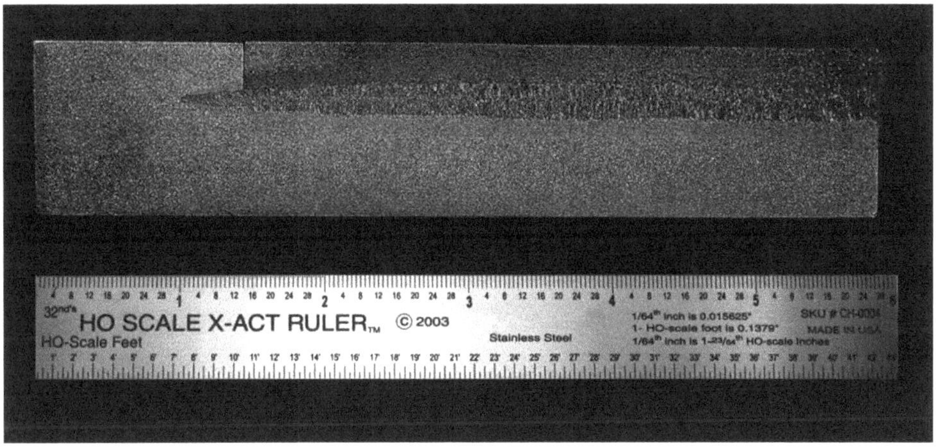

Fig. 7. Photograph illustrating an electron beam's deep, high aspect ratio weld capability.

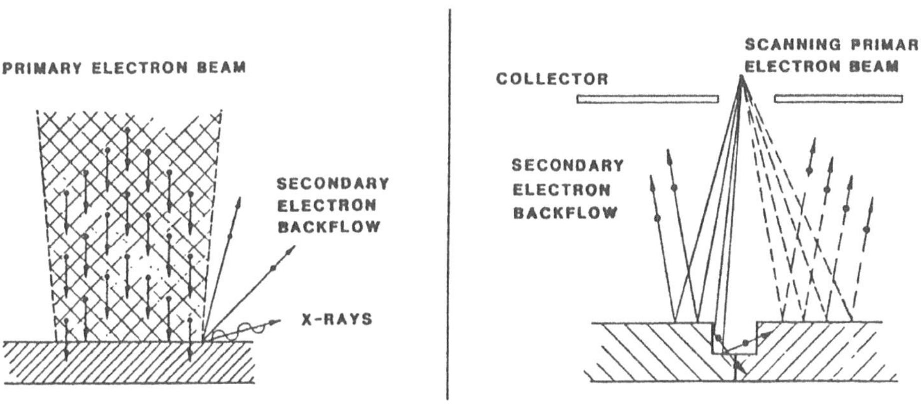

Fig. 8. SEES (Secondary Electron Emission Sensing) seam tracking principle.

This was done with ~54 kW of beam power and a weld speed on the order of 0.25 m/min. Each 2.8 m diameter circumferential weld produced requires about 45 min of continuous welding. During this time, the part's motion (x, y, and rotary) and the beam's operating parameters (V, I, focus, etc.) are all automatically controlled in a continuous interpolating fashion by the system's CNC-based controller.

As outlined in Ref. 22, due to the extent of the weld (~9 m long) performed at the beam power and weld speeds outlined above, a seam tracking system called SEES (Secondary Electron Emission Sensing) was provided on this machine. Figure 8 is a schematic representation of how this SEES system works. The system monitors the secondary electron emission produced by momentarily (in a matter of only 1 or 2 ms) scanning the electron beam side-to-side in front of the weld impingement point and monitoring the signal produced to provide a direct correlation between the position of the beam impingement point and the upcoming seam path's location. When this information is fed to the CNC control unit, it ensures that the beam stays properly aligned to the weld seam regardless of any part distortion resulting from thermal effects during the 45 min weld time. In addition, a circular weld-spot oscillatory motion is provided during welding to both help ensure that any seam gap present is entirely covered by the beam impingement spot and to provide a "stirring" effect to the weld melt zone, as a means of reducing gas entrapment pockets that could result in the formation of voids in the final weld during the solidification phase.[23]

3.2. *High efficiency impellers*

Another example of the application of HVEBW is shown in Fig. 9.[24] In this application, EBW is used to simultaneously weld and braze. Figure 9(a) illustrates this hybrid process for which the term EBraze was coined a decade ago by Dresser–Rand Corporation, the company that first utilized it. They are still using the process today to join impellers in high efficiency turbo-compressors.

An EB tee-weld is used with a special shim inserted between the faying surfaces of the tee-joint formed by the cover and impeller blade. In this operation, the beam simultaneously produces a welded and a brazed joint — one where the brazed segment helps to eliminate any joint gap that could subsequently precipitate a fatigue failure during operation. The variety of complex, multi-bladed impeller parts to be EBrazed necessitates a fairly large EBW system with 5 axes of CNC capacity to control part motion in conjunction with an even greater number of beam parameters. A system supplied for this application is shown in Fig. 9(b).

3.3. *Speed gears*

Figure 10 is an example of a partial vacuum EB welding (PVEBW) application where a gear and synchronizer ring have been EB welded under partial vacuum conditions to provide a finished speed gear. Also shown in this figure is a magnified cross section of the ∼6 mm weld. This cross section clearly shows that the PVEBW method can produce the relatively narrow, parallel-sided weld needed for such a part while minimizing part distortion during welding. In this and other similar applications, it is common practice to

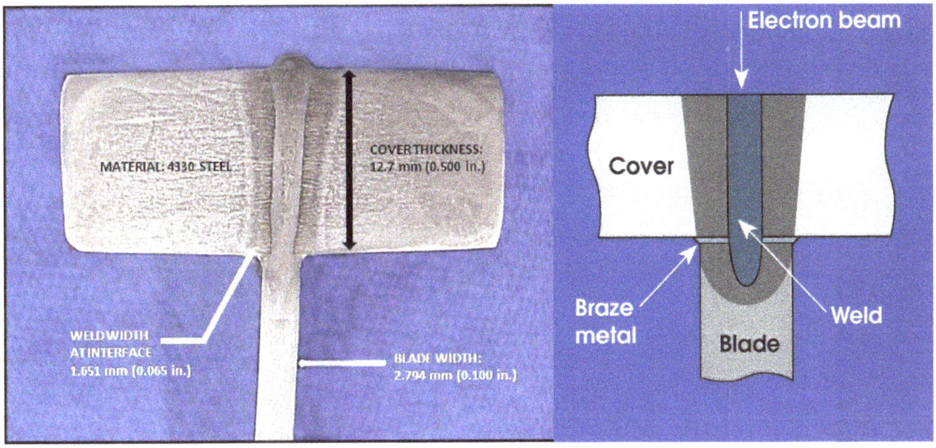

(a)

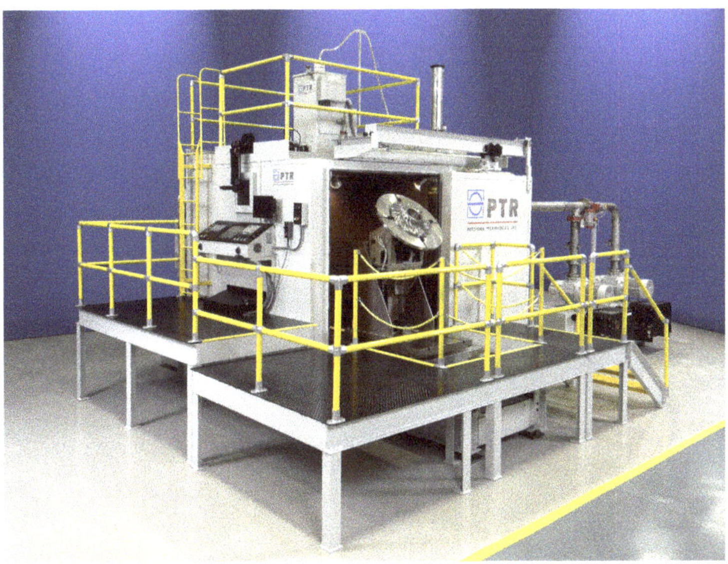

(b)

Fig. 9. (a) Thick cover to thin blade EBraze tee-joint weld profile; (b) large chamber, multi-function HVEBW unit.

utilize a sharply focused beam with some form of dynamic beam deflection in order to enhance the overall characteristics of the final weld produced.

The significant production advantage of the PVEBW method is that it provides users with the ability to minimize the total time needed to produce a welded part.[26] By using a weld chamber sized to accommodate a

Fig. 10. Photographs showing a finished welded speed gear and enlarged view of the final weld produced.[25]

single part (or depending on the part size, possibly a few parts), the small chamber volumes are quickly (in 3–6 s) evacuated to the partial vacuum level desired — as opposed to the longer (on the order of minutes) pump-down times required for HVEBW. Once the required partial vacuum level is achieved, the beam is turned on, and the workpiece is either moved under the beam or the beam is moved over the workpiece to complete the weld task.[18] Operating in this fashion allows parts like the speed gear to be welded at a production rate of ~200 parts/h. PVEBW is an ideal joining means for industries seeking to produce parts requiring vacuum-type EB welds at high part-throughput rates.

Figure 11 shows a typical PVEBW unit: a two-station load lock unit capable of processing three parts per cycle. Finished parts are unloaded and new parts loaded in one station. The loading station is then pre-pumped to the desired welding vacuum level while the parts in the other station are going through the welding process.

3.4. *Drive rings*

The non-vacuum EB welding (NVEBW) method provides another means of achieving high part-throughput EB welds. Since the welding is performed at atmospheric pressure, the time-consuming process of evacuating a work chamber is eliminated. In this method, the electron beam is generated in a high vacuum region and then passes through a series of differentially-pumped stages that are at successively higher pressures and separated by apertures,

Fig. 11. Photograph of a PVEBW unit.[25]

until it reaches the workpiece at atmospheric pressure. Due to the type of beam diffusion that results when applying the EB in atmosphere (as exhibited earlier in Fig. 6), the workpiece must be fairly accessible to the limited area of travel over which the beam is still able to provide a high enough power density to accomplish welding. Consequently, NVEBW is generally only used when the exit nozzle (the point at which the beam leaves vacuum and enters atmosphere) can be positioned so that the standoff distance (the distance from exit nozzle to top surface of joint) is in the range of 6–25 mm. Even though it restricts the use of NVEBW to this limited standoff distance, the beam diffusion effect provides the advantage of being able to accommodate joint fit-up tolerances that are far less stringent than those required for in-vacuum EB or out-of-vacuum LB (laser beam) welding.[27]

Figure 12 shows an NVEBW weld task that involves joining a drive ring onto the turbine bowl assembly of an automotive torque converter. An enlarged view of the lap weld being produced is also shown. For this application, the beam is directed onto the joint at a 45° angle, using a standoff distance of ∼12 mm, and welding is accomplished at weld speeds on the order 80 mm/s. This allows the welds on these ∼300 mm diameter parts to be completed in roughly 11 s, providing the capacity to cyclically produce these large diameter parts at a rate of ∼200 parts/h.

Figure 13 shows a non-vacuum EB performing a similar type of weld utilizing the beam at a 45° angle at a standoff distance of ∼12 mm.[19] This figure provides a fairly good view of the non-vacuum EB weld process being

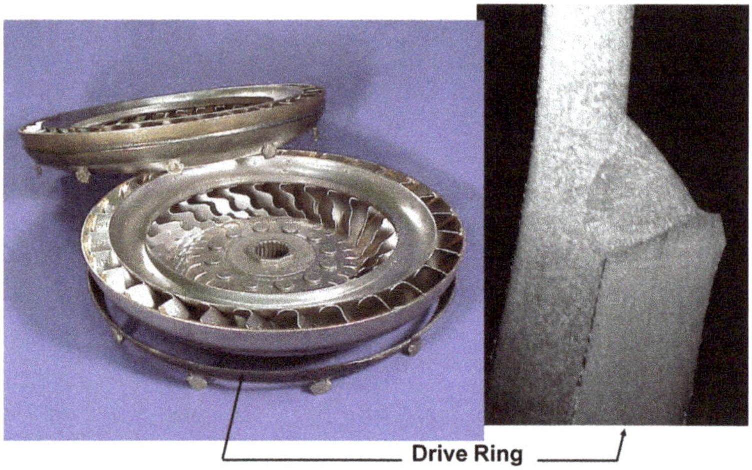

Fig. 12. Non-vacuum EB-welded drive ring and weld being produced.[25]

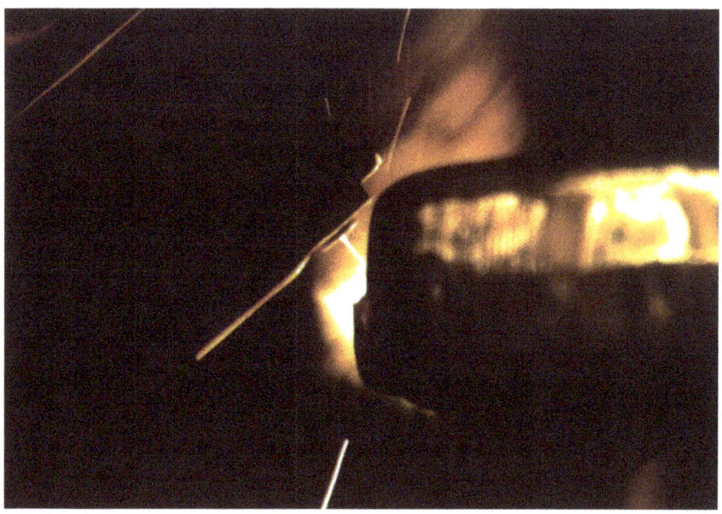

Fig. 13. Non-vacuum EB weld in the process of being applied.

applied. Figure 14 is a photograph of a NVEBW unit used for this type of out-of-vacuum EB welding application.[28] As seen in Fig. 14, tooling consists of a 4-station rotary index table totally surrounded by a radiation-tight lead enclosure with a single opening at the operator's station. In operation, the operator will unload a welded part and load a new part through that opening, and then push the palm-buttons. This action causes the index table to rotate one position, moving the previously loaded part into the

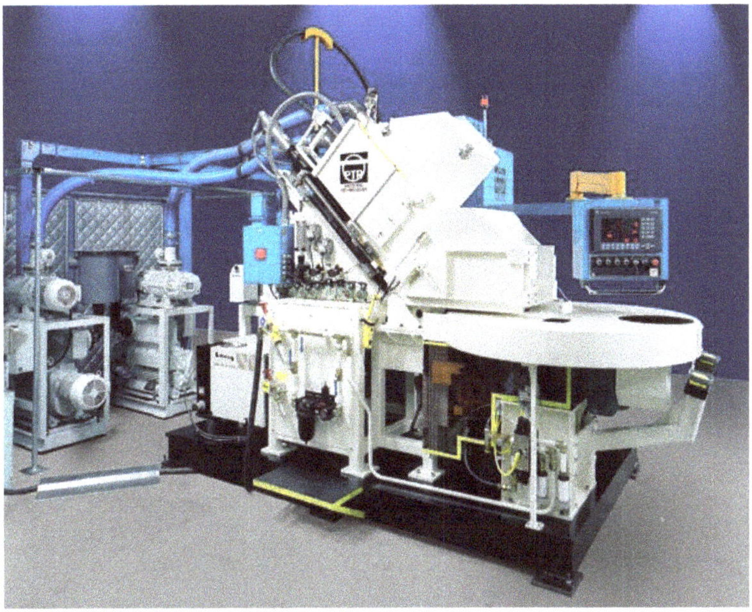

Fig. 14. Photograph of a typical NVEBW welding system.

#2 weld station where the weld occurs and the part just loaded into the #1 dwell station. Simultaneously, the part just welded moves into the #3 dwell station, and the previously welded part moves to the #4 station (the operator station) for unloading. Each cycle lasts approximately 14 s (10 s for welding, during which time the operator unloads/loads parts at station #4, and 4 s to transfer the next part to be welded into position), thus providing a capacity for welding ~250 parts/h.

Figure 15 shows a more complex-shaped part that is presently produced using the NVEBW process.[20] This part involves a 3D weld path that requires two axes (x and y) of part motion and one axis (z) of beam generator motion be applied simultaneously. These ~1 m-long aluminum crossbar parts for automobiles require two joint-welds per part. Welding is performed at speeds in excess of 10 m/min, providing a finished-welded part throughput rate of >100 parts/h. Because of the size of the part being welded, the complexity of the tooling used to hold it, and the requirement that both part and beam generator be moved to accomplish the two welds, this application is performed in a roughly 3 m^3 room-style, radiation-tight enclosure employing automated parts-transfer devices to manipulate the part through each weld cycle.

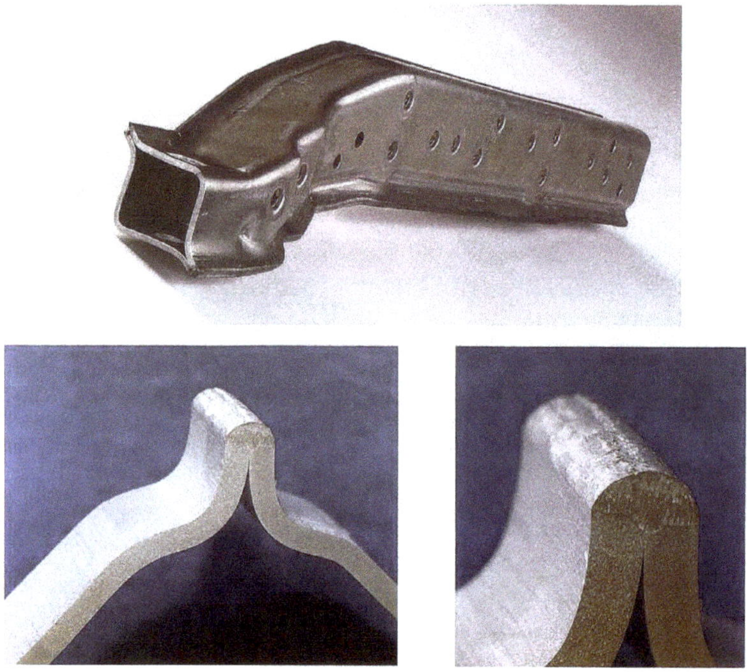

Fig. 15. Non-vacuum EB welded aluminum crossbar piece.

4. EB Cutting and Drilling

An electron beam's versatility for use in vacuum as a machining (EBM) tool lies in its capacity for full parameter control and precisely repeatable results.[29] By adjusting the beam power density parameters (beam current, voltage, and focused spot size), an extremely intense EB spot approximately two orders of magnitude greater in intensity than that utilized in EBW can be produced at the workpiece. When this high beam density (10^7–10^9 W/cm^2) is applied to a material surface, the intense local energy delivery causes the material at the beam's point of impact with the workpiece to be constantly removed via a thermal erosion process. This process can then be used for cutting or drilling holes in the material.

EB cutting is limited in use because there are more economical techniques available (laser, water jet, or plasma arc cutting). However, EB drilling (EBD) is a more widespread technique in which the deposition of the electron beam's energy is used to form a hole, as schematically shown in Fig. 16(a). This figure also illustrates a key technique used in this process that ensures consistency in the diameter of the holes being drilled. A backing material capable of producing a large volume of gas upon interaction with the beam

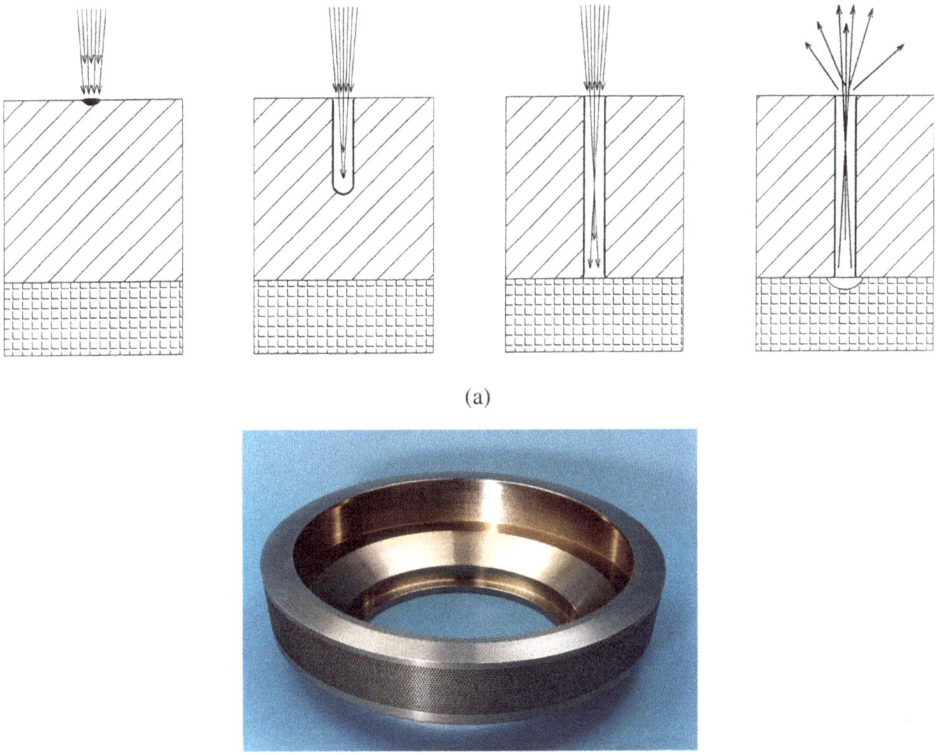

(a)

(b)

Fig. 16. (a) Schematic representation of EB drilling process; (b) photograph of a part containing ~25 000 EB-drilled holes.[25]

is placed on the material to be drilled. This gas then expands up through the hole to eject any residual molten material from it. By utilizing the capability of modern day CNC control systems to quickly interpolate programmed part positioning and beam operating level commands, as well as synchronize the part motion with the beam output current level, the drilling of each hole can be done extremely rapidly. The beam is then momentarily (in a matter of only tenths of milliseconds) switched off while the part is repositioned to the location of the next hole, and then switched back on in a pulsing manner. This cycle is repeated literally "on the fly" numerous times until the entire hole pattern is completed.[30]

Although many material drilling applications are now accomplished under atmospheric pressure conditions using a laser, EBD's flexibility, high quality holes, and faster drilling rates make it more suitable for use in certain drilling tasks. This is due to the ability of EBD to precisely control both the shape (height, width, and slope rate) and frequency of each repetitive energy pulse being delivered, as well as the capability to simultaneously control both the motion of the workpiece (x, y, and rotary) and the operational parameters of the beam (pulsing, spot location via deflection, etc.). These features greatly enhance the ability of the process to dramatically increase the rate at which exact size holes (depth & profile) can be accurately and reliably produced.

Figure 16(b) is an example of EBD's production drilling capability. The spinner shown in the figure is used in the manufacture of fiberglass insulation and contains ∼25 000 EB drilled holes, all 0.55 mm in diameter. These holes are serially produced in a specific pattern at rates on the order of hundreds of holes/s under CNC control that simultaneously governs both the beam pulsing and high-speed deflection functions in conjunction with the workpiece motion commands. Much higher rates in the range of thousands of holes/s can be achieved when producing smaller diameter holes and using thinner materials.

In addition to the "on-the-fly" drilling speeds and precise drilling/positioning tolerances possible with EBD, it offers other advantages such as eliminating tool wear associated with mechanical drilling. Even so, EBD is a highly specialized technique and the number of units in operation is only a few percent of the total number of industrial EB processing units in use today.

5. EB Heat Treating

Because EB processing is a fast, efficient, and highly versatile means for delivering energy to a workpiece, it is often utilized for a number of thermal surface-type industrial processes.[31] However, like EBD, EB heat treating is a very small segment of the EB processing market with the number of units in operation being only a few percent of the total.

In EB heat treating applications, the beam is used as the source of heat energy that initiates a physical change to the surface characteristics. In one industrial application, the beam is used for surface hardening, either the conventional rapid heating and quenching phase-transformation-type hardening or a rapid melting and re-solidifying "glazing" type hardening.

For conventional phase-transformation hardening, a rapidly scanning (either continuous or pulsed) beam is used to locally heat a specific area to a temperature above the transformation temperature (but below the melting point) of the workpiece. When the beam is extinguished, this heated area is quickly self-quenched by the still cool bulk portion of the workpiece, producing a hardened surface segment. For glazing, the beam's heat input is used to quickly melt a specific area of the workpiece surface that rapidly solidifies to form the desired hardening effect when the beam is extinguished.

With phase-transformation hardening, the level of hardness achieved depends on the percentage of carbon content in the treated material and the depth of hardness attained by adjustment of the beam's energy input and interaction time in a highly controllable fashion. With glazing, the beam input energy and reaction time are both carefully controlled to rapidly melt and re-solidify a thin layer of the workpiece surface, forming a "metallic-glass"-like surface layer with an extremely high hardness value. It can be utilized for surface hardening of materials with a carbon or non-carbon content.

In both cases, the EB can quickly deposit the exact amount of energy needed to heat the treated area to the exact temperature required in a very precisely controlled fashion. Both methods require some form of vacuum environment, either the same order of magnitude as that required for beam generation or one that has a higher ambient pressure level than is required for generating the beam.

Figure 17 depicts an EB being used to simultaneously heat treat several lobes on a cam shaft by employing both high-speed and multi-axis beam

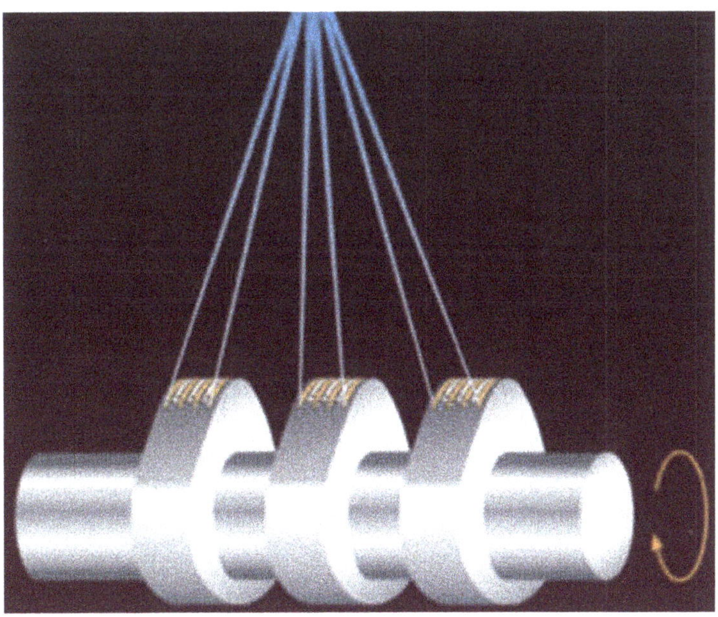

Fig. 17. Simultaneous EB heat treatment of several lobes on a cam shaft.

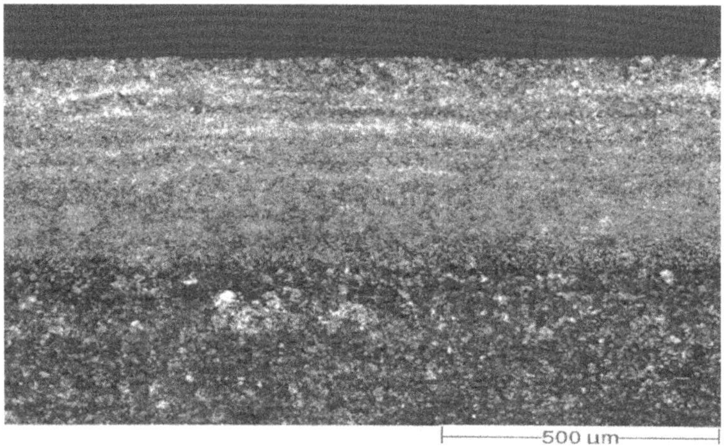

Fig. 18. Magnified view of an EB hardened area profile.

deflection control in a combined fashion.[32] Figure 18 shows a micrograph profiling the type of hard martensite surface that is achieved by the application of the EB transformation hardening process.[33]

6. EB Melting and Casting

The final segment of the industrial use of EB thermal processing described in this chapter is EB melting and casting. While many other areas of EB applications in industrial processing exist (such as cross-linking of plastics and rubber, curing of coatings, food irradiation, and sterilization of medical products), most of them are non-thermal in nature and are described in Chapter 3. Irradiation processes rely on chemical and physical changes in the material caused by the electrons themselves, not on thermal deposition.

Even though the industrial use of EB melting and casting is not as widespread as the other EB thermal applications discussed in this chapter, this area does merit some degree of mention because it employs the highest power industrial electron beams used in industry.[34] First used during the early 1900s to melt titanium, EB melting utilizes EB guns designed specifically for low voltage (on the order of 25 kV), high amperage (on the order of amps) operation with a capacity for delivering beam powers in the range of 0.75–1.25 MW per gun to the part being processed. Applied in a controlled vacuum environment, this high amount of thermal energy provides a melting and refining operation that permits distillation of volatile elements, including vaporization of suboxides, evolution and removal of gasses, and vaporization of metallic impurities with vapor pressures above that of the

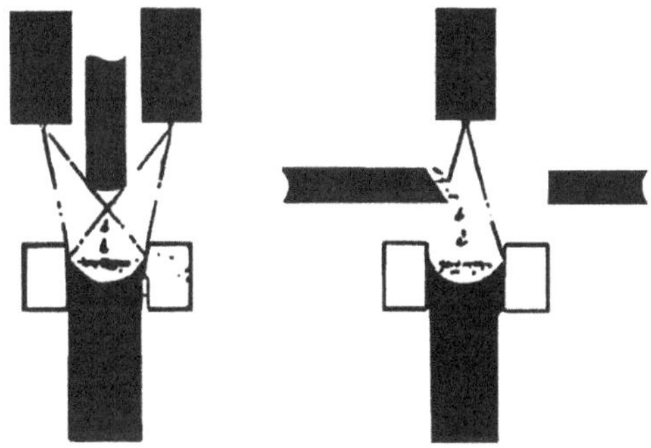

(a) (b)

Fig. 19. Representation of different electron beam drip melting techniques.

processed material. In all cases, the processed material is cast in water-cooled copper containers (i.e., cold molds) to prevent contamination of the ingot from occurring.[35]

Two basic operational modes prevail in electron beam melting technology today: drip melting and hearth melting. Figure 19 schematically illustrates two methods of employing the drip melting technique.[36] Figure 19(a) is classic drip melting using a center vertical material feed. Two electron guns are shown, but furnaces utilizing this concept may utilize from one to six guns. Figure 19(b) shows another option for the drip melting mode: a single electron gun with side-feeding of the stock to be melted. In all drip melting processes, the ingot is formed in a water-cooled mold to assure that the melted metal remains free from mold contamination. The drip melting technique has been and will remain the most prevalent mode used in the processing of refractive and reactive metals.

In the hearth melting mode, the metal is melted in a water-cooled hearth (again no contamination) before casting in a cold mold. In addition to the removal of gases and higher volatility impurities, hearth melting removes heavier-than-melt inclusions by allowing these to settle to the bottom of the hearth. Lighter components either melt or float to the top for removal prior to the melted material flowing into the water-cooled mold. Figure 20 provides a schematic illustration of the hearth melting mode and shows multiple guns being employed as a means for not only providing the melt but also for controlling it in both the water-cooled hearth and mold.[36]

There are numerous equipment providers (e.g., ALD Vacuum Technologies, von Ardenne Anlagentechnik, and Verdi Metallurgical Corporation) and industrial users (e.g., energy, medical, aerospace, and optical coating)

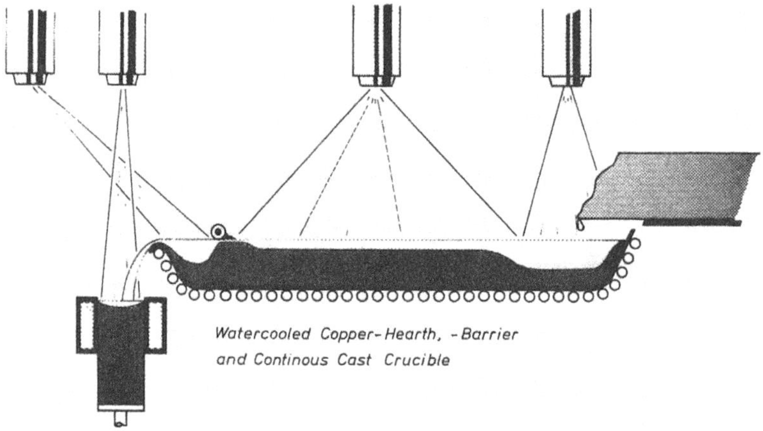

Fig. 20. Representation of a multi-gun hearth melting/casting process.

of the two EB melting modes worldwide. It is estimated that the EB melting furnaces presently in operation throughout the world have a sum total melting power capacity that is well in excess of 35 MW.

7. Summary and Future Trends

The industrial use of electron beams for welding, drilling, and heat treating applications began in the 1950s and continues today on a worldwide basis. Starting as a relatively unique laboratory tool and initially of interest to only the fairly technical nuclear and aeronautical industries, EB processing has steadily grown to become a highly sophisticated production tool used by almost all segments of industry today.[37] Among the many innovative new applications, several with significant industrial potential are worth mentioning for future trends: additive manufacturing, surface texturing, simulated multi-beam operation, and large-area electron beam surface machining.

Additive manufacturing (originally called EB free form fabrication and rapid prototyping/manufacturing) is a relatively new process that relies on EB melting. In this technique, a 3-axis computer-controlled table is used to move a base or a part under an electron beam that is melting a wire of material being fed into it, as depicted in Fig. 21. The melted material can then be precisely deposited to produce a three dimensional part. What makes this application particularly interesting is that the US NASA is considering it for lengthy manned space missions (e.g., to Mars) as a means for astronauts to rapidly produce parts they might need.[38]

EB surface texturing is a relatively new technique that utilizes high bandwidth beam deflection to sequentially produce many "beamlets" that are

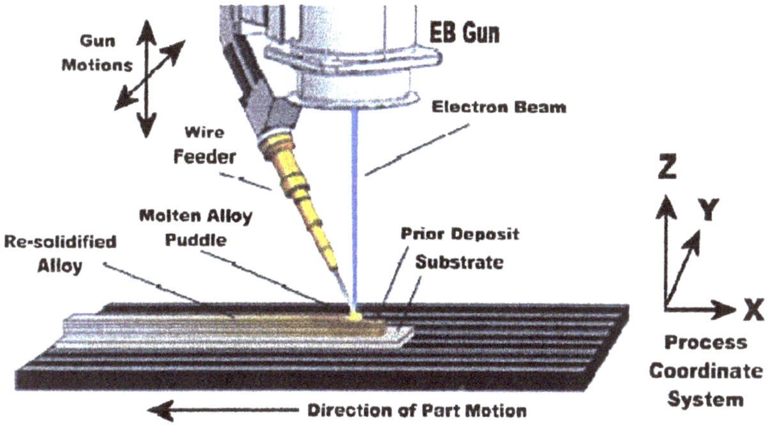

Fig. 21. A simplified view of the EB additive manufacturing process.

Fig. 22. Scanning electron microscope image of a surface produced by an EB surface texturing process.

then used to rapidly blast away material to form surface hills and valleys, as seen in Fig. 22, in order to provide an enhanced surface for such applications as lubricant retention or bonding.[39] The process is currently being used by several aircraft manufacturers to "glue" composite materials to Al or Ti aircraft fuselage structures.

Simulated multi-beam operation also uses high-speed beam deflection techniques to form many seemingly simultaneous beams, as depicted in

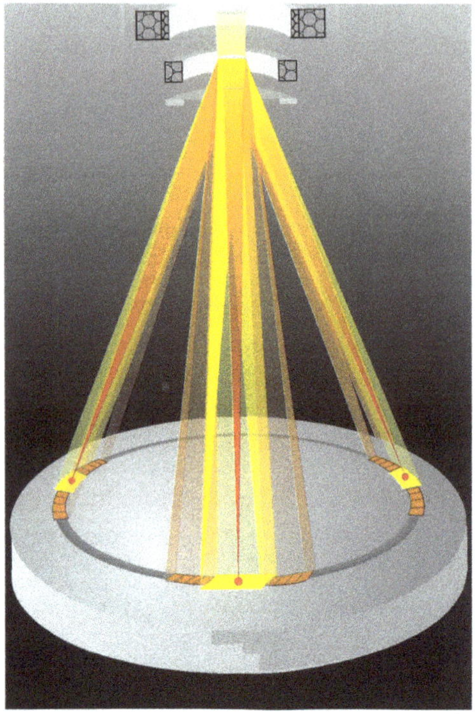

Fig. 23. Graphic illustration of an EB simulated multi-beam process.

Fig. 23, for welding applications. These can then be used for such purposes as performing pre- and post-weld treatment concurrently with welding, or for reducing part distortion by simultaneously welding at different locations.[40]

The use of plasma-generated large-area electron beams to provide a surface smoothing effect is a relatively new EB surface machining technique that has been investigated for the rapid final polishing of hard-to-polish metal surfaces such as cast titanium parts (like those used in removable dentures), hard die and tool steels, and the inside surfaces of irregular metal molds. Other uses include fast removal of burrs, milling marks, or defective layers from metal surfaces. Recent studies have shown that surface gloss and smoothness comparable to or better than conventional mechanical or hand polishing can be achieved in just a few minutes of irradiation time.[41–43] These studies also report that the EB-irradiated surfaces exhibit a much higher degree of corrosion resistance compared to non-irradiated parts. Unlike the traditional highly-focused electron beams that use a heated filament as the source of electrons (as described in this chapter), these large-area beams are formed by applying a low-voltage (~25 kV) pulsed negative bias to an inert gas plasma (usually argon) producing a Penning discharge effect that causes the electrons to move toward a ring-shaped anode. This then allows a high current, large-diameter beam of electrons to be extracted from the plasma source. Polishing of the irradiated parts is accomplished by the repeated rapid melting and re-solidifying of the surface due to heat and surface tension effects.

At least two companies (both in Japan) have developed industrial large-area EB units that generate short pulses of high energy-density electron beams up to 6 cm in diameter to irradiate workpieces placed in a vacuum chamber. An optional 5-axis control mechanism is offered to handle complex parts. Indications are that as many as 100 of these units have been delivered worldwide.

In summary, the ability of electron beam thermal processing to perform tasks of the very shallow surface nature to the very deep penetration type makes it a unique and invaluable tool in a broad spectrum of industrial applications. Continued development of equipment, along with advances in beam deflection and part-handling techniques, is expected to lead to even more widespread use of electron beam processing well into the future.

Acknowledgements

The author would like to thank Mary Cosgrove, Dale Butrymowicz, and Derek Mayers of PTR-Precision Technologies, Inc., all of whom assisted with the preparation of this chapter.

References

1. M. von Pirani, *US Pat. No. 848,600* (1907).
2. H. B. Cary and S. C. Helzer, *Modern Welding Tech.* (Pearson Education, 2005), p. 202.
3. K. H. Steigerwald, *Physikalesche Verhandlungen* **4**, 123 (1953).
4. H. A. James, *J. Vac. Sci. & Tech.* **7**(6), 539 (1970).
5. J. A. Stohr, in *Nov. 1957 Fuel Element Conf., Paris, Proc*, TID 7456, Book J (US Atomic Energy Commission, 0–17, 1958).
6. *Proc. Alloyd Electron Beam Symposiums, 1959–1963* (Alloyd Electronics Corp., Cambridge, MA).
7. G. Purtan *et al.*, *American Machinist* **23**, 95 (1959).
8. D. E. Powers, *AWS High Energy Beam Welding & Materials Processing Conf. Proceedings*, 187 (1992).
9. S. Schiller, U. Heisig and S. Panzer, *Electron Beam Technology* (Wiley & Sons, New York, 1982).
10. G. R. LaFlamme and D. E. Powers, *Welding Journal* **70**(10), 33 (1991).

11. J. W. Elmer, A. T. Teruya and D. W. O'Brien, *Welding Journal* **72**(11), 493s (1993).
12. D. E. Powers, in *Proc. 1st Int. Electron Beam Conf. (IEBW)*, Session 3, Paper 2 (International Institute of Welding, 2009), on CD available from iiwelding.org.
13. Y. Arata, *Development of Ultra High Energy Density Heat Sources and their Application to Heat Processing* (Osaka University JWRI Publication, 1984).
14. D. E. Powers, in *Proc. Int. Power Beam Conf.*, 8804-6 (ASM International, OH, 1988).
15. F. Samuelson, in *OSU Electron Beam Welding Symp. Proc.*, 157 (Ohio State University, 1966).
16. E. Gajdusek, *Welding Journal* **59**(7), 17 (1980).
17. *Recommended Practices for Electron Beam Welding* (American Welding Society, Miami, 2004), Doc. C7.1M/C7.1:2004.
18. G. Schubert and D. E. Powers, *Practical Welding Today* **32** (July/August 2003).
19. K.-R. Schulze and D. E. Powers, in *Colloquium on Welding and Melting by Electron and Laser Beams, CISFFEL 6 Proc.* (French Atomic Energy Commission Institute for Welding, France, 1998).
20. G. Schubert and D. E. Powers, in *Proc. Global Powertrain Congress (GPC2000)*, 115 (Detroit, 2000).
21. D. E. Powers, *Welding Journal* **86**(12), 32 (2007).
22. M. J. Carroll and D. E. Powers, *Welding Journal* **64**(8), 34 (1985).
23. R. Mayer, W. Dietrich and D. Sundermeyer, *Welding Journal* **53**(6), 35 (1977).
24. G. LaFlamme, J. Rugh, S. MacWilliams and R. Hendryx, *Welding Journal* **85**(1), 44 (2006).
25. *AWS Welding Handbook, 9th Ed.*, Vol. 3, Chapter 13 (American Welding Society, Miami, 2007).
26. J. F. Hinrichs, P. W. Ramsey, R. L. Ciofoni and T. M. Mustaleski, *Welding Journal* **53**(8), 488 (1974).
27. D. Powers and G. Schubert, in *Proc. Global Powertrain Congress (GPC2000)*, 110 (Detroit, 2000).
28. D. Powers *et al.*, in *Proc. IIW 54th Annual Assembly — Ljubljanu, Slovenia*, Paper IV-795-01 (International Institute of Welding, July 2001).
29. E. J. Weller, Ed., *Nontraditional Machining Processes, 2nd Edition* (Society of Manufacturing Engineers, Dearborn MI, 1984).
30. D. Dobeneck and A. Parella, in *Proc. Electrochem. Soc. 1974 San Francisco Mtg.* (Electrochemical Society, 1974).
31. D. E. Powers and W. Dietrich, in *1st Nat'l. Heat Treatment Conf. (Sydney, Australia) Proc.*, 44 (1984).
32. W. J. Farrell and J. D. Ferrario, *Welding Journal* **66**(10), 41 (1987).
33. J. E. Jenkins, *Metal Progress*, 38 (July 1981).
34. R. Bakish, *Industrial Heating*, 12 (August 1985).

35. W. Dietrich, in *Proc. 2nd Int. Congr. Int. Org. for Vac. Sci. & Tech., Washington DC* (American Vacuum Society, 1961).
36. R. Bakish, Ed., *Proc. Electron Beam Melting and Refining State of the Art Conferences, 1983–1987* (Bakish Materials Corp., Englewood NJ, 1983–1987).
37. D. E. Powers, *Welding Journal* **90**(1), 2 (2011).
38. K. M. B. Taminger, R. A. Hafley and D. L. Dicus, in *Proc. 2002 International. Conf. on Metal Powder Deposition for Rapid Manuf.*, 51–60 (Metal Powder Industries Federation, 2002).
39. F. Smith, *DVS Welding & Cutting*, Issue 4/2005, Specialist Article Section (2005).
40. M. Mücke and C. Scheiblich, in *Proc. 1st Int. Electron Beam Conf. (IEBC)*, Session 2, Paper 6 (International Institute of Welding, 2009), on CD available from iiwelding.org.
41. J. Tokunga *et al.*, *Dental Materials Journal* **28**(5), 571 (2009).
42. Y. Uno *et al.*, *J. Materials Processing Tech.* **187/188**, 77 (2007).
43. Z. Yu, Z. G. Wang, K. Yamazanki and S. Sano, *J. Materials Processing Tech.* **180**, 246 (2006).

Chapter 3

ELECTRON BEAM MATERIALS IRRADIATORS

Marshall R. Cleland
IBA Industrial, Inc.
151 Heartland Blvd., Edgewood, NY 11717, USA
Marshall.Cleland@iba-group.com

Radiation processing is a well established method for enhancing the properties of materials and commercial products by treating them with ionizing energy in the form of high-energy electrons, X-rays, and gamma rays. Beneficial effects include polymerizing, cross-linking, grafting and degrading plastics, sterilizing single-use medical devices, disinfecting and disinfesting fresh foods, purifying drinking water, treating wastewater and other toxic waste materials that harm the environment, and many other applications that are still being evaluated. Industrial electron accelerators of several types have been developed and are being used for these applications. More than 1800 electron accelerators are presently installed in facilities worldwide for these purposes.

1. Introduction

The irradiation of materials and commercial products with ionizing energy in the form of accelerated electrons and high energy X-rays emitted by energetic electrons, as well as with gamma rays emitted by radioactive nuclides, has been evolving for more than fifty years. These methods, usually called radiation processing, can modify the physical, chemical, or biological properties of materials. In many applications, these modifications can improve the utility and increase the value of commercial products. Some typical applications include:

- The conversion of liquid low-molecular-weight monomers into solid high-molecular-weight polymers to cure inks, coatings, and adhesives without using volatile solvents, to graft monomers onto solid polymers in order to modify their surface properties, and to make strong fiber-reinforced composite materials.
- The cross-linking of solid polymers to produce strong, tough plastic and elastomeric materials and products that will not melt at elevated temperatures or dissolve in organic solvents.

- The reduction in molecular weight of solid polymers, such as cellulose and polytetrafluoroethylene, to make them more useful for some applications.
- The sterilization of medical devices as an alternative to the use of heat or toxic gases.
- The disinfection or pasteurization of fresh foods to reduce the risk of food-borne diseases.
- The disinfestation of insects in fresh fruits and grains to preserve these food supplies and to avoid the exportation or importation of insects to or from other countries.
- The degradation of toxic compounds in waste materials to reduce their deleterious effects on the environment.

A few radiation effects on materials were discovered during the 1920s and 1930s, but serious interest in this field did not materialize until the late 1940s and early 1950s when high-intensity radiation sources were developed for research in the applications of atomic energy. The earliest industrial applications were for cross-linking of plastic materials and products, sterilization of medical devices, and disinfection of foods. Cross-linking applications have since evolved into the largest market for industrial electron accelerators. Curing (polymerizing and cross-linking) of inks, coatings, and adhesives developed later, as did the treatment of waste materials for environmental protection.

Many different types of electron accelerators have been developed to serve these industrial applications. These electron beam (EB) accelerators are reliable electrical equipment that produce ionizing radiation without radioactive isotopes — they can be turned on or off like other industrial equipment. Thus, EB units avoid the security, transportation, and disposal issues associated with long-lived gamma ray emitting radioisotopes such as cobalt-60 that are used mainly for medical device sterilization. Market surveys indicate that there are more than 1800 high-current EB units installed worldwide for industrial use, providing an estimated added value to numerous products in excess of 80 billion USD.[1–4] There are about nine times more industrial EB units in worldwide operation than gamma ray irradiators.[5] This number of commercial EB installations does not include the nearly 1000 low-current accelerators used for research purposes or the EB systems covered in Chapters 2 and 7.

The energy and beam current or beam power ratings for electron accelerators are each selected to satisfy the requirements of a particular irradiation application. The penetration of energetic electrons into materials increases with their kinetic energy, which can be as low as 75–80 keV or as high as

10–12 MeV for radiation processing applications. Lower energies are difficult to use because of the very short electron ranges in solids, liquids, and gases; higher energies are usually avoided because of the possibility of inducing radioactivity in some materials. The processing rate increases with the electron beam current, which may be as low as a few mA or as high as several hundred mA, and with the EB power, which may extend from a few kW to as high as 700 kW.

With the recent development of high-current, high-energy EB accelerators, conversion of electron beam power to X-radiation is now a commercially viable alternative to the industrial use of gamma rays.[6–9] X-rays are emitted when energetic electrons strike any kind of material and are deflected by the atomic nuclei in the material. The forward X-ray intensity increases with the EB current, the kinetic energy of the electrons, and the atomic number of the target material. In order to give sufficient intensity and penetration to compete with large gamma ray sources, industrial X-ray generators intended for radiation processing produce electrons with energies in the range of 3–7 MeV and beam powers in the range of 100–700 kW. X-ray energies higher than 7.5 MeV are not allowed by the United States Food and Drug Administration (USFDA) for food irradiation in order to avoid causing nuclear reactions in a tantalum or gold X-ray target and in the mineral constituents of the food.

Many publications on radiation chemistry, the practical applications of ionizing energy, and the technology of industrial electron accelerators are now available. Numerous papers on these topics have appeared in the 15 proceedings of the International Meetings on Radiation Processing (IMRP).[10] Extensive information on radiation chemistry and applications of radiation processing is also available in many books and patents.[11–24]

A comprehensive overview of this field is presented in a booklet to be published by the International Atomic Energy Agency (IAEA).[25] This chapter summarizes some of the information that is presented in more detail in the IAEA booklet (the present author being one of the primary authors of that work), as well as material from other papers written or presented by the author and his associates, as referenced.

2. Physical Properties of High-Energy Electrons and X-Rays

2.1. *High-energy electrons*

The depth of penetration, or range, of high-energy electrons in irradiated materials not only increases in proportion to their kinetic energy but also depends on the density and atomic composition of the material. This energy is transferred mainly through collisions with atomic electrons in the material. Materials with higher electron densities will have higher absorbed doses (defined as the energy deposited per unit mass; see Sec. 2.3) near the entrance surface but lower electron ranges. Normal hydrogen has more electrons per unit mass than any other element because of its lack of neutrons. Thus materials with a higher percentage of hydrogen content, such as polyethylene $(H_4C_2)_n$, polypropylene $(H_6C_3)_n$, and water (H_2O), will receive higher surface doses per unit electron fluence (defined as the total incident electron flux integrated over time) but will have shorter electron ranges than other plastic materials like polystyrene $(H_8C_8)_n$, polycarbonate $(H_{14}C_{16}O_3)_n$, and polytetrafluoroethylene $(F_4C_2)_n$.

At incident energies above 150 keV, the absorbed dose tends to rise with increasing depth to ~50% of the maximum electron range and then decrease to a relatively low value near the end of the range. This initial increase is due to the production of multiple secondary electrons by collisions with atomic electrons. Several useful depth quantities can be obtained from the depth-dose curves[26] as illustrated in Fig. 1:

(i) R_{opt} — the depth where the dose at the exit surface equals the dose D_e at the entrance surface.
(ii) R_{50} — the depth where the exit dose equals half of the maximum dose.

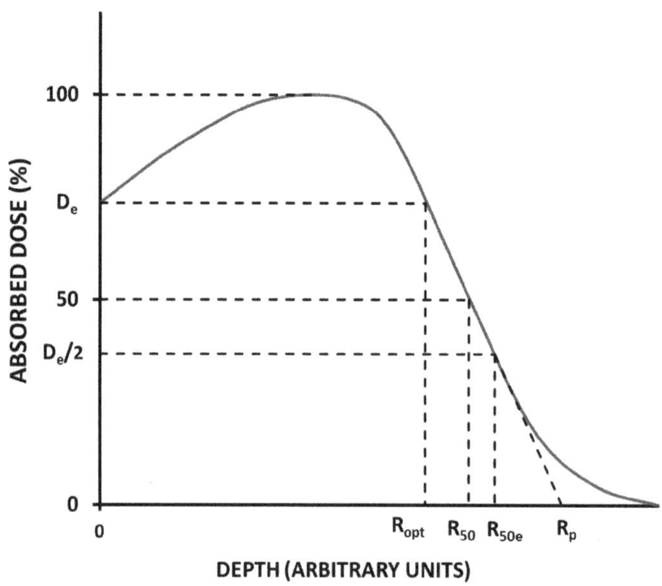

Fig. 1. An electron depth-dose curve showing various useful depth quantities.

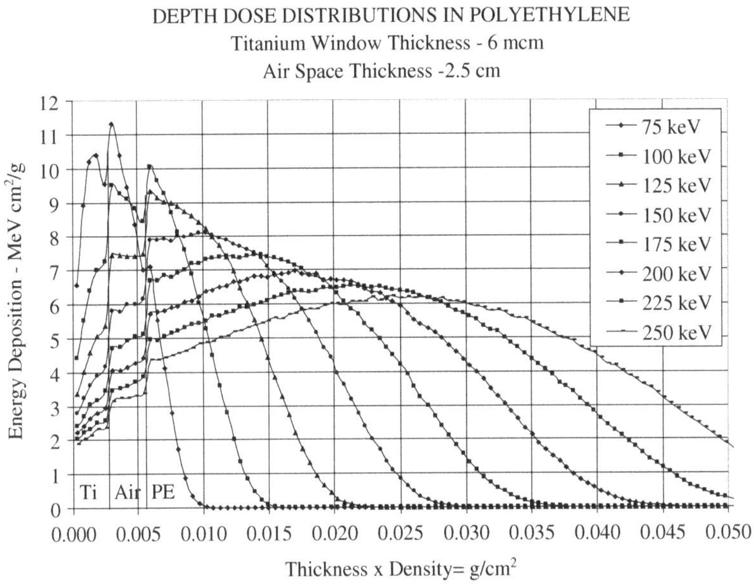

Fig. 2. Depth-dose curves in HDPE for electron beam energies from 75 to 250 keV showing the energy deposited in the beam window (Ti), the air space (Air), and the material (PE).

(iii) R_{50e} — the depth where the exit dose equals half of the entrance dose.
(iv) R_p — the depth where the tangent line at the inflection point of the decreasing depth-dose curve intersects the depth axis.

To illustrate the behavior of depth dose for low energy electron beams, Fig. 2 shows the theoretical depth-dose curves in high density polyethylene (HDPE) for beam energies from 75 to 250 keV. These curves were generated by Monte Carlo calculations using the one-dimensional TIGER code from the Integrated Tiger Series ITS3.[27] In reality, the electron beams are generated in a vacuum and first pass through a vacuum window and air, depositing some of their energy in each, before entering the material being irradiated. This must be taken into account in the calculations. In this example, the beam is assumed to pass through a titanium window with a thickness of only 6 μm and then through 2.5 cm of air, which gives the air a weight per unit area nearly the same as the beam window.

It can be seen that the penetration in the HDPE increases with increasing incident electron energy, and that the entrance dose and the maximum dose decrease with increasing electron energy. The decreases in the entrance and maximum doses are caused by decreases in the electron energy deposition, or stopping power, as the incident electron energy increases. Similar curves for

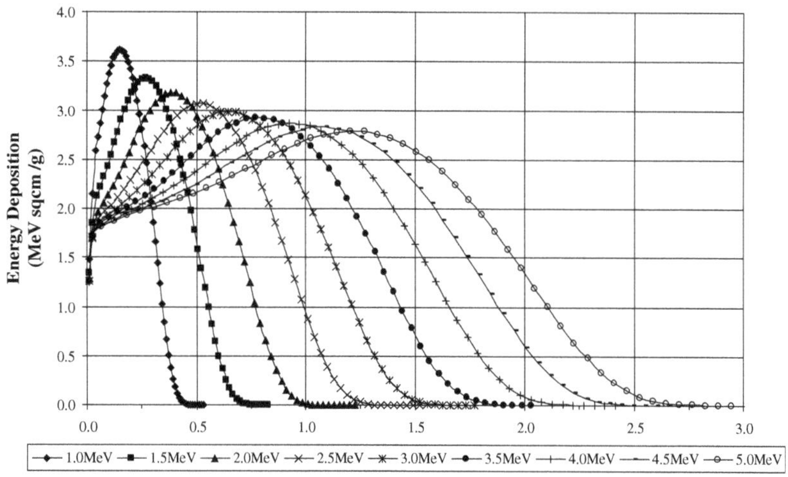

Fig. 3. Depth-dose distributions for electrons in water with beam energies from 1.0 to 5.0 MeV. The depth axis is in cm or g/cm^2.

polypropylene and other low-density plastics will be nearly the same because the units for the depth axis are given in the combined units of thickness in cm multiplied by the volume density in g/cm^3. The depth axis units are then g/cm^2 (i.e., weight per unit area). This quantity is also called the area or areal density. For materials with different volume densities, the depth values in cm can be obtained by dividing the areal density by the density of the material in gm/cm^3.

A similar set of theoretical depth-dose curves in water for higher electron energies from 1.0 to 5.0 MeV is shown in Fig. 3.[28] In this case, the thickness of the titanium beam window is assumed to be 40 μm and the air space is 15 cm, which are typical values for applications in this energy range. It is obvious from these calculations that the effect of the window and air on the entrance dose is far less than for lower energy beams. In these curves, the depth values in cm are the same as the weight per unit area in g/cm^2 because the density of water is assumed to be 1.0 g/cm^3.

In the extended energy range from ~1 to 10 MeV, electron beam penetration is often expressed in terms of the optimum depth parameter R_{opt} in material of unit density. The nearly linear relationship between this optimum thickness and the incident electron energy is shown in Fig. 4. A similar relationship exists for lower energies, but the optimum penetration is much less, ranging from ~10 to 400 microns for unit density materials.

When a flat sheet of material is irradiated from opposite sides, the thickness can be increased by a factor of ~2.4 over the thickness of

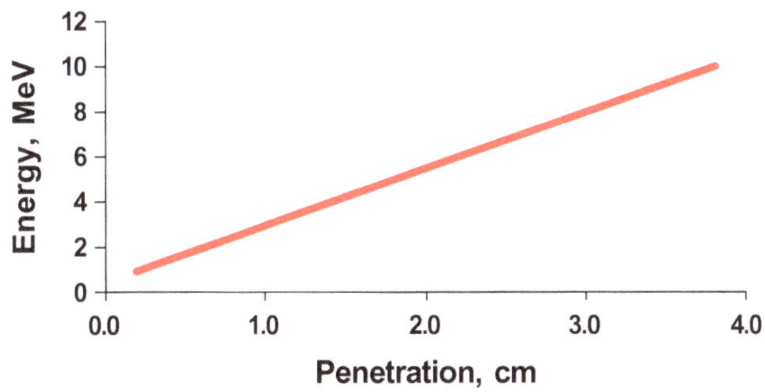

Fig. 4. Optimum electron beam penetration R_{opt} in cm of water (unit density) as a function of incident electron energy for mid- to high-energy beams.

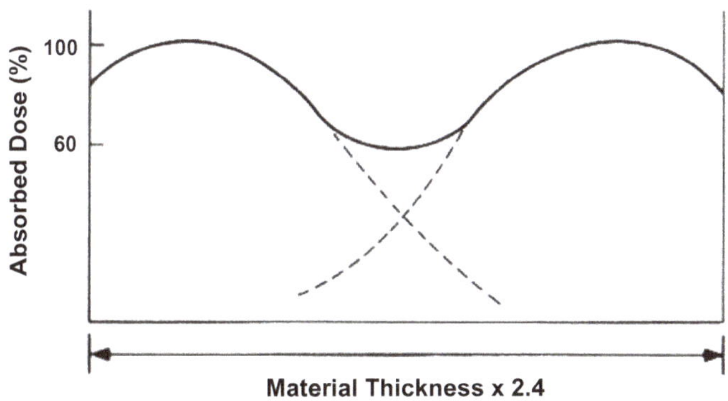

Fig. 5. Depth-dose distribution for EB irradiation from opposite sides of flat materials.

material irradiated from a single side because the decreasing portions of the depth-dose curves are then effectively combined in the middle of the sheet, as shown in Fig. 5.[15,16] This method is often used to sterilize large packages of low-density medical products.

2.2. *High-energy X-rays*

When an energetic electron strikes any material, electromagnetic radiation is emitted in the form of X-rays. Characteristic monoenergetic X-ray photons are emitted when atomic electrons are excited by the incident electrons, and bremsstrahlung photons are produced by the interaction of incident electrons with the nucleus of an atom. Because high-energy bremsstrahlung X-rays are a penetrating form of ionizing radiation, thick shielding is needed around high-energy X-ray irradiation equipment to prevent worker exposure to high levels of radiation. The required shielding thickness increases with the maximum X-ray energy, which equals the energy of the incident electrons. Low-energy EB equipment from 75–800 keV can be shielded with a combination of steel and lead panels. However, concerns over the costs and safety of lead can be alleviated by using thicker steel shielding without lead. Higher energy X-ray facilities are usually shielded using thick concrete walls.[1] Entrance labyrinths with three or four corners are designed to attenuate scattered X-rays from the irradiation area so that the radiation intensity outside of this area is negligible.[29]

In industrial X-ray generators, the X-rays are produced by striking a metal target with the electron beam from an EB accelerator. The forward X-ray intensity increases with the electron current, electron energy, and atomic number Z of the target material.[30–32] Tungsten is used in many X-ray generators for other applications (see Chapter 7), but water-cooled tantalum is the practical choice for large-area targets. The optimum X-ray yield is achieved when the target thickness is ~40% of the maximum electron range in the target material. The water cooling channel and stainless steel backing plate must add enough material to stop the incident electrons in the target assembly. The efficiency for converting incident electron power to X-ray power in the forward direction depends on accelerator energy and target Z-value. Efficiencies ranging from 8–12% have been calculated for electron energies between 5.0 and 7.5 MeV on a tantalum target with a cooling channel. The angular distribution of the X-rays is peaked in the forward direction, and the dose rate in the treated material can be controlled by a combination of beam current, product conveyor speed, and distance from the target. Very-high-power, high-energy electron accelerators (e.g., 300 kW at 5.0 MeV or 700 kW at 7.0 MeV) can compensate for the inefficiencies of X-ray conversion.[33,34]

Figure 6 illustrates the forward emission of 5.0 MeV X-rays from a tantalum target. This is very different from the isotropic emission of gamma ray sources and is a distinct advantage in the treatment of pallet loads of products. Figure 7 shows that X-ray penetration is much greater than the highest energy electrons (10 MeV) from industrial EB systems and comparable to the penetration of cobalt-60 gamma rays. X-ray dose rates are at least one order of magnitude higher than in a typical gamma ray facility but significantly less than EB dose rates. Product throughput rates for X-ray processing with high-power, high-energy EB equipment can exceed the EB throughput rates of low- to medium-power, high-energy (10 MeV) industrial linacs.[35] Thus, with the advent of very-high-current, high-energy accelerators, X-rays are now a viable alternative to gamma rays from radioactive

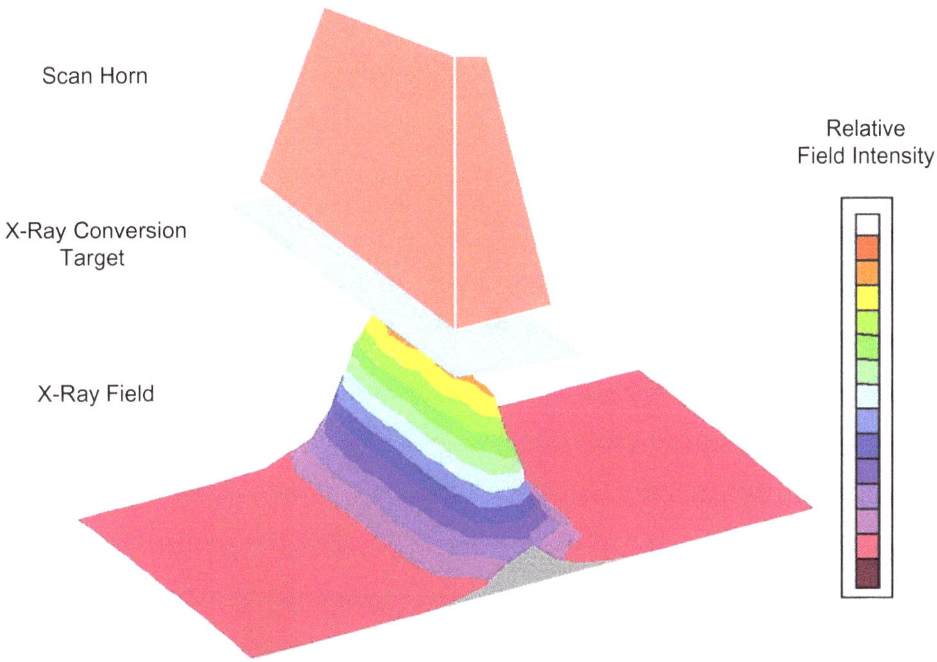

Fig. 6. Angular distribution of 5.0 MeV X-rays in a stack of plastic plates. (Diagram courtesy of IBA Industrial, Inc.)

isotopes in treatment facilities needing more than ~1.5–2 MCi of cobalt-60 for applications such as medical device sterilization and food irradiation.[6–8,36] The main advantages of X-ray processing for irradiation applications can be summarized as follows:

- Greater penetration for treating large volumes of products, such as pallet loads of packaged medical devices or food.
- Controllable dose rate that can facilitate dose-rate sensitive processes such as monomer polymerization.
- Low-temperature, non-thermal process that eliminates adverse heat-related effects on materials.

2.3. *Radiation dosimetry*

The beneficial effects in materials treated with ionizing radiation are related to the absorbed dose, which is defined as the energy absorbed per unit mass of material. The international unit of dose is the gray (Gy), which is defined as the absorption of one joule per kilogram (J/kg). A more common unit for

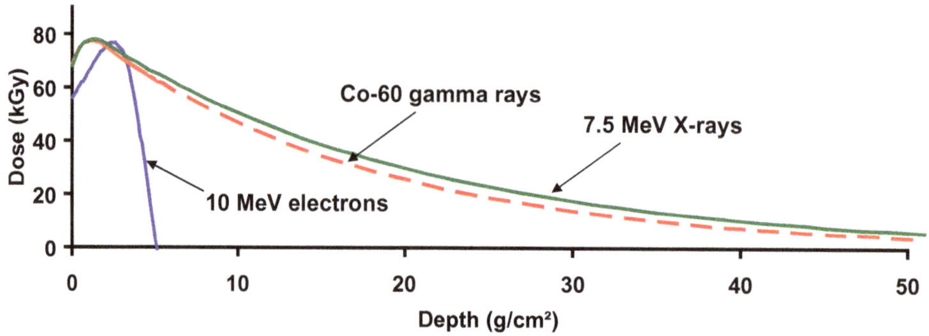

Fig. 7. Comparison of dose and penetration in water of high energy electrons, cobalt-60 gamma rays, and high energy X-rays.

most radiation processing applications is the kGy (1 kGy = 1 kJ/kg or 1 J/g). The gray has replaced the rad, which is defined as the absorption of 100 ergs per gram (or 10^{-5} J/g). Therefore, 100 rad is equivalent to 10^{-3} J/g, or 1.0 J/kg, or 1.0 Gy. Even though the rad is now obsolete, many irradiation processes are still specified in rad, krad, or Mrad, with 1.0 Mrad being equivalent to 10 kGy.

For doses above 1.0 kGy, the average absorbed dose can be measured by the temperature rise in the irradiated material using Eq. (1):

$$D_{ave} = c\,(T_2 - T_1), \tag{1}$$

where D_{ave} is the absorbed dose in kGy, T_2–T_1 is the temperature rise in °C, and c is the thermal capacity in J/g/°C. This relationship is used by standards laboratories to measure the dose in materials that have known thermal capacities such as water, graphite, or plastics.[37] Other dosimeters can then be calibrated by exposure to the same radiation source in the laboratory. This calibration procedure is usually done with gamma radiation from cobalt-60 or cesium-137 sources. The value of c for water at normal temperature and pressure is 1.0 calorie/g/°C or 4.18 J/g/°C. Therefore, a dose of 1.0 kGy increases the temperature by only 1.0/4.18 = 0.239°C.

Commercial irradiation facilities employ a variety of dose indicators that are more convenient to use than the thermal or calorimetry method described above. Among these may be several different kinds of plastic films containing dyes that change the optical density of the film after being irradiated. These films are called dosimeters, but unlike a calorimeter, they do not measure the dose directly. Another method utilized is the detection of a long-lived free radical in a small piece or thin film made of alanine, an amino acid, using an electron spin resonance (ESR) spectrometer. This procedure is more accurate and reproducible because alanine is less influenced by temperature, humidity, ambient ultraviolet radiation, or dose rate than the dyed films. However, an ESR spectrometer is much more expensive than an optical densitometer. A third method is detecting the unsaturated carbon double bonds in polyethylene film. These transvinylene segments are formed by hydrogen abstraction during the irradiation process. They can be detected using an infrared spectrometer, which is much less expensive than an ESR spectrometer. Polyethylene film also has the attractive feature of being less affected by extraneous conditions than the dyed films.[38]

2.4. *Dose versus electron beam power*

The absorbed dose defined above can be related to the emitted power of the radiation source, the treatment time, and the mass of the irradiated material in several forms, as follows:

$$D_{ave} = F_p(PT/M), \tag{2}$$

$$D_{ave} = F_p\,P/(M/T), \tag{3}$$

$$D_{ave}/T = F_p\,P/M, \tag{4}$$

where D_{ave} is the average dose in the irradiated material in kGy, F_p is the fraction of emitted radiation power absorbed by the material, P is the power emitted by the radiation source in kW, T is the treatment time in s, and M is the mass of the material in kg. The quantity $P \cdot T$ is then the emitted energy in kJ, M/T is the mass throughput rate of the process in kg/s, and D_{ave}/T is the average dose rate in kGy/s.[39] Values of F_p can be estimated from the depth-dose distribution curves or calculated more accurately with a suitable Monte Carlo code and can range from 0.25–0.75 depending on the size and shape of the material.

2.5. *Dose versus electron beam current*

The relationship between the absorbed dose, the electron beam current, the treatment time, and the area of the irradiated material is often more convenient to use than the dose–beam power relations given in Eqs. (2) to (4) because it avoids the estimation or calculation of the power absorption factor F_p. This relationship can also be expressed in various forms as follows:

$$D_z = D(e,z)\, F_i\, I\, T\,/(10\,A), \tag{5}$$

$$D_z = D(e,z)\, F_i\, I/(10\,A/T), \tag{6}$$

$$D_z/T = D(e,z)\, F_i\, I/(10\,A), \tag{7}$$

where D_z is the dose in kGy at depth z in the material, $D(e,z)$ is the energy deposition per electron per unit area density (thickness times volume density) in MeV/(g/cm^2) at depth z, F_i is the fraction of emitted electron beam current intercepted by the material, I is the emitted beam current in mA, T is the treatment time in s, and A is the area of the material in m^2 at depth z. The quantity A/T is then the area throughput rate of the process in m^2/s and D_z/T is the dose rate in kGy/s.[39] The appropriate value of F_i can usually be estimated with sufficient accuracy by the ratio of the area of the material to the total area irradiated by the electron beam. The units of energy deposition are often written as MeV $\cdot$ cm^2/g, which is less obvious to understand. Values of $D(e,z)$ can be obtained from depth dose curves or calculated with a suitable Monte Carlo code. A typical value at the entrance surface of water or polyethylene for electron energies from 2.0–10 MeV is ~1.8–2.0 MeV/(g/cm^2), as can be seen from Fig. 3. It is higher for lower energies, increasing to about 10 MeV/(g/cm^2) at 100 keV (see Fig. 2).

3. Industrial Electron Accelerators

Industrial electron beam accelerators can be classified by their maximum electron energies. Systems rated as "low energy" usually range from 75–300 keV, with the upper limit sometimes extended to 500 keV. The "medium-energy" range usually means from 300 or 500 keV to 5 MeV, and the "high-energy" range is from 5 MeV to 10 or 12 MeV. These energy ranges reflect the material penetrations required for different types of applications, as listed in Table 1.

At least 20 different business entities in various countries around the world now offer high-current electron beam accelerators over these energy ranges. These companies utilize a variety of different accelerator designs to cover the known as well as emerging irradiation applications.

3.1. *Low-energy accelerators*

In the low-energy range (below 300 keV), EB accelerators usually employ a long linear cathode or multiple cathodes to distribute the electrons over a wide web of material that is moving through the beam. The acceleration

Table 1. EB energy range and material penetration for several irradiation market segments.

Market segment	Electron energy	Maximum penetration
Surface Curing	80–300 keV	0.4 mm
Shrink Film	300–800 keV	1.6 mm
Wire & Cable	0.4–3 MeV	11 mm
Sterilization	3–10 MeV	40 mm

is done inside a long evacuated tube connected to a direct current (dc) high-voltage generator. Most low-energy acceleration tubes employ a single gap between a thermionic cathode (the source of electrons) and an anode (a thin metallic window at ground potential). This window must be thin enough to allow the electrons to exit the acceleration tube and irradiate materials or products in air. The high voltage is typically generated using a low-frequency, iron-core transformer and a rectifier circuit to produce a dc potential. The high-voltage power supply is usually connected to the accelerator with a flexible high-voltage cable. Low-energy accelerators can be locally self-shielded with high density metal, usually lead and steel, but steel alone can also be used. The major suppliers of these low-energy, self-shielded systems are Energy Sciences Incorporated (ESI), now a subsidiary of Iwasaki Electric Company (Japan); BroadBeam Equipment, a division of PCT Product and Manufacturing, LLC (US); Advanced Electron Beams (AEB, US); and NHV Corporation, a subsidiary of Nissin Electric Company (Japan).

In recent times, the fastest growing market for industrial accelerators has been in this low-energy sector. Applications involve the curing (without the use of solvents) of inks, coatings, and adhesives, all of which consist of liquid reactive monomers that can be polymerized and cross-linked by electron beam irradiation. An important advantage of this type of curing has been the elimination of air pollutants — mainly solvents that are emitted by conventional printing and coating processes. Also, it has been found that the overall energy efficiency of these radiation processes is higher than other methods such as forced air drying. Most of the low-energy EB accelerators in this category are integrated into printing, coating, or similar continuous-web processes. When compared to other process equipment like conventional printing presses, coating systems, and drying ovens, an EB unit takes up a very modest total amount of space in the process line. When the web of material is moved under the beam, entering and exiting at an angle (as depicted

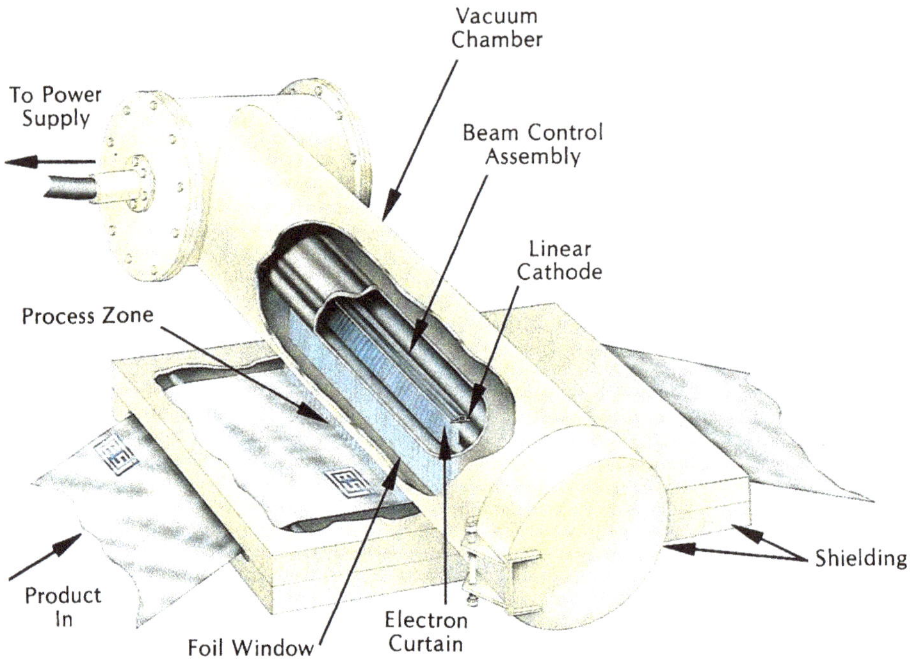

Fig. 8. Low energy EB accelerator design using a single long linear cathode filament. (Courtesy of Energy Sciences, Inc.)

in Fig. 8), the amount of shielding required can be minimized. However, a low-energy EB unit can also be arranged with a longer entrance structure for treating flat materials such as wooden panels.

Figure 8 is a diagram of the first commercial low-energy EB device to use a long linear cathode filament to form a uniform density electron "curtain" for irradiating wide swaths of materials. Units based on this concept take up little plant floor space and can be easily fitted into coating and conversion process lines. In the early 1970s, Energy Sciences Inc. (ESI) developed the Electrocurtain™ system shown in Fig. 9 based on this patented design.[40] ESI has since improved and downsized the technology through several generations of products and now offers a variety of EB processing systems. This includes the EZCure™ product line of compact, less expensive units, one of which is shown in Fig. 10.

A disadvantage of using a single linear filament is that when it burns out, the electron beam chamber has to be opened to replace it. An alternative scheme was developed at RPC Industries, which was later acquired by BroadBeam Equipment (now a division of PCT Products and Manufacturing). It is a segmented cathode design that links short filament segments

Fig. 9. Electrocurtain low-energy EB unit with a 3.6 m wide by 1.5 m deep footprint. The power supply (not shown) occupies slightly more than 1 m^3 of space. (Courtesy of Energy Sciences, Inc.)

in parallel.[41] When a single filament burns out in a segmented cathode, the adjacent filaments can still provide enough surface dose uniformity to continue operating without disrupting the treatment process. The segmented filament design is used in 100–300 keV self-shielded equipment. Figure 11 shows such a filament assembly in an accelerator with the titanium beam window removed. Much like ESI and other EB equipment manufacturers, PCT/BroadBeam has produced downsized versions of low-energy irradiators in its LE BroadBeam series. A segmented filament is also used by NHV in its Curetron™ low-energy system.

A third approach to reduce the cost of low-energy electron accelerators has been developed by Advanced Electron Beams (AEB). While most low-energy irradiator systems are made to meet the customer's requirements, AEB produces only two standard designs: a circular EB unit with a 26 cm end window and a longer smaller-diameter unit with a 41 cm side window. Unlike other low-energy accelerators, the AEB units are evacuated and sealed at the factory and designed for plug-and-use operation. They do not need to be continuously evacuated. Figure 12 is a diagram of the

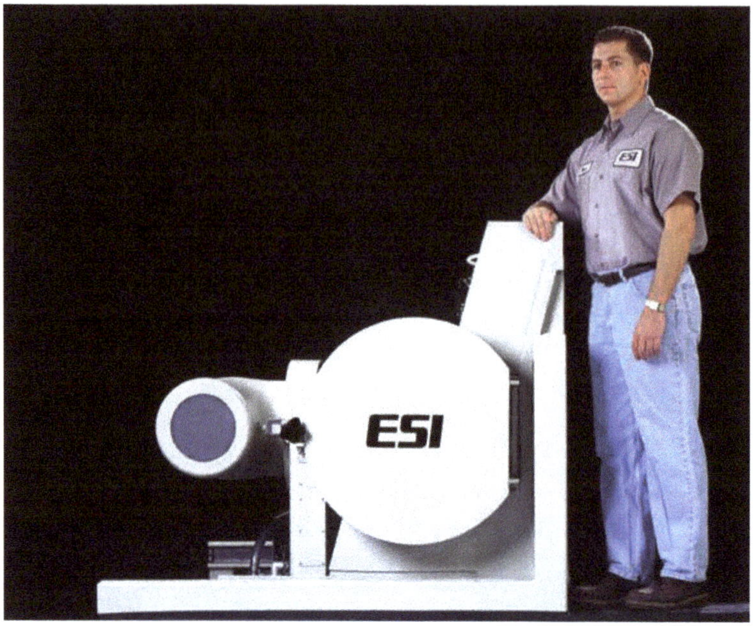

Fig. 10. EZCure-I^{TM} system with a person for size perspective. (Courtesy of Energy Sciences, Inc.)

Fig. 11. Photograph of a segmented low-energy EB filament array with the beam exit window removed. (Equipment designed by RPC Industries, now Broadbeam Equipment.)

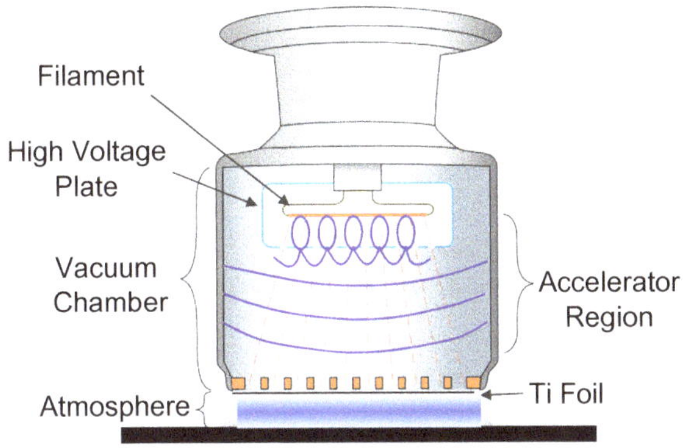

Fig. 12. Diagram of a modular compact low-energy factory-sealed EB accelerator. This unit weighs ~15 kG, is 27 cm in diameter and 33 cm in height. (Courtesy of Advanced Electron Beams).

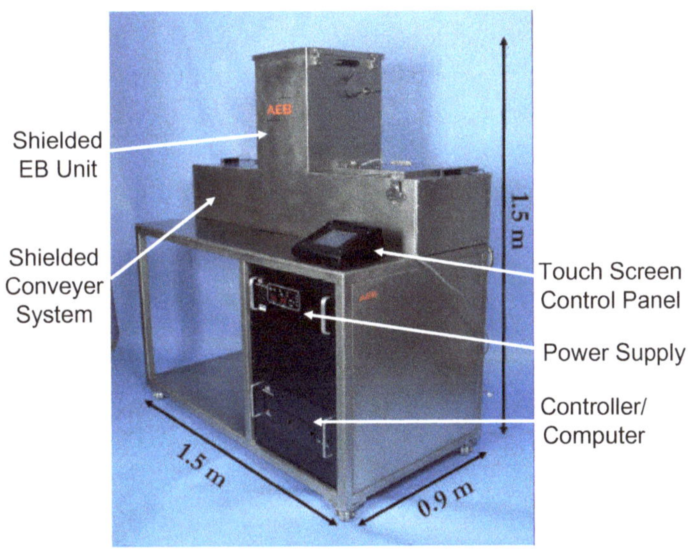

Fig. 13. Self-shielded low-energy application development unit. (Courtesy of Advanced Electron Beams.)

AEB circular EB module. Multiple modules can be placed in a variety of configurations to irradiate a wider area, treat both sides of a web in-line, or treat products in three dimensions.

AEB also makes a low-cost, self-shielded laboratory unit shown with major components labeled in Fig. 13. This economical Application Development Unit operates at 80–120 keV and enables research on materials of sufficiently low gauge thickness (e.g., thin films and liquids for inks, coatings, or composite matrices, and even gases) to be conducted within any laboratory. It provides sufficient beam current to simulate industrial processes.[42] Laboratory units are also offered by ESI and NHV.

The major end-user of low-energy self-shielded EB systems, the Cryovac division of Sealed Air Corporation, produces its own 500 keV accelerators for irradiating heat-shrinkable multi-layer plastic films for food packaging. The design for these ICT systems was acquired from the former supplier, High Voltage Engineering Company, when HVEC exited the accelerator business. Cryovac has reportedly built more than 130 units for use in its own production facilities. A description of this type of accelerator is given in the next section.

3.2. *Medium-energy accelerators*

In the medium-energy range (300 keV to ~5 MeV), most of the industrial EB accelerators in use are high current dc potential-drop devices, although pulsed radio frequency (RF) linacs (discussed in the high-energy accelerator section) are also employed. One example is the ILU family of linear accelerators developed in Russia at the Budker Institute of Nuclear Physics (BINP) for medium energy applications. These are pulsed RF resonant cavity systems operating at 100–200 MHz that produce electron energies from 0.6–4 MeV with beam powers up to 50 kW at 4 MeV.

For the dc accelerators in this energy range, the use of a flexible high-voltage cable between the generator and the accelerator is not practical at higher voltages, so systems operating at voltages above 800 keV combine the high-voltage generator and the acceleration tube within the same enclosure. Compressed sulfur hexafluoride gas at pressures of ~6–7 atm is usually provided for insulation.

The electron acceleration tubes must employ multiple electrodes, often called dynodes, with small apertures to distribute the voltage and avoid internal sparking. Therefore, the electron source must be a small single cathode that produces a small-diameter electron beam. After acceleration, the beam is usually scanned rapidly with a deflection magnet energized with an alternating current (ac) to spread the electrons over a wide web of material or over other types of product handling systems for electrical wire or plastic tubing. The scanning frequency is usually about 100–200 Hz. The beam then exits to atmosphere through a triangular-shaped "scan horn". The high-voltage generator for these accelerators usually consists of a multi-stage cascaded rectifier circuit. This can be energized with medium-frequency or high-frequency ac power, which is coupled either capacitively or magnetically to the rectifiers.

Electron accelerators used in routine production applications must have high beam currents, e.g., tens of mA, in order to provide the high dose rates required. Research equipment in the same mid-energy range, such as Van de Graaff generators (>550 units built) or Pelletron^{TM} systems (an electrostatic device using insulated chains instead of belts to build up charge; >150 units built) do not attain such high currents and high dose rates. A few other types of low-current, mid-energy electron accelerators have been developed, but like the Van de Graaff and Pelletron systems, these are also limited to research functions.

The major types of dc accelerator structures used to obtain high beam powers at medium energy are:

- The magnetic, series-coupled Insulating Core Transformer (ICT). The ICT was originally developed in the US by High Voltage Engineering Corporation (HVEC) and uses series coupling between disc-shaped transformer cores to transmit low frequency ac power from the primary windings at ground potential to all of the secondary windings at successively higher dc voltages. Thin insulating sheets are inserted between the cores to minimize high-voltage breakdown. Rectifiers and capacitors are used to produce the dc voltage at each stage, with the stages all connected in series between the high-voltage terminal containing the electron gun and ground. The major manufacturers of ICTs are Vivirad SA (France) and Wasik Associates, Inc. (US). A variety of ICTs with voltages from 300 kV to 3 MV at power outputs up to 100 kW have been produced by HVEC, Wasik, and Vivirad. As mentioned earlier in the low-energy section, Cryovac builds ICT systems for its own use. These were mentioned in the low-energy category primarily because they are used for processing thin-film heat-shrink material, which is typically a low-energy application. In addition, several companies in China now build ICTs from 300 kV to 1.2 MV, mainly for use in that country.
- The magnetic, parallel-coupled Electron Transformer-Rectifier (ELV) developed at the Budker Institute of Nuclear Physics (BINP), Russia. The ELV uses parallel coupling with the primary winding extending the full length of the high voltage assembly. A variety of these systems have been built ranging from 400 kV to 2.5 MV with output power up to 400 kW at 1 MV and 100 kW at 2.5 MV. ELV systems are available from BINP and EB Tech Co., Ltd (Korea) under a joint-venture with BINP.

- The transformer-driven dc accelerators from the D. V. Efremov Scientific Research Institute of Electrophysical Apparatus (NIIEFA), Russia. These systems are similar to BINP ELVs in that the conical primary winding extends the full length of the accelerator. The latest systems have been extended up to 750 kV.
- The capacitive, series-coupled Cockcroft–Walton system from NHV Corporation (Japan). This balanced, series-cascade Cockcroft–Walton system is energized at a drive frequency of 3 kHz and does not use the precision components required in the accelerators of this type originally developed for physics research. The NHV Cockcroft–Walton system can attain 5.0 MeV at 150 kW and can be used for X-ray generation.[43,44]
- The capacitive, parallel-coupled Dynamitron® from IBA Industrial, Inc. (US). The Dynamitron was originally patented in 1959[45] and then developed at Radiation Dynamics, Inc., which was later acquired by IBA. These systems transmit high frequency (100 kHz) ac power to all of the rectifier stages of the generator in parallel using semi-circular electrodes that surround the high voltage column. The electrodes are connected to a center-grounded, iron-free, high-Q transformer to form a balanced resonant circuit that is energized with a triode oscillator circuit. The electrodes induce ac voltages that are rectified by the series stages between the high voltage terminal and ground. Figure 14 shows a Dynamitron high voltage column surrounded by the electrodes and mounted inside the pressure vessel.

Developed during the late 1950s and early 1960s, the ICT and the Dynamitron were the earliest of the mid-energy EB accelerators and are the most widely used in the category. The ICT and ELV mid-energy accelerator designs are limited to electron energies of 2.5–3.0 MeV. Figure 15 is a photograph of a new 550 keV ICT system built by Wasik for irradiation of wire. The Dynamitron can attain high beam power, e.g., 150 kW at 3.0 MeV and 300 kW at 5.0 MeV; hence, it can also be used as a high-energy, high-power X-ray generator. Figure 16 is a cutaway illustration of a 5 MeV, 300 kW IBA Industrial Dynamitron. During the last 20 years, ICT and Dynamitron-type systems have also been built by a number of vendors and institutes in China, mainly for use in that country. These systems were developed at national laboratories in the 1990s, and the technology was transferred to at least six commercial Chinese companies for production.

High-current, mid-voltage EB units are mainly used for cross-linking of wire and cable insulation and heat-shrinkable tubing, and for partial curing of rubber components in the tire industry where EB processing is a

Fig. 14. Photograph of the Dynamitron developed by Radiation Dynamics, Inc.

Fig. 15. The 550 keV ICT system from Wasik Associates with the cabinet door open. (Courtesy of Wasik Associates, Inc.)

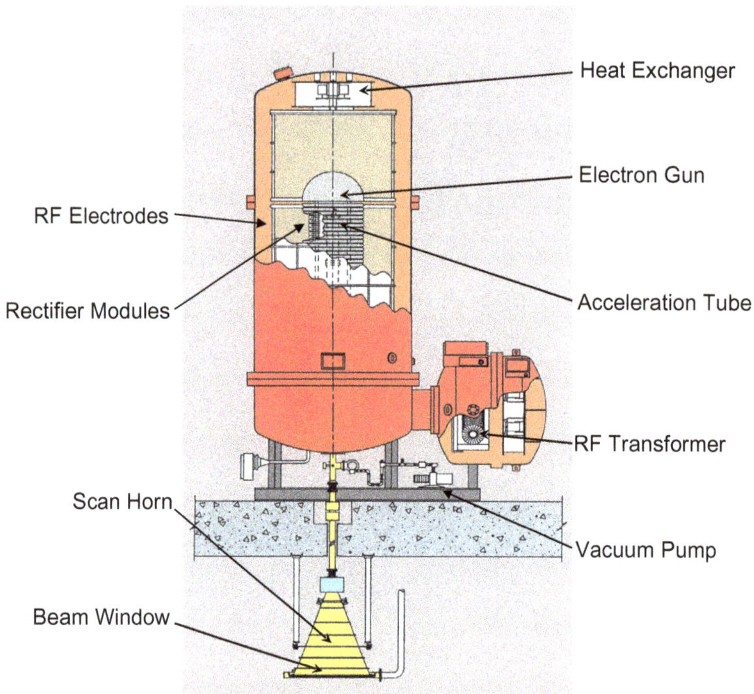

Fig. 16. Cutaway diagram of a 5 MeV 300 kW Dynamitron. (Courtesy of IBA Industrial, Inc.)

well-accepted industrial manufacturing practice. They can also be used in the X-ray mode for sterilizing medical devices, irradiating fresh foods, and curing thick composite products.

3.3. *High-energy accelerators*

In the high-energy range (5–10 MeV), most industrial EB accelerators use electric fields generated by radio frequency (RF) power to accelerate electrons. High-power, high-voltage dc generators could be made for energies greater than 5 MeV, but they would be much larger in size and more expensive to make and house than RF accelerators.

With the exception of the Russian ILU linacs that are based on a single large-diameter toroidal cavity resonating at a VHF frequency of around 176 MHz, most of the electron linacs used for high-power EB irradiation consist of a linear series of small coupled cavities resonating at microwave frequencies — the smaller the cavity, the higher the frequency. These microwave linacs are usually designated by the frequency band into which they fall: L-band for 1–2 GHz, S-band for 2–4 GHz, and C-band for 4–8 GHz.

The cavities are energized by a microwave generator, either a magnetron or a klystron, and must all be carefully tuned to the same resonant frequency. In general, the L-band systems can provide the highest beam powers (up to 100 kW) but are the largest in size. The smaller S-band linacs are the most prevalent systems for EB irradiation but are usually limited to beam powers of ~30 kW. The lower frequency ILU linacs are capable of producing beam powers up to 100 kW, but these are much larger systems than the microwave linacs. In contrast to the linacs, Rhodotrons accelerate electrons by circulating the beam through repeated passes in a single coaxial resonant cavity to achieve high energies in a relatively small-footprint structure. Rhodotrons operate in the VHF frequency range (100–200 MHz) and can produce beam powers of up to 700 kW.

The electron energy from a linac is not limited to the 5–10 MeV range. Linacs can produce electron energies from 2–16 MeV for radiographic inspection (see Chapter 7) and even higher energies for cancer therapy. Because of the extensive use of lower power linacs in X-ray radiography and in medical therapy, these must be distinguished from the high-power linacs that are suitable for material irradiation. Linacs for radiation processing of materials are normally limited to a peak energy of 10 MeV to minimize induced radioactivity in the products, but they can also be downsized to energies of a few MeV.[46] However, high duty factors and high dose rates are required for most material irradiation applications. Linacs are typically operated in a pulsed mode with low duty factors from a few tenths of a percent up to about one percent. The duty factor is the fraction of beam operating time (the beam pulse width times its repetition rate) during which the beam is on. Linacs have relatively modest energy conversion efficiencies, with about 30% of input ac power being converted to output beam power. Radiographic, medical, and research linacs operate at low duty factors and also with relatively low average dose rates compared to industrial irradiation linacs. In addition to determining the product throughput rate, the dose rate can influence some of the effects that electron beams have on materials and products.

One of the earliest linacs for industrial irradiation was a 7 MeV, 5 kW unit made by HVEC in 1957 for the Ethicon Division of Johnson & Johnson. It was used to sterilize absorbable sutures made from animal tissues. Another early linac was a 10 MeV, 5 kW unit made in 1960 by Varian Associates for the Riso National Laboratory in Denmark.

Since then, many other more powerful linacs have been made by several companies for applications in radiation processing. Such linacs have been produced for more than four decades in France by Getinge–La Calhene and its antecedents, Linac Technologies, MeV Industrie SA, and CGR–MeV. In 1967, CGR–MeV installed a 10 MeV, 10 kW linac in an irradiation facility in France. In 1985, MeV Industrie installed a 10 MeV, 20 kW CIRCE^{TM} linac for food irradiation, and since 1987, several CIRCE linacs have been installed for medical device sterilization.[47,48] One of these accelerators has also been installed for use in cross-linking the matrix of carbon fiber composites at the European aerospace consortium Aerospatiale's facility in the Bordeaux region of France (1991). This is a 10 MeV, 20 kW accelerator with X-ray capabilities. Industrial irradiation linacs can be positioned vertically or horizontally or can use beam deflection magnets to arrange the accelerator and its scan system to meet application requirements. For 10 MeV, industrial linacs can be as long as 4 m. Figure 17 shows a 10 MeV CIRCE linac showing the scan horn pointing upward in a vertical position.

In the late 1980s, Atomic Energy of Canada, Ltd. (AECL) developed the Impela linac, which was the most powerful 10 MeV system available at the time. This L-band linac operates in an unusual long-pulse, high duty factor mode to achieve average beam powers of 50 kW. The prototype system was installed at AECL's Chalk River Laboratories, but the company exited the

Fig. 17. CIRCE 10 MeV 20 kW accelerator assembled in the manufacturer's workshop for testing. (Photo courtesy of Getinge–La Calhene, previously Linac Technologies.)

linac business in 1998 after building only a few commercial systems — one was installed in the New Jersey facility of E-Beam Services, Inc. and another at the Vancouver facility of Iotron Industries Canada Inc. Impela® systems are still available from Iotron who acquired the rights to the technology from AECL.

The Mevex Corporation in Canada has also been designing and building a variety of industrial EB irradiation linacs since the late 1980s. Mevex now offers compact 3–8 kW systems in the 3–6 MeV energy range that are suitable for small products irradiation and has developed a 10 MeV, 30 kW system for high-power applications.

In Japan, Mitsubishi Heavy Industries offers a variety of industrial EB linacs ranging in energy from 3 to 10 MeV. These systems can produce beam powers from 2–6 kW for the small C-band and medium size S-band linacs up to 31 kW for their most powerful S-band 10 MeV linac.

In the US, L-3 Services Inc., Pulse Sciences (L-3 PS) produces 5 and 10 MeV high-power electron linacs and incorporates these into turn-key systems for both food irradiation and medical device sterilization. The 10 MeV systems are based on compact S-band linacs that can produce up to 18 kW of beam power. Figure 18 shows two of these linacs, each with 15 kW average beam power, installed side by side in an EB medical products sterilization facility. The larger but more powerful 5 MeV L-band linac produces up to 100 kW of beam power. These 5 MeV linacs can be used for either high-power EB irradiation applications or for the generation of X-rays for food irradiation.

Another high-power, high-energy linac is being developed by the Budker Institute (BINP) in Russia. The basic resonant cavity accelerating structure is the same as used in the medium energy ILU systems, but the ILU-14 will reportedly attain energies from 7.5–10 MeV and beam powers up to 100 kW by coupling together as many as 6 modules of resonant cavities.[49] In contrast to the higher frequency linacs described above, these cavities operate at 176 MHz. The lower frequency results in a larger accelerator, but it can be energized by efficient triode RF power tubes. A photograph of a prototype ILU-14 system is shown in Fig. 19.

The other type of accelerator now commonly used for high-energy EB radiation processing is the IBA Rhodotron. A Rhodotron accelerates electrons by passing the beam repeatedly through a single coaxial resonant cavity. The underlying accelerating principle was conceived and demonstrated in the late 1980s by a group of researchers from the French Atomic Energy Agency (CEA). This patented design was licensed to IBA in the

Fig. 18. Two 15 kW, 10 MeV electron linac systems installed side by side in a medical products EB sterilization facility. (Courtesy of L-3 Services Inc., Pulse Sciences.)

Fig. 19. Prototype ILU-14 system. (Photograph courtesy of Budker Institute of Nuclear Physics.)

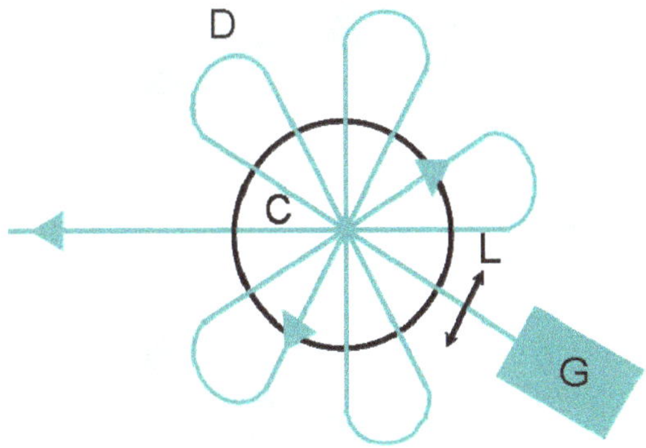

Fig. 20. Beam trajectory in a six-pass Rhodotron accelerator. Labeled components are G: electron gun, L: magnetic lens, C: accelerating cavity, D: bending magnet. (Courtesy of IBA Industrial, Inc.)

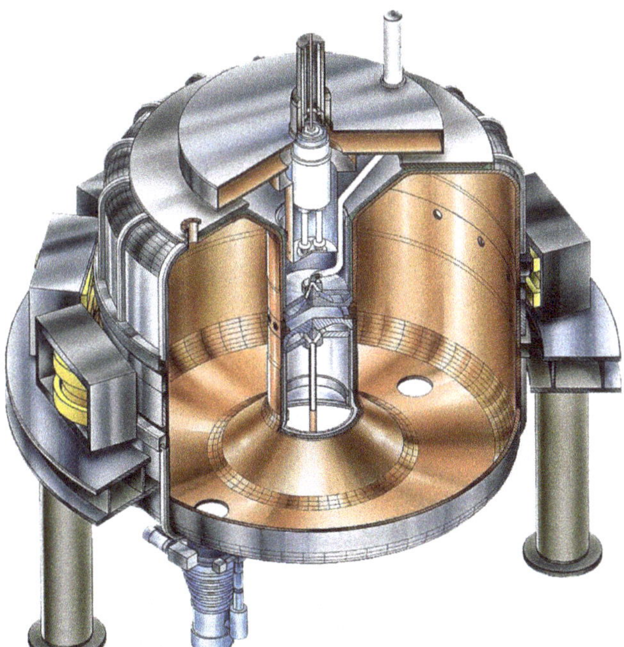

Fig. 21. Cutaway view of a Rhodotron resonant cavity with beam bending magnets. (Courtesy of IBA Industrial, Inc.)

early 1990s where it was subsequently developed into the commercial Rhodotron systems built by IBA today. Figure 20 illustrates the basic accelerator concept and Fig. 21 is a cutaway drawing of a Rhodotron resonant cavity.

Whereas radio frequency linacs typically operate with less than 100 kW of average beam power, Rhodotrons can produce up to 200 kW at 10 MeV and 700 kW at 7.0 MeV. With such high beam power, the generation of X-rays has become a viable option for radiation processing.[6] The recirculating beam in a Rhodotron also gives it the capability of having multiple beamlines with different electron energies drawn off the same accelerator (but not concurrently).

High-power Rhodotrons have already demonstrated the industrial viability of X-ray processing. The US Postal Service is using a 130 kW Rhodotron installed in a facility in New Jersey to sanitize the mail for critical US Federal government departments and agencies in Washington, DC. A 10 MeV electron beamline from this accelerator is used for irradiating trays of flat mail and a 5.0 MeV line with a tantalum X-ray target is used for treating bulk mail in sacks with X-rays. This facility clearly illustrates the versatility of the Rhodotron, with the 10 MeV beamline being bent downward to a lower vault for electron beam processing and two other beamlines at 5.0 and 7.0 MeV being directed horizontally toward X-ray targets. Only one beamline can be used at a time.

4. Major Applications of Industrial EB Irradiators

4.1. *Cross-linking of materials*

Cross-linking of materials is the largest application of electron beam irradiator systems. The major categories of cross-linking applications are shown according to relative market size in Fig. 22. These applications cover a

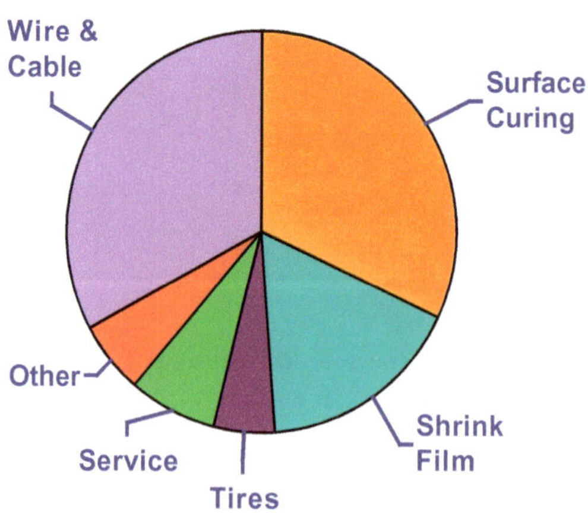

Fig. 22. Major end-uses of EB radiation cross-linking.

wide range of accelerator energies to accommodate the required electron beam penetration in the various products irradiated in these diverse market segments. Different end-use areas also tend to use different beam handling systems, depending on the size and shape of the materials or products being processed. The Service segment includes contract irradiation work performed at various facilities for outside companies. These services cover many of the applications included in the other segments. This section will give a brief overview of some of the most prevalent industrial uses of EB irradiation cross-linking.

4.1.1. *Wire and cable insulation*

Cross-linking of the insulation on electrical wires and cables was one of the first commercial applications of radiation cross-linking and is now one of the most widely used applications of EB processing. Cross-linking keeps the insulation from melting when the conductor is overheated and also makes it more resistant to a fire. This is particularly important for wires used near the engines in automobiles: when they are exposed to flames, the jacketing will not burn away or melt. Such wires usually meet the US Underwriters Laboratory flame retardant standards. Cross-linking also increases the toughness and abrasion resistance of the insulation and makes it insoluble in organic solvents. Polyethylene (PE) or a blend of PE and ethylene-propylene rubber is normally used as the basic polymer, but other materials can be added such as aluminum trihydrate (hydral), a common ingredient that makes the insulation flame retardant. Processing aids, antioxidants, and cross-linking enhancers such as trimethylolpropane triacrylate (TMPTA) are usually added as well, but most wire manufacturers will not disclose the actual ingredients in their proprietary coating compounds.

The cross-linking is accomplished by moving the wire back and forth many times through a scanning electron beam. Multiple-drum fixtures and guides control the positions of the wires as they pass through the beam. The divergence of the scanning beam and the slight twisting of the wires as they pass between the drums cause the dose distribution around the wires to be sufficiently uniform. The minimum electron energy depends on the diameter of the wire and the thickness of the insulation. It can be as low as 0.5 MeV for low-voltage wires in automobiles and electronic appliances and as high as 3 MeV for the jackets on multiconductor cables. Processing speeds can be as high as several hundred meters per minute with high-current electron accelerators.

4.1.2. *Heat-shrinkable plastic tubing and film*

The Raychem Corporation was founded in 1957 to manufacture and sell radiation-cross-linked wire systems. They also developed a variety of new products using the elastic recovery effect shown by cross-linked polyethylene (PE) when it is heated above its melt transition temperature. Heat-shrinkable plastic tubing can be irradiated using the same kind of under-beam fixtures that are used for insulated wire. The cross-linked tubing is then expanded using a process that was patented in 1963 (US Patent No. 3,086,242). Polyethylene consists of a mixture of amorphous and crystalline regions, but cross-linking occurs mainly in the amorphous regions. When cross-linked PE is heated above its melt transition temperature, the crystalline domains melt and the material behaves like a weak rubber. Irradiated PE tubing in this condition can then be expanded to a larger diameter by using compressed air. When the tubing is cooled in the expanded shape, the crystalline regions re-form and hold the tubing in the larger dimension. When it is reheated, the crystalline domains melt again and the tubing contracts to its original dimension. This unique characteristic of PE is being used to make heat-shrinkable products to cover bundles of cables, cable connectors, pipes, etc. Short pieces of tubing are used to cover wire connections, with the tubing being similar to the wire jacketing. An adhesive or sealant can be applied inside the tubing to make a waterproof seal over the connection. When it is heated again, the tubing contracts and conforms to the connector inside. In addition to tubing, specially shaped heat-shrinkable enclosures are used to cover a variety of other products.

Raychem's development of heat-shrinkable polyethylene tubing prompted the Cryovac division of W. R. Grace & Co. (now part of the Sealed Air Corporation) to develop an irradiation process for producing heat-shrinkable films for food packaging. The film is usually irradiated in the form of a thin-walled, large-diameter tube that is then heated and expanded with compressed air and cooled in the expanded dimension.

The concept was to pass an extruded tubular form back and forth through a beam, thereby absorbing most of the beam power, and then blow the irradiated and heated material into the desired film dimensions. This became known as the "double-bubble" process, one bubble being the extrudate and the other the blown film.[50,51] Sometimes the film is irradiated as a flat sheet using low-energy (300 keV) EB units. It is then heated and stretched with a tenter as is commonly done in the plastic film industry to orient films. Packages wrapped with such films are exposed briefly to hot air in order to shrink the films and make a tight enclosure. Uncooked frozen turkeys for Thanksgiving dinners are often wrapped in this way. Heat-shrinkable food films usually have at least five layers: an inner food contact layer, a tie coat, a gas barrier film, another tie coat, and an outer layer for abrasion resistance and printing. Electron energies in the range of 0.3–0.5 MeV are usually sufficient for this process. The exact energy used depends on the film thickness and whether the irradiation is done in a single pass or multiple passes through the beam.

4.1.3. *Curing of inks, coatings, and adhesives*

Inks, coatings, and adhesives formulated with reactive monomers and oligomers can be cured more rapidly with electron beams than with ultraviolet (UV) radiation or thermal curing. Electron beam curing involves a combination of polymerizing and cross-linking. Electron beams are often used with wide-web presses for high volume production and for printed items that require fine graphics and color highlights (electrons can penetrate pigments whereas UV cannot). An advantage of EB curing and cross-linking of coatings is that the pigmentation does not interfere with the cross-linking process as it does with UV. Also, EB ink formulations tend to be considerably less complex than UV formulations. Because the temperature rise with EB curing is not as high as with heat curing, EB can be used to cure coatings on heat-sensitive substrates, such as plastic films. In contrast to UV curing, toxic initiators are not needed with EB curing, so such coatings can be used for applications where they are in contact with foods. EB curing consumes less energy than UV or thermal curing. It also eliminates the volatile solvents that are used in thermal curing, which can pollute the atmosphere. The minimal temperature rise during EB treatment allows this method to cure the adhesive layer between films that have different coefficients of thermal expansion without producing the interfacial strains that are caused by thermal curing.[52]

The suppliers of monomers and oligomers for inks, coatings, and adhesives have had to address the issues of toxicity, Clean Air Act compliance, food contact, and many other areas of concern in contemporary manufacturing. These have been major factors in changing the portfolio of materials available to formulators and users over the past several decades. The energy required to cure formulations of inks, coatings, or adhesives using electron beams is significantly less than that used with alternative drying systems, even with what are called "high solids" content products. Thus, besides eliminating volatile organic compounds (VOCs), EB curing also lessens potential greenhouse gas emissions.

Low-energy EB processing is also used in making thin film laminates or thin film overlays. Higher energy EB or even X-rays can be used to cure the adhesive bonds between thicker substrates. Materials with different coefficients of thermal expansion can be bonded with EB-curable adhesives without creating as much interfacial strain as that produced when using thermal curing. Pressure sensitive adhesives (PSAs) can also be cured using electron beams.

4.1.4. *Automobile tires*

In 1933, the B. F. Goodrich Company obtained a patent (US Patent No. 1,906,402) on the industrial use of electron beams to vulcanize natural rubber without heat. Today, several of the components of automobile tires are given low doses of electron beam radiation before they are assembled into a complete tire. This process is referred to as partial cross-linking or partial curing. First, the tire components are extruded and then EB-irradiated to change them to a gel condition. The irradiation stabilizes the thicknesses of the components and prevents the migration of the tire cord and steel belt through the supporting material during the final thermal cure of the tire. Partial curing also maintains the tacky surface condition needed to allow the different layers to stick together. This technique results in more uniform and better balanced tires that can be thinner and more flexible than tires made without irradiation. The finished tire is fused together during a thermal molding process that determines the tread pattern, appearance, shape, and other features of the tire.[53]

Different elastomeric materials are used for various functions in a tire. Formulated halogenated butyl rubbers (as BIIR) are used for inner liners. Side-walls are made from ethylene-propylene rubbers (EPDM) because of the inherent ozone resistance of this polymer. Chafer strips are also partially EB cured. Figure 23 shows where these components are placed in a finished partially-cross-linked, thermally-cured tire. Since tires are made in such large quantities, the use of less rubber can recover the cost of an EB accelerator in about one year. Many US patents provide detailed information on various aspects of this application. These have all expired so their disclosures are now in the public domain.[54]

4.1.5. *Polyethylene foam*

Cross-linked plastic foams, typically made from polyethylene, are used in a wide variety of familiar items, including the insulation and cushioning in furniture, medical products, sports helmets, athletic safety pads, and

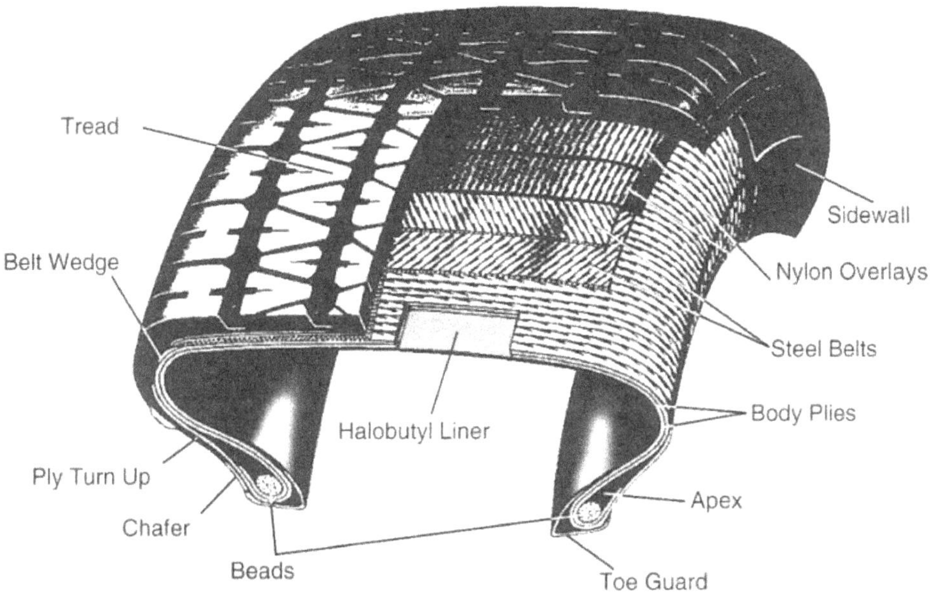

Fig. 23. Cross section of a modern radial ply tire.

automobiles. Plastic foam is made by incorporating a blowing agent into the plastic and then heating the mixture to convert the blowing agent to a gas, resulting in bubbles being formed inside the plastic material. The heating process must be carefully controlled to preserve the bubbles. The plastic/blowing agent mixture can be cross-linked with thermo-chemical reactions, but this risks premature activation of the blowing agent, which is by nature thermally unstable. Since electron beam processing causes only a small rise in temperature, the mixture can be cross-linked with EB irradiation without activating the blowing agent.

The type of PE used, the amount of blowing agent, the radiation dose, and the blowing process all combine to make a well-defined closed-cell foam.[55] Significant among the many uses of these foams is its use in automobiles for safety and protection, particularly as cushioning under the interior header. Figure 24 shows some of the automobile interior panels that use closed-cell PE foams. Another example is the use of PE foam as backing material for medical products.

4.2. *Radiation sterilization of medical devices*

The Johnson & Johnson Corporation (J&J) was the first industrial organization to sterilize medical devices with ionizing radiation. The Ethicon

Fig. 24. Examples of automotive use of EB-cross-linked, closed-cell polyethylene foams.

Table 2. Some common medical disposable articles sterilized by ionizing radiation.

Syringes	Absorbents
Catheters	Gloves
Drains	Surgical gowns and drapes
Tubing	Hand towels
Urine bags	Beakers and labware
Drain pouches	Petri dishes
Bandages	Culture tubes

Division of J&J began using electron beam processing in 1956 to sterilize absorbable sutures made with animal tissues. A wide variety of medical devices are now being sterilized using EB processing,[17–19] several examples of which are listed in Table 2. Three issues must be considered when planning to use electron beams or X-rays derived from electron beams for medical device sterilization: (1) the materials used in the manufacture of the device, (2) the manufacturing stage at which the device will be sterilized, and (3) the radiation exposure or dose needed to attain sterility assurance levels. The studies that have been used in the past to determine sterility assurance levels were done mainly with gamma ray irradiation. X-rays are a viable alternative to radioactive gamma ray sources since they can provide higher dose rates — although not nearly as high a dose rate as electron beams can provide. Studies involving EB sterilization have shown a dose rate dependence of the lethality of radiation exposure: higher dose rates can cause greater lethality. Hence, the possibility exists for a future reduction in the sterilizing dose with EB and perhaps with X-ray processing as well.[25]

Radiation sterilization is also used for non-disposable medical items such as hip and other joint replacements. These implants are made from combinations of metals and plastics. The biocompatibility of various materials and the use of radiation to enhance such devices are not discussed in this chapter, but they are described briefly in Chapter 1.

The compatibility of the different components has to be considered when ionizing radiation is used for sterilizing medical devices. This form of energy not only kills micro-organisms but can also affect the properties of some materials. Medical devices are made of many different materials (some of which are metals that are not affected by ionizing radiation). However, most materials are non-metallic, such as formed polymers, composite structures, and even ceramics. If metals are part of a device, such as the needle of a syringe, then there must be sufficient EB energy to penetrate this higher-density material, or the product orientation under the beam must be chosen to irradiate the entire article.

Plastics are commonly used to replace glass in some medical products. However, there are concerns over the discoloration of polyvinyl chloride (PVC) when exposed to ionizing radiation and over the leaching into the blood of plasticizers used to impart flexibility in PVC formulations. Alternatives have been developed to replace the use of PVC in medical devices. These newer materials are based on polyethylene blends that include the optically clear metallocene catalyzed polyethylenes (mPE) or polypropylenes (mPP) as well as laminates of these materials. Such products remain clear when sterilized by electron or X-ray beams. A properly formulated PE material will also be enhanced by radiation exposure while not losing its flexibility.[1,9,25]

Rigid transparent medical devices can be molded from radiation tolerant plastics based on polycyclic (ring) molecules such as polystyrene (PS), polycarbonate (PC), and polyethylene terephalate (PET). The manufacturers of PC and PET have developed specialty grades that minimize the discoloration from radiation exposure. Flexible tubing and other products for medical devices can also be made out of silicone elastomers. In general, materials like these are radiation tolerant.

Electron beams were used at the very outset of radiation sterilization going back to the mid 1950s. High-energy (10 MeV) accelerators have been used since the 1960s for sterilizing packaged medical devices, both at in-house manufacturing facilities and service centers. Some mid-energy accelerators at 5.0 MeV and below are also used for medical device sterilization. Many packaged devices have a low bulk density so that the penetration of electrons is sufficient. If needed, packages can be irradiated from opposite sides thereby increasing the beam penetration to about 2.4 times more than that for single-sided treatment (see Fig. 5). Larger packages of medical devices have historically been sterilized using large gamma ray sources such as cobalt-60, but the use of X-rays has recently emerged as an alternative (X-rays have the same effective penetration as gamma rays, as was shown in Fig. 7).[6] In addition, X-rays can have somewhat higher dose rates than gamma rays and might not be as deleterious to some polymers, formulated PP, for example, as gamma ray radiation.[1,9,25]

To attain more uniform dose distributions in large pallet loads of medium and high-density products, a turntable can be used to rotate the pallet in the X-ray beam. Such a system, the Palletron™, was originally conceived and patented by MDS Nordion (part of MDS Inc., Canada) and then licensed to IBA. Figure 25 illustrates a Palletron in front of a long X-ray target equipped with thick steel collimators on each side to limit the width of the X-ray beam.[56,57]

The dose distribution in a pallet load rotating in a scanning collimated X-ray beam has been calculated with a Monte Carlo code. This indicates energy deposition levels so that one can estimate the differences between the maximum dose (D_{max}) and the minimum dose (D_{min}) and attain an

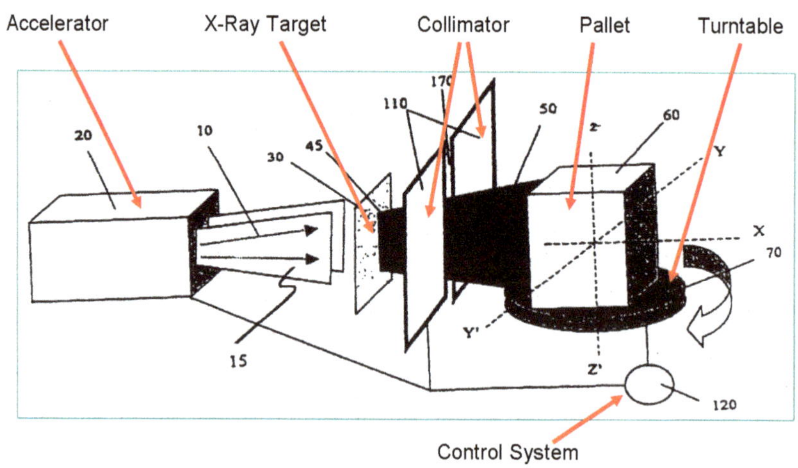

Fig. 25. Diagram of a Palletron showing rotation of pallets in front of a collimated X-ray beam. (Courtesy of IBA Industrial.)

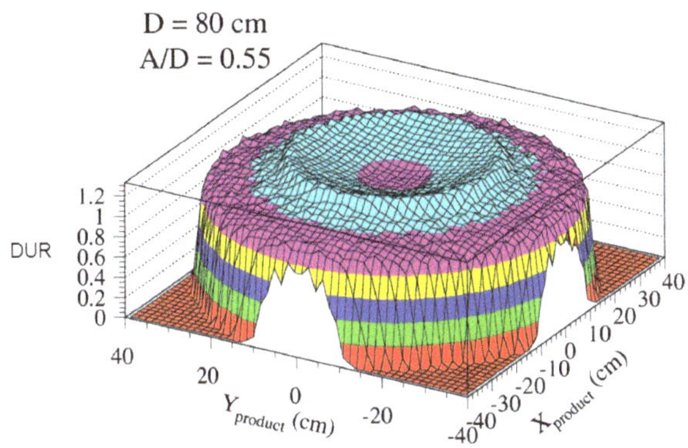

Fig. 26. Monte Carlo illustration of uniform dose distribution for rotating product.

acceptable dose uniformity ratio (DUR). Figure 26 shows the results of the three-dimensional Monte Carlo calculation of DUR for a cylindrical object rotating in front of a collimated tantalum X-ray target intercepting a 5 MeV electron beam. In this example, the diameter of the cylinder is assumed to be 80 cm and the average product density is 0.8, which is more appropriate for food products than medical devices. With a proper choice of the collimator width, the surface dose can be reduced to the dose in the middle of the rotating container, thereby obtaining a low value for the dose uniformity ratio (DUR). Parameters such as electron energy and beam current, distance from the X-ray target, width of the collimator opening, and size and density of the product load can be adjusted to obtain the required dose and dose uniformity ratio.[56,57]

Dose setting is a specific feature of the radiation sterilization process to obtain the required sterility assurance level (SAL), which is 10^{-6} in most countries. This value means that the probability of having live organisms on sterilized products would be less than one in a million devices. Many medical device manufacturers already employ the US Good Manufacturing Practices (GMP), in particular to document control, inspection and testing, validation of the process, commissioning of the plant, and microbiological monitoring. Basic requirements are directly related to quality management (quality assurance and quality control), personnel (qualification, training, and hygiene), equipment and premises (production area, storage area, quality control area, and ancillary areas), documentation (specifications, processing and packing instructions, and procedures and records), production, contract manufacture and analysis, complaints and product recall, and self-inspection.[17–19,58]

Fig. 27. SterBox in-line sterilization tunnel with one 5 MeV, 5 kW, MeVAC accelerator, shown in open position for service access. (Photo courtesy of Getinge–La Calhene, previously Linac Technologies.)

In 2008, Getinge–La Calhene (formerly Linac Technologies) developed a medium-energy (3–5 MeV, 5 kW) self-shielded, in-line EB system that provides sterilization of packaged medical products or sterile transfer of packages into an aseptic product filling system. The entire STERBOX™ system (except for the electrical cabinet and cooling system) is installed inside a stainless steel housing, which encases the lead shielding and has a footprint less than 20 m^2, as shown in Fig. 27. This permits the entire unit to be located inside a GMP clean room. In-line sterilization is accomplished at the same speed as the production and packaging processes, which reduces the logistics and storage costs associated with trans-shipment of products to outside sources for sterilization.[59,60] An installed STERBOX system has been validated by the US Food and Drug Administration (USFDA). For sterilization of higher density products, the STERBOX Twin contains two 5 MeV, 5 kW accelerators that enable two-sided treatment of products to be conducted in a single pass.

4.3. *Irradiation of foods*

The possibility of using X-rays to kill microorganisms was investigated soon after Wilhelm Roentgen discovered this form of radiation in 1895. In 1896,

F. Minck published a paper describing his unsuccessful attempt to inactivate microorganisms with X-rays.[61] In 1905, Hugo Lieber received a US patent on the use of ionizing radiation to preserve food, even though the chemical and biological effects of irradiation were not yet understood.[62] In 1906, Joseph Appleby and Arthur John Banks obtained a British patent for treating food with X-rays.[63] In 1921, the US Department of Agriculture (USDA) demonstrated that X-rays could kill *Trichinella spiralis* in pork,[64] and they investigated the preservation of ground beef with ionizing radiation during World War II. In 1953, the US started a major effort on food irradiation.[65] Since then there has been an evolution in the use of radiation processing for disinfecting pathogens from food and for disinfesting insects from fresh fruits and vegetables. However, despite the possibility of reducing the labeling requirements established by the USFDA and acceptance by members of the public who have had the opportunity to buy and consume irradiated food, the major food processors have been reluctant to adopt this technology.

The industrial high-energy electron accelerators (10 MeV) described earlier for medical sterilization have also been used to successfully irradiate food and foodstuffs to eliminate pathogens.[47,48,65] The processing systems for attaining acceptable dose distributions with packaged medical products can also be used for food irradiation. In addition, X-rays derived from 5.0 or 7.5 MeV beams can be used because their penetration is similar to that of gamma ray sources.

As an example, the National Center for Electron Beam Research at Texas A&M University (www.ebeam.tamu.edu) that performs EB and X-ray sterilization research and commercial food irradiation has both types of equipment operating in either EB mode or X-ray mode: two opposing, vertically mounted 10 MeV, 18 kW linacs for electrons and one horizontally mounted 5 MeV, 15 kW linac for X-rays. The two 10 MeV electron beams can be used simultaneously to obtain a uniform dose distribution in thick or dense products such as packages of ground beef. The facility has been USDA–FSIS (Food Safety and Inspection Service) approved for commercial food irradiation with the same equipment and conveyance system as is used for handling research materials. There are strict standard operating procedures in place to accommodate both research and commercial processing.[66]

The effectiveness, the minimal effect on nutritive value, and the general safety of food irradiation have been demonstrated over and over again, and the World Health Organization (WHO) has been supporting this method for processing food for many years.[67,68] Some of the food items that can be and have been irradiated include red meat, poultry, and spices. Irradiated spices are widely used in processed foods because these ingredients, which are used for flavor enhancement, are less affected by irradiation than by other forms of pasteurization such as thermal or chemical processes.

A joint expert committee convened by the IAEA, WHO, and Food and Agriculture Organization (FAO) of the United Nations on the wholesomeness of irradiated food announced in 1980 that any food irradiated with a dose below 10 kGy is wholesome. The Codex Alimentarius Commission approved a regulation to accept foods irradiated lower than 10 kGy in 1983 and further approved foods irradiated at higher than 10 kGy if so warranted in 2003. This is contained in the FAO WHO Codex General Standard for Irradiated Foods: Codex Stan 106–1093, rev.1–2003, which also calls for the proper labeling of irradiated food.

There are variations regarding the regulatory compliance of food irradiation in different countries, and universal acceptance of the recommendations of the WHO and FAO has not been achieved. In the US, for example, the USFDA has established specific regulations for the maximum absorbed dose for irradiation of different foods: (i) 4.5 kGy for fresh red meat products, (ii) 3.0 kGy for poultry, (iii) 1.0 kGy for fresh fruits and vegetables, and (iv) 30 kGy for spices.[69]

In Japan, the irradiation of potatoes after harvest for sprout inhibition was approved in 1972, and the commercial irradiation of potatoes has been conducted since 1974, now at $\sim 10^5$ tons per year. Nonetheless, the Ministry of Health and Welfare of Japan has not approved any other irradiated food items for public consumption, and its regulations are not harmonized with the Codex Alimentarius. In the European Union, there are directives in place to guide member states in formulating laws, rules, and regulations on food irradiation, such as Directives 1999/2/EC and 1999/3/EC, and Directives 89/397/EEC and 93/99/EEC. Even so, there remain country-by-country variations as to which particular irradiated foods can be consumed by the public.

Given the outbreak of illnesses in consumers caused by food contaminated with *Escherichia coli* 0157:H7, *Salmonella enteridis*, and other pathogens, and the consequent food product recalls of meat and vegetables, the irradiation of food within the US, for example, is in a state of flux. The reluctance of the major food processors and distributors to adopt the radiation process, more than any consumer resistance, remains a market barrier.

5. Other EB Irradiation Applications

In addition to the examples of major end-use applications described above, there are many other existing or emerging applications of EB irradiation

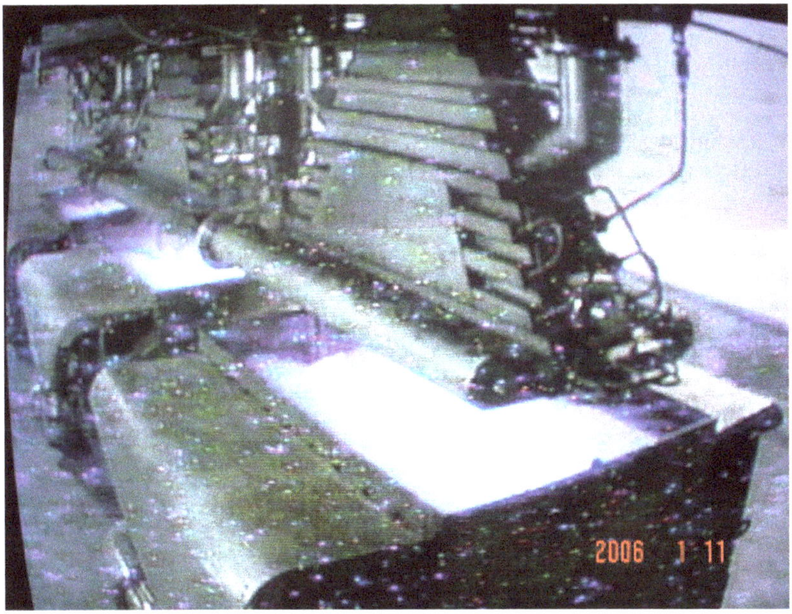

Fig. 28. EB treatment of waste water with dye residual contaminants.

processing. These are proven and effective industrial EB processes, but most are limited by the size of a given market or by limited commercial acceptance. A few examples of such applications with significant industrial potential are given below.

5.1. *Treatment of waste materials*

As with food irradiation, the pathogenic microorganisms in sewage sludge can be killed by EB treatment.[70–72] In addition, toxic organic compounds in wastewater can be decomposed by EB irradiation.[73–75] Full scale demonstration facilities have been constructed as well as transportable EB systems mounted in vans.[76] While transportable systems are not yet being used commercially, there is renewed interest in using EB systems to break down pollutants in wastewater contaminated by cleaning refinery vessels. An upward-flowing wastewater system was developed to use all of the beam power for treating the water.[77] A similar system is being used in a full scale EB water treatment facility in Korea where a 1 MeV 400 kW ELV type accelerator with three beamlines irradiates under-beam troughs to eliminate the residual dyes from a textile manufacturing plant.[78–80] Figure 28 shows these beams in operation.

5.2. *Cleaning of stack gases*

Experiments in Japan in the early 1970s showed that EB processing can be used to eliminate both of the acid-forming gases sulfur dioxide (SO_2) and nitrogen oxides (NO_x) from the exhaust gases of fossil-fueled electric power plants.[81,82] The EB treatment process is unique in that it can eliminate both of these combustion gases. Alternative systems like wet scrubbers can remove SO_2 but not NO_x. When ammonia is injected into the gas being treated, the result is a powdery precipitate of mixed sulfur and ammonium nitrates that can be collected and used as fertilizer.[83–86] Such a system is currently being used in Poland, Japan, and China, and one is planned for Saudi Arabia.

The use of EB irradiation to eliminate contaminants in other gases, such as the exhaust emissions from highway tunnels or volatile organic compounds (VOCs) from industrial factories, has been explored. Some of these processes remain in the pilot or feasibility stage.[87]

5.3. *Curing of composite materials*

EB irradiation has been used to cure the polymeric matrix materials in carbon fiber composites. Recent studies have shown that such matrices can also be cured in a mold to produce shaped articles such as vehicle fenders using X-rays. The X-rays penetrate through the molds and the complex shapes of the products. EB and X-ray curing are not thermal processes, and they cause only a slight rise in temperature. Therefore, residual internal strains within a composite structure caused by differences in thermal expansion coefficients are minimized. Also, the exposure times at the high temperatures needed in the thermo-chemical processes can be reduced.[35,88] Figures 29 and 30

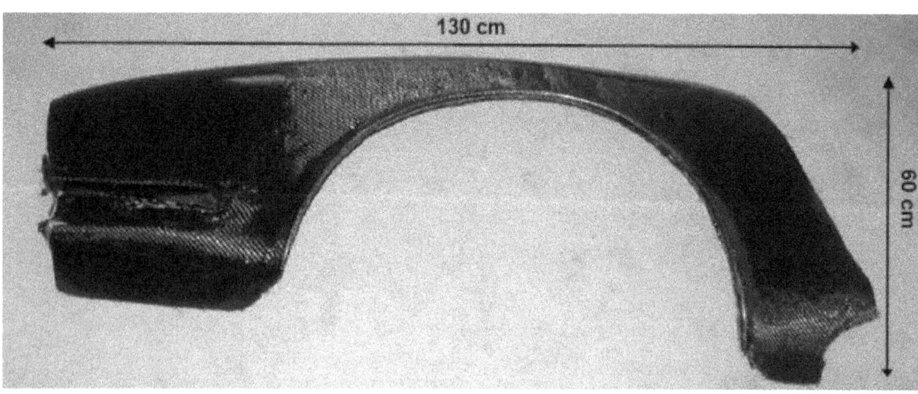

Fig. 29. X-ray cured carbon-fiber composite sports car fender.

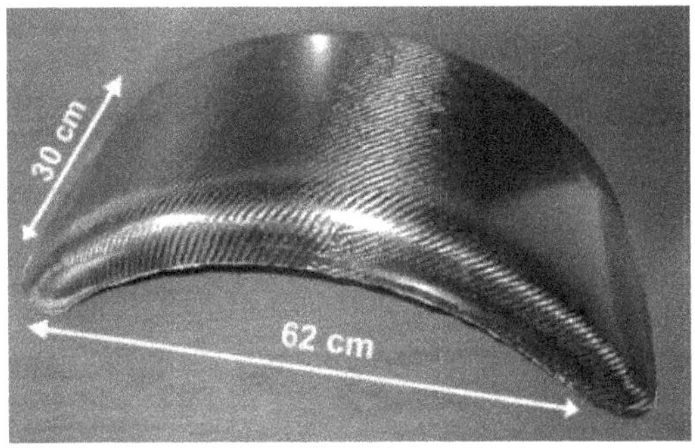

Fig. 30. X-ray cured carbon-fiber composite motorcycle fender.

show a carbon-fiber sports car fender that was X-ray cured in a vacuum-bagged polyester mold and a fender for a wide-wheeled motorcycle that was also X-ray cured in a mold. Both exhibited Class A surfaces, excellent fiber wetting with the matrix material, and satisfactory properties of the final composite products.

Carbon fibers used in composite manufacturing have also been treated with electron beams in order to enhance the adhesion of the matrix to the fiber. Improvements in mechanical properties of cured composites have been observed irrespective of the initial sizing on the fiber.[89–92]

5.4. *Silicon-carbide fiber manufacturing*

Silicon-carbide (SiC) fiber is of interest in space applications. Polycarbosilane is first extruded into fiber strands and then EB irradiated to cross-link the fiber.[93] The conventional process involves heating the polycarbosilane. This produces a fiber with lower heat resistance due to the presence of oxygen. EB cross-linked SiC fiber can maintain high tensile strength up to 1700 °C, while thermally cross-linked SiC fiber keeps its strength only up to 1200 °C.

5.5. *Production of fuel cells*

Proton exchange membranes (PEM) used in fuel cells are being developed based on radiation grafting of styrene onto polymer films such as PTFE (polytetrafluoroethylene), PVDF (polyvinylidene fluoride), PE (polyethylene), and PP (polypropylene) using low-energy EB processing. The grafted surface is then sulfonated. These films have excellent mechanical properties and the desired proton conductivity.[94–96]

Platinum (Pt) and ruthenium (Ru) nanoparticles are used as catalysts in fuel cells. They are prepared by exposing water/ethylene glycol solutions containing Pt(IV) and Ru(III) ions to EB radiation and then depositing the nanoparticles onto carbon particles for support. These PtRu/C catalysts perform very well in fuel cells.[97,98]

5.6. *Cross-linking of PTFE and rubber sheeting*

PTFE can be cross-linked by electron beam irradiation at high temperatures slightly above the melt transition (330–340 °C) while immersed in an inert gas. Increased mechanical properties and wear resistance make the cross-linked PTFE suitable for sliding parts, rollers, and bearings. Commercial quantities of this material are being produced in Japan.[99]

Wide sheets of calendared EB-cross-linked rubber are used for industrial building roofing, for pond and water basin linings, and as a material to prevent leakage from landfills. This type of sheeting is commonly made from EB-cross-linkable polyolefins, most notably formulated ethylene-propylene-diene rubber (EPDM), which responds well to EB processing.[25]

5.7. *Seed and soil disinfestation*

EB treatment is an alternative to the conventional treatment of seeds to control insects with toxic fumigants that can leave trace residues. In this process, seeds flow by gravity between two oppositely directed low-energy (105–145 keV) EB units so that only the surface of the seed is exposed to ionizing irradiation. Tests have shown that treating some seeds with an EB process results in an 11% increase in crop yield — more than double the yield from seeds treated by chemical disinfestation.[100]

EB treatment of soils with doses of ~50 kGy eliminates pathogens and sterilizes the soil. Such sterilized soils can then be inoculated with nitrogen-fixing bacteria, such as *Bradyrhizobium* and *Rhizobium*, to enhance yields for crops such as soy beans.[101,102]

5.8. *Human tissue sterilization*

Radiation sterilization has been approved for tissue banking because it is a non-thermal process. Soft tissues, such as skin to be used for grafting onto patients, are not adversely affected by exposure to sterilizing doses of ionizing radiation, so their mechanical properties remain intact.[103,104]

5.9. *Direct food contact coatings*

The Food Packaging Alliance of the industrial association RadTech International North America has obtained broad USFDA clearance for the use of a variety of acrylate monomers that can be used in formulating direct food contact coatings. Such coatings can be used instead of film laminates to simplify the printing and conversion of food packaging materials by enabling a printer to apply an over-print coating during the printing process. Low-energy EB irradiation can then cure the printing itself as well as the food contact coating.[105]

6. Summary

Materials irradiation is currently the largest application of electron beam accelerators, and the future growth of this industry looks quite robust. As has been described in this chapter, these systems are now widely used in a number of industrial processes and are being developed for even more. Cross-linking, which has been the largest application to date, should find even more new applications as the use of these systems spreads throughout the world. The new high power electron beam accelerators that have been developed specifically for this application can also be used for the sterilization of medical products, which has traditionally been done with radioactive sources.

The potential of food irradiation has barely been tapped due partly to fear of radiation and partly to the slow regulatory process in the food industry. Fortunately, these impediments are being diminished today. The pasteurization and extension of food shelf life are slowly being developed in a regulated environment. With the large volumes of food that could be processed, this may become a huge application of irradiation. The results would lead to a significant improvement in the quality and amount of food available worldwide, particularly in developing countries where food losses are significant because refrigeration is not widely available.

Other EB irradiation applications have been proven to be technically feasible but have not been exploited by industry to any significant degree. These include the treatment of waste materials for degradation or sterilization and the production of composite materials. There are competing technologies in some of these areas, and others are just emerging as industrial processes. However, as the cost of producing and using radiation decreases, many of these applications should mature.

In summary, the future of materials irradiation with electron beam accelerator technology appears very bright. The recent development of new electron beam systems specifically for these applications, as well as the expanded knowledge of irradiation processes, has resulted in numerous new opportunities for near term markets as well as long term developments.

Acknowledgements

The author would like to thank his long-time colleague Anthony J. Berejka for his tireless efforts to teach an old nuclear physicist enough radiation chemistry to understand the chemical processes involved in many of the EB polymerization, cross-linking, and sterilization applications briefly described in this chapter. The substantial contributions of the editors, Robert and Marianne Hamm, to the content and style are also appreicated.

References

1. A. J. Berejka, in *Emerging Applications of Radiation Processing*, IAEA-TECDOC-1386 (International Atomic Energy Agency, Vienna, 2004), pp. 65–72.
2. S. Tagawa *et al.*, *J. Nucl. Sci. Tech.* **39**(9), 1002 (2002).
3. Y. Lin, in *Proc. Second Asian Part. Acc. Conf.* 852 (2001), available at http://accelconf.web.cern.ch/AccelConf/a01/PDF/FRAM01.pdf.
4. S. Machi, presented at *Topical Meeting AccApp09, IAEA Satellite Meeting Appl. Electron Accel.*, paper SM-EB-04 (International Atomic Energy Agency, Vienna, 2009), available at http://www-pub.iaea.org/MTCD/publications/PDF/P1433_CD/datasets/abstracts/sm_eb-04.html.
5. A. G. Chmielewski and A. J. Berejka, Radiation Sterilization Centers Worldwide, in *Trends in Radiation Sterilization of Health Care Products* (International Atomic Energy Agency, Vienna, 2008), pp. 49–62.
6. *Radiation Source Use and Replacement*, National Research Council of the National Academies (The National Academies Press, Washington DC, 2008).
7. J. Meissner *et al.*, *Rad. Phys. Chem.* **57**(3–6), 647 (2000).
8. Y. Jongen *et al.*, in *Emerging Applications of Radiation Processing*, IAEA-TECDOC-1386, (International Atomic Energy Agency, Vienna, 2004), pp. 44–54.
9. A. J. Berejka, *Nucl. Instr. Meth. Phys. Res.* **B261**, 86 (2007).
10. Papers in Proc. Int. Meetings on Radiation Processing, *Rad. Phys. Chem.* **9**(1–3, 4–6) (1977), **14**(1–2, 3–6) (1979), **18**(1–2, 3–4, 5–6) (1981), **22**(1–2, 3–5) (1983), **25**(1–3, 4–6) (1985), **31**(1–3, 4–6) (1988), **35**(1–3, 4–6) (1990), **42**(1–3, 4–6) (1993), **46**(4–6), Part 1 and 2 (1995), **52**(1–6) (1998), **57**(3–6) (2000), **63**(3–6) (2002), **71**(1–2) (2004), **76**(11–12) (2007), **78**(7–8) (2009) (Elsevier Ltd., Oxford).

11. J. W. T. Spinks and R. J. Woods, *An Introduction to Radiation Chemistry*, 3rd Edition (Wiley-Interscience, New York, 1990).
12. A. Singh and J. Silverman, Eds., *Radiation Processing of Polymers* (Oxford University Press, Hanser Publishers, New York, 1992).
13. J. G. Drobny, *Radiation Technology for Polymers* (CRC Press, Boca Raton, 2003).
14. R. Bradley, *Radiation Technology Handbook* (Marcel Dekker, New York, 1984).
15. J. H. Bly, *Electron Beam Processing* (International Information Associates, Yardley, 1988).
16. R. J. Woods and A. K. Pikaev, *Applied Radiation Chemistry: Radiation Processing* (Wiley-Interscience, New York, 1994).
17. M. R. Cleland and J. A. Beck, in *Encyclopedia of Pharmaceutical Technology*, Volume 5, Eds. J. Swarbrick and J. C. Bolan (Marcel Dekker, New York, 1992), p. 105.
18. M. R. Cleland, M. T. O'Neill and C. C. Thompson, in *Sterilization Technology*, Chapter 9, Eds. R. F. Morrissey and G. B. Phillips (Van Nostrand Reinhold, New York, 1993), p. 218.
19. *Trends in Radiation Sterilization of Health Care Products*, (International Atomic Energy Agency, Vienna, 2008).
20. P. Loaharanu and P. Thomas, Eds., *Irradiation for Food Safety and Quality* (Technomic Publishing Co., Lancaster, 2001).
21. V. Komolprasert and K. M. Morehouse, *Irradiation of Food and Packaging*, (American Chemical Society, Washington DC, 2004).
22. C. H. Sommers and X. Fan, Eds., *Food Irradiation Research and Technology* (IFT Press, Blackwell Publishing Ltd., Oxford, 2006).
23. E. A. Abramyan, *Industrial Electron Accelerators and Applications* (Hemisphere Publishing Corporation, New York, 1988).
24. W. H. Scharf, *Particle Accelerators — Applications in Technology and Research* (Research Studies Press Ltd., Taunton, 1989).
25. *Industrial Radiation Processing with Electron Beams and X-rays*, Booklet in press (International Atomic Energy Agency, Vienna, 2011).
26. *Standard Practice for Dosimetry in an Electron-Beam Facility for Radiation Processing at Energies Between 300 keV and 25 MeV*, ASTM INTERNATIONAL ISO/ASTM 51649, Nuclear (II), Solar, and Geothermal Energy, Vol. **12.02**.
27. M. R. Cleland, R. A. Galloway and A. J. Berejka, *Nucl. Instr. Meth. Phys. Res.* **B261**, 94 (2007).
28. M. R. Cleland, T. F. Lisanti, and R. A. Galloway, *RDI-IBA Technical Information Series* TIS 01556 (IBA Industrial, Inc., Edgewood, NY, 2003).
29. *Radiation Protection for Particle Accelerator Facilities*, NCRP Report No. 144 (National Council on Radiation Protection and Measurements Washington DC, 2003).

30. J. P. Farrell, *Rad. Phys. Chem.* **14**(3–6), 377 (1979).
31. J. P. Farrell, S. M. Seltzer and J. Silverman, *Rad. Phys. Chem.* **22**(3–5), 469 (1983).
32. S. M. Seltzer, J. P. Farrell and J. Silverman, *IEEE Trans. Nucl. Sci.* **NS-30**(2), 1629 (1983).
33. R. A. Galloway, T. F. Lisanti and M. R. Cleland, *Rad. Phys. Chem.* **71**(1–2), 551 (2004).
34. M. Abs, Y. Jongen, E. Poncelet and J.-L. Bol, *Rad. Phys. Chem.* **71**(1–2), 287 (2004).
35. A. J. Berejka, M. R. Cleland, R. Galloway and O. Gregoire, *Nucl. Instr. Meth. Phys. Res.* **B241**, 847 (2005).
36. M. R. Cleland in *Advances in Radiation Chemistry of Polymers*, IAEA-TEC-DOC-1420 (International Atomic Energy Agency, Vienna, 2002), pp. 111–123.
37. *Standard Practice for Use of Calorimetric Dosimeter Systems for Electron Beam Dose Measurements and Routine Dosimeter Calibration*, ISO/ASTM51631:2003(E) Vol. 12.02 (ASTM International, West Conshohocken, 2011).
38. A. J. Berejka, M. Driscoll and D. Montoney, in *Proc. International Polyolefins Conference 2009*, Society of Plastics Engineers (Curan Associates, New York, 2009).
39. M. R. Cleland, R. Galloway, F. Genin and M. Lindholm, *Rad. Phys. Chem.* **63**(3–6), 729 (2002).
40. B. S. Quintal, US Patent No. 3,702,412 (1972).
41. S. R. Farrell, L. J. Demeter and P. G. Wood, US Patent No. 3,863,163 (1975).
42. X. Coqueret, B. DeFoort, J. M. Dupilier and G. Larnac, in *Radtech Europe Conf. Papers Archive 2003,* http://www.radtech-europe.com/files_content/december%202004%20papers/coqueretpaper december2004.pdf.
43. S. Uehara *et al.*, *Rad. Phys. Chem.* **42**(1–3), 515 (1993).
44. K. Mizusawa, M. Kashiwagi and Y. Hoshi, *Rad. Phys. Chem.* **52**(1–6), 475 (1998).
45. M. R. Cleland, US Patent No. 2,875,394 (1959).
46. R. R. Smith and S. R. Farrell, in *14*th *Int. Conf. Appl. Acc. Res. Ind.*, AIP Conference Proceedings **392** (American Institute of Physics, Melville, 1997), pp. 1093–1098.
47. C.-L. Gallien, J. Paquin and T. Sadat, *Rad. Phys. Chem.* **22**(3–5), 759 (1983).
48. T. Sadat, *Rad. Phys. Chem.* **25**(1–3), 81 (1985).
49. V. S. Podobaev *et al.*, in *Proc. RuPAC-2010* (JACoW, 2010), pp. 411–413, available at http://accelconf.web.cern.ch/accelconf/r10/papers/frchb03.pdf.
50. W. G. Baird, Jr. *et al.*, US Patent 3,022,543 (1962).
51. W. G. Baird, Jr., *Rad. Phys. Chem.* **9**(1–3), 225 (1977).
52. J. V. Koleske, *Radiation Curing of Coatings* (ASTM International, West Conshohocken, 2002).

53. J. D. Hunt and G. Alliger, *Rad. Phys. Chem.* **14**(1–2), 39 (1979).
54. US Patents on Tire Irradiation: 3,933,553 (1976), 3,933,566 (1976), 4,089,360 (1978), 4,102,761 (1978), 4,108,749 (1978), 4,122,137 (1978), 4,139,405 (1979), 4,166,883 (1979), 4,176,702 (1979), 4,202,717 (1980), 4,221,253 (1980), 4,230,649 (1980), 4,756,782 (1988), 4,851,063 (1989).
55. D. A. Trageser, *Rad. Phys. Chem.* **9**(1–3), 261 (1977).
56. J. Kotler and J. Borsa, US Patent 6,504,898 (2003).
57. F. Stichelbaut *et al.*, *Rad. Phys. Chem.* **71**, 291 (2004).
58. I. Kaluska and Z. Zimek, in *Proc. Symp. Techniques for High Dose Dosimetry in Industry, Agriculture and Medicine*, IAEA-TEC-DOC-1017, IAEA-SM-356/33 (International Atomic Energy Agency, Vienna, 1999).
59. A. G. Chmielewski, T. Sadat and Z. Zimek, in *Trends in Radiation Sterilization of Health Care Products* (International Atomic Energy Agency, Vienna, 2008), pp. 40–41.
60. D. Morisseau, P. Fontcuberta and F. Malcolm, in *Proc. Int. Topical Meeting on Nucl. Res. Appl. and Utilization of Accelerators* (International Atomic Energy Agency, Vienna, 2009), SM/EB-08.
61. F. Minck, *Muenchener Medicinische Wochenschrift* **43**(5), 101 (1896).
62. H. Lieber, US Patent 788,480 (1905).
63. J. Appleby and A. J. Banks, British patent GB 1609 (1906).
64. B. Schwartz, *J Agri. Res.* **20**, 845 (1921).
65. J. F. Diehl, *Rad. Phys. Chem.* **62**(3–6), 211 (2002).
66. Dr. Suresh Pillai, private communication, National Center for Electron Beam Research, Texas AM University.
67. *Report of a Joint FOA/IAEA/Who Expert Committee: Wholesomeness of Irradiated Food*, WHO Technical Report Series 659 (World Health Organization, Geneva, 1981).
68. *Food Irradiation: In Point of Fact* **40** (World Health Organization, Geneva, 1987).
69. *Fourth Report on Needs in Ionizing Radiation Measurements and Standards* (Council on Ionizing Radiation Measurements and Standards, Duluth, 2004), pp. 86–92, CD-ROM.
70. J. G. Trump, US Patent 3,901,807 (1975).
71. J. G. Trump, E. M. Merrill and K. A. Wright, *Rad. Phys. Chem.* **24**(1), 55 (1984).
72. E. H. Bryan *et al.*, in *Radiation Energy Treatment of Water, Wastewater and Sludge* (American Society of Civil Engineers, New York, 1992).
73. W. J. Cooper, R. D. Curry and K. E. O'Shea, Eds., *Environmental Applications of Ionizing Radiation* (John Wiley & Sons, New York, 1998).
74. M. H. O. Sampa, P. R. Rela and C. L. Duarte, in *Environmental Applications of Ionizing Radiation* (John Wiley & Sons, New York, 1998), pp. 521–530.

75. C. Kurucz, in *Proc. Work. Appl. Ion. Rad. Decon. Env. Resources* (National Science Foundation, Environmental Engineering Program, Washington DC, 1994), pp. 79–89.
76. M. N. Schuetz and D. A. Vroom in *Environmental Applications of Ionizing Radiation* (John Wiley & Sons, New York, 1998), pp. 63–82.
77. P. R. Rela *et al.*, *Rad. Phys. Chem.* **57**, 657 (2000).
78. B. Han *et al.*, in *Radiation Treatment of Gaseous and Liquid Effluents for Contaminant Removal*, IAEA-TECDOC-1473 (International Atomic Energy Agency, Vienna, 2005), pp. 101–110.
79. D. Kim *et al.*, US Patent 6,121,507 (2000).
80. B. Han *et al.*, in *Proc RuPAC 2006* (JACoW, 2006), pp. 123–125, available at http://accelconf.web.cern.ch/accelconf/r06/PAPERS/THLO02.PDF.
81. S. Machi *et al.*, *Rad. Phys. Chem.* **9**, 371–388 (1977).
82. S. Machi, *Rad. Phys. Chem.* **22**, 91–97 (1983).
83. A. G. Chmielewski *et al.*, *Modern Power Systems*, 53 (2002).
84. A. G. Chmielewski *et al.*, *Rad. Phys. Chem.* **71**(1–2), 441 (2004).
85. A. G. Chmielewski, *Rad. Phys. Chem.* **76**(8–9), 1480 (2007).
86. L. Genli *et al.*, in *Environmental Applications of Ionizing Radiation* (John Wiley & Sons, New York, 1998), pp. 113–121.
87. O. Tokunaga, in *Environmental Applications of Ionizing Radiation* (John Wiley & Sons, New York, 1998), p. 108.
88. A. J. Berejka, in *Proc. 2007 SAMPE Conf. Baltimore, MD* (Society for the Advancement of Material and Process Engineering, Baltimore, 2007), available at http://www.sampe.org/store/paper.aspx?pid=4573#p4573.
89. C. Giovedi *et al.*, *Nucl. Instr. Meth. Phys. Res.* **B236**, 526 (2005).
90. C. Giovedi *et al.*, in *Proc. 2nd Inter. Symp. Utilization of Accelerators*, IAEA-CN-115-29 (International Atomic Energy Agency, Vienna, 2005).
91. E. S. Pino, L. D. B. Machado and C. Giovedi, *Nucl. Sci. Techn.* **18**(1), 39 (2007).
92. C. Giovedi, L. D. B. Machado and E. S. Pino, in *Proc. Inter. Nuclear Atlantic Conf. INAC 2007*, available at http://www.ipen.br/biblioteca/2007/inac/11960.pdf.
93. M. Sugimoto, T. Shimoo, K. Okamura and T. Segucki, *J. American Ceramic Soc.* **78**(4), 1013 (1995).
94. M. Yoshida *et al.*, Japanese Patent, filed no. 2008-049319 (2008).
95. J. Chen *et al.*, *Rad. Phys. Chem.* **76**, 1367 (2007).
96. F. Muto *et al.*, *Nucl. Instr. Meth. Phys. Res.* **B265**(1), 162 (2007).
97. D. F. Silva *et al.*, *Mat. Res.* **10**, 367 (2007).
98. D. F. Silva *et al.*, in *Proc. Inter. Nuclear Atlantic Conf. INAC 2007*, available at http://www.ipen.br/biblioteca/2007/inac/11959.pdf.
99. T. Seguchi, *Rad. Phys. Chem.* **57**, 367 (2000).
100. C. Zago and P. R. Rela, in *Proc. Inter. Nuclear Atlantic Conf. INAC 2007*, available at www.ipen.br/biblioteca/2007/inac/12065.pdf.

101. D. Tsai and P. R. Rela, Master of Science Dissertation, São Paulo University and Institute for Nuclear Energy Research (2006).
102. S. M. Tsai *et al.*, *Visão Agrícola* **5**(Ano 3), 31 (2006).
103. S. C. Bourroul, M. R. Herson, E. S. Pino and M. B. Mathor, *Cell. Mol. Biol.* **48**(7), 803 (2002).
104. A. Dziedzic-Goclawska *et al.*, *Cell and Tissue Banking* **6**, 201 (2005).
105. Adtech International North America Food Packaging Alliance: see www.radtech.org/Industry/food.htm.

Chapter 4

ACCELERATOR PRODUCTION OF RADIONUCLIDES

David J. Schlyer
Medical Department, Brookhaven National Laboratory
Upton, NY 11973, USA
schlyer@bnl.gov

Thomas J. Ruth
TRIUMF and BC Cancer Agency, Vancouver, BC, Canada
truth@triumf.ca

While many radioactive isotopes in use today are found in nature, many more are artificially produced by irradiating target materials with nuclear particles. Two different technologies can provide the energetic particles needed: nuclear reactors, which produce a flux of neutrons, and particle accelerators, which produce a flux of charged particles. This chapter will deal with the important aspects of the production of radionuclides with accelerators, along with some details on their applications, commercially-available accelerator systems used for this purpose, and the size of the equipment business.

1. Introduction

The early history of the practical use of radioactive isotopes (commonly called radionuclides or radioisotopes) and the birth of accelerators capable of producing them are intertwined with many pivotal events and discoveries in nuclear physics and nuclear chemistry in the early 1900s. While the existence of naturally-occurring radioactive substances had been known since the 1890s, the first known practical application of a radionuclide as a tracer was made in 1911 by George de Hevesy (considered the father of nuclear medicine).[1] At the time, de Hevesy was a young Hungarian researcher working with naturally-occurring radioactive materials in the Manchester laboratory (UK) of Ernest Rutherford. Not having much money, he lived in a boardinghouse and ate his meals there. He began to suspect that some of the meals might be made from leftovers from the preceding days or even weeks, but he could never be sure. To confirm his suspicions, de Hevesy put a small amount of radioactive material into the remains of a meal. Several days later when the same dish was served again, he used a simple radiation detector (a gold leaf electroscope) to check if the food was radioactive. It was, and his suspicions were confirmed. (According to the story, the landlady was furious and threw him out of the boardinghouse.) De Hevesy went on to perform pioneering work on radioactive tracers, and in 1923 he was the first to use them to study processes in living systems. He won the Nobel Prize in chemistry in 1943 for his work on the use of isotopes as tracers in the study of chemical processes.

A key discovery was made in 1919 by Ernest Rutherford (who had already won the Nobel Prize in chemistry in 1908 for his research on radioactivity) when he was the first to recognize that one element could be transformed into another by artificial means.[2] He showed that when nitrogen gas was bombarded with α particles (the nucleus of a ^{4}He atom) from a radioactive source, an α was occasionally stopped, and at the same time a proton was emitted with a great deal of kinetic energy. This was actually the first demonstration of a man-made nuclear reaction: he had formed oxygen-17 — a naturally-occurring stable isotope of oxygen unknown at that time — via the ^{14}N(α, p)^{17}O reaction. This production of a stable isotope using atomic particles led the way to the production of artificial radioactive isotopes a few years later.

Rutherford's early research on nuclear reactions was limited because he was constrained to using small numbers of low energy α particles from radioactive sources for his projectiles. In 1927, he announced that what was needed was a reasonably sized apparatus that could produce a large supply of energetic particles and thereby launched a race of sorts among researchers at that time to conceive and demonstrate a device that could accelerate particles to energies of many MeV. This work quickly led to the invention of the Cockcroft–Walton and Van de Graaff high voltage electrostatic devices, but these were limited in the energy or beam current they could produce. Concepts for linear accelerators also emerged, but these were either impractical to build or needed a source of RF power that was not available at the time. The real breakthrough came in 1929–1930 with the invention of the cyclotron by Ernest O. Lawrence after he saw a paper on Rolf Wideröe's concept of linearly accelerating particles in many small steps using an alternating voltage. Even though he could read little or no German, he understood the schematic. He knew that charged particles could be bent with a magnetic field and from the paper he realized that those particles could be accelerated by a small incremental voltage.[3] Neither concept was new, but Lawrence was the first to put the two together. In 1930, the first cyclotron models were produced by his students Nels Edlefsen and M. Stanley Livingston at the University of California in Berkeley, successfully demonstrating the concept.[4,5] Bigger and more powerful versions were built and installed over the next several years at the Radiation Laboratory at Berkeley (dubbed the "Rad Lab"). Lawrence received the Nobel Prize in physics in 1939 for his invention. (An excellent treatment of these and other key events in the history of accelerators can be found in *Engines of Discovery – A Century of Particle Accelerators*.[6])

The final key ingredient was the discovery that radioactive elements could be artificially produced. In 1932, Irene Curie and Frederic Joliot observed that the emission of energy they had been studying by bombarding boron with α particles (from radium) continued even after the source was removed. They subsequently proved (by chemical means) that an unstable isotope of nitrogen — ^{13}N, which they dubbed "radio-nitrogen" — had been created. In 1934, they reported this first discovery of the production of "artificial radioactivity"[7] and won the Nobel Prize for their work in 1935. They further suggested that these substances might well be produced using other projectiles such as protons, deuterons, or neutrons.

After learning of the Curie–Joliot discovery, Lawrence realized that he and his fellow researchers had already been creating radioactive isotopes in their studies of deuteron-induced disintegrations of various target elements using the Rad Lab cyclotrons. In fact, the high energy charged particles and high beam currents from those and even larger cyclotrons built soon after would make it possible to produce a wide variety of radioactive isotopes, thereby ushering in their wide-spread use in research, industry, and medicine.

One early example of this was the production and use of radioactive carbon isotopes by Samuel Ruben and Martin Kamen to study the pathways in photosynthesis. They had been using ^{11}C produced by bombarding ^{10}B with deuterons from Lawrence's 37-inch cyclotron to map the pathways of carbon dioxide in plants, but the half-life was too short. They then decided to try to make a new isotope of carbon that had been suggested to exist by Curie in 1934[8] by continuously irradiating graphite with deuterons in the cyclotron for 120 hours. At 8 a.m. on February 27, 1940, Kamen rushed into Ruben's office with the sample just out of the cyclotron. Ruben analyzed it and found they had produced just enough ^{14}C to conclusively identify it. As nuclear reactors became available after World War II, they produced ^{14}C from neutron capture, which allowed them to make much greater quantities and led to the use of ^{14}C as a tracer to examine a huge number of pathways in biology and medicine.

Even though there are other examples of the early use of accelerator-produced radionuclides as tracers, it took several decades before they began to play a major role as radiotracers and radiopharmaceuticals. In the 1950s, the production of radionuclides for medical use started to grow, due in large part to the discovery that ^{201}Tl could be used as an ideal tracer for studying blood flow in the heart.[9] The formulation of ^{18}F-tagged glucose analog (FDG) in the mid-1970s and the demonstration of its use for studying metabolism was a major breakthrough leading to the development of the now widely-used nuclear medicine imaging modality known as Positron Emission Tomography (PET).[10]

The cyclotrons that were used for many years to produce radionuclides were initially designed as physics research machines capable of accelerating protons, deuterons, $^3He^{+2}$, and ^{4}He (α particles). However, most of the radionuclides in routine use today are produced with beams of protons. The simplicity of design for proton-only cyclotrons has resulted in systems that are capable of generating two or more simultaneous beams of varying energies and intensities.[11] While cyclotrons are by far the dominant type of accelerator used in the commercial production of radionuclides, modern RF linear accelerators have also been developed for this purpose. The radio frequency linear accelerator (RF linac) is limited to a single beam, but the reduced shielding required and the operational stability of linacs offer some advantages over cyclotrons. Both modern cyclotrons and linacs are completely computer-controlled and can be in continuous operation for extended periods with only minimal attention.

There are three major reasons why accelerator-produced radionuclides are so widely used and are becoming ever more popular: (1) accelerator-produced radionuclides have more favorable decay characteristics (particle emission, half-life, gamma rays, etc.) compared to those produced by reactors; (2) radionuclides cannot usually be produced in a reactor with high specific activity (the activity of a given radionuclide per unit mass); and (3) perhaps most important is that access to a reactor is often very limited. The number of reactors available for radionuclide production has diminished while the number of accelerators used for that purpose has grown significantly.

The remainder of this chapter will give an overview of some of the radionuclides that are routinely produced by accelerators, their applications, and a description of the accelerators commonly used to produce them. In addition, some of the basic principles of radionuclide production with an accelerator will be presented, including some of the important physics, chemistry, and engineering issues involved in creating practical radionuclide production targets.

2. Applications of Radionuclides

Radioactive materials are pervasive in modern society, with the largest use by far being in nuclear medicine. While there are many non-medical industrial uses, the majority of them employ reactor-produced radionuclides. A few examples of these industrial uses are:

- Many large manufacturers use radionuclides in the production of metals, chemicals, plastics, paper, pharmaceuticals, rubber, glass and clay products, food, tobacco, textiles, and other products.
- Various industries use radiotracers to study mixing efficiency and gas or fluid flow rates and patterns in columns and fractionation towers.
- Radioactive sources are used in the oil, gas, and geothermal industries.
- Tiny ^{241}Am sources are incorporated into many smoke detectors, and photocopiers use small amounts of radioactive material to eliminate static and prevent paper jams.

More than 50 known radionuclides are now produced with accelerators for use in nuclear medicine to diagnose disease and treat cancer. Some of the most widely used are listed in Table 1 along with their half-lives and areas of application. Many other examples can be found in the literature and in the recently published IAEA Technical Report TRS 468.[12] For more in-depth information on these and other accelerator-produced radionuclides, the interested reader should consult that document and the numerous references cited therein.

2.1. *Radiotracers*

The term radiotracer refers to a radioactive species that is used to follow (i.e., trace) the uptake into or function of an organ system in a living plant or animal or in a physical/chemical process. In most applications of radiotracers, the radionuclides are used in "trace" (i.e., very small) concentrations and are often incorporated into a biologically- or chemically-significant molecule or compound. By definition, the radionuclides and the compounds to which they are attached must obey three basic tracer principles:

(1) The tracer behaves or interacts with the system to be probed in a known and reproducible fashion.

Table 1. Some common accelerator-produced radionuclides.

Radionuclide	Half-life	Primary uses
Astatine-211 (^{211}At)	7.2 h	Radioimmunotherapy
Cadmium-109 (^{109}Cd)	461.4 d	Industrial quantitative analysis, environmental and biomedical tracer studies, and some interest in nuclear medicine as a generator of ^{109m}Ag
Carbon-11 (^{11}C)	20.4 min	PET imaging and biomedical studies
Cobalt-57 (^{57}Co)	271.8 d	Nuclear medicine
Copper-64 (^{64}Cu)	12.7 h	PET imaging and radioimmunotherapy
Copper-67 (^{67}Cu)	62 h	Radioimmunotherapy
Fluorine-18 (^{18}F)	109.8 min	PET imaging
Gallium-67 (^{67}Ga)	3.26 d	SPECT imaging
Germanium-68 (^{68}Ge)	270.8 d	Calibration of PET cameras and use as a ^{68}Ga generator for biomedical studies
Indium-111 (^{111}In)	2.83 d	SPECT imaging
Iodine-123 (^{123}I)	13.2 h	SPECT imaging
Iodine-124 (^{124}I)	4.2 d	PET imaging and radiotherapy
Nitrogen-13 (^{13}N)	10 min	PET imaging
Oxygen-15 (^{15}O)	122 s	PET imaging
Palladium-103 (^{103}Pd)	17 d	Interstitial seeds for cancer brachytherapy
Sodium-22 (^{22}Na)	2.6 a	Calibration of PET cameras and ion chambers
Thallium-201 (^{201}Tl)	73.06 h	SPECT imaging
Yttrium-86 (^{86}Y)	14.7 h	Targeted therapy imaging

(2) The tracer does not alter or perturb the system in any measurable fashion.
(3) The concentration of the tracer can be measured.

A notable exception is in the use of radionuclides for internal radiotherapy to treat cancer and other diseases, where the second principle is in a strict sense broken since the point of delivering the radiotoxic substance is to deliberately cause damage to the targeted tumor or tissues. However, in order for the radiotoxic substance to localize, it must still follow a known chemical behavior without perturbing the pathway that leads to the localization.

Radioactive isotopes of naturally occurring elements or chemical congeners of these elements were first used as radiotracers and continue to be used to the present day. Examples of these include radioiodine used to monitor iodine uptake in the thyroid and to label a variety of compounds. Iodine-123 has been of interest for nearly three decades because of its unique chemistry that makes it possible to attach it to a wide variety of molecules and its γ-ray energy (159 keV) that is well matched to SPECT (Single Photon Emission Computed Tomography) cameras. The ability to produce this isotope in high purity from enriched ^{124}Xe targets makes it possible to ship ^{123}I over long distances and still have enough specific activity to use for SPECT.

2.2. *Nuclear medicine imaging*

Nuclear medicine imaging differs from other radiological imaging modalities such as X-ray and computed tomography (CT) in that radiotracers are used to map out the function of an organ system or metabolic pathway. Thus, imaging the concentration of these agents in the body can reveal the integrity of these systems or pathways. This is the basis for the unique information that a nuclear medicine scan can provide for a wide variety of organ/functional systems of the body. Research into these biological processes is also greatly aided by the use of radiotracers to elucidate pathways and biodistribution.

Two widely used types of radionuclide imaging are SPECT and PET. The principles involved in each are similar: a radioactive tracer is injected into the body and the subsequent gamma radiation emitted is detected outside of the body. This is in contrast to X-ray or CT imaging where the source of the radiation is outside of the body and the information provided is mainly anatomical. SPECT generally uses longer-lived radiotracers that can be stored on site, while PET mainly utilizes shorter-lived positron (β^+) emitters that must either be made in close proximity to the use site or delivered from a production facility no more than a few hours away (as in the case of ^{18}F-FDG).[13,14]

SPECT imaging is accomplished using several rings of detectors that are equipped with heavy metal collimators to determine the position within the body of the source of gamma rays coming from the radiotracer given to the patient. Gamma rays that make it through the collimator come along a straight line from a volume within the body. This information is then used to reconstruct a three dimensional image of the distribution of radioactivity. Smaller collimators result in better image resolution, but the penalty is that fewer gamma rays enter the detector and the scan time must be longer to get a good image. The basic principle of SPECT imaging is illustrated in Fig. 1.

At the present time, SPECT imaging is much more widely used than PET imaging, but PET is growing at a more rapid rate. It is estimated that in 2010

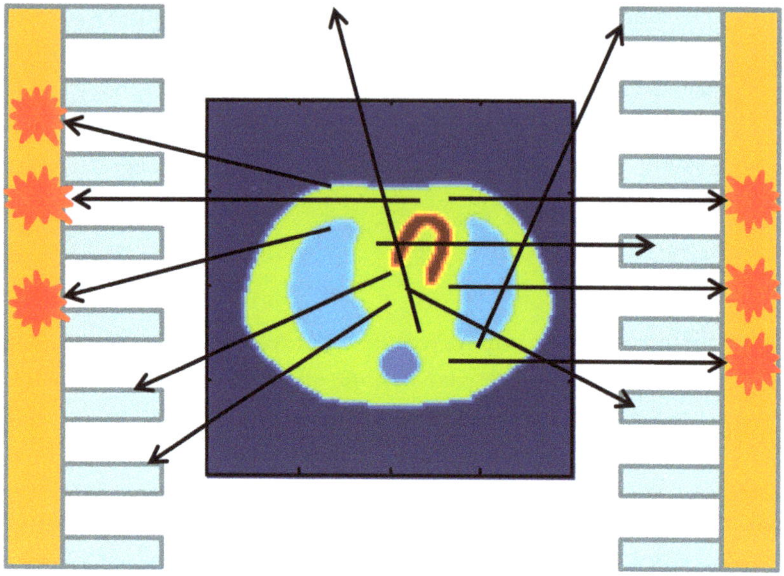

Fig. 1. Illustration of SPECT imaging showing the gamma rays that are used to reconstruct the image of the radiation.

more than 26 million SPECT scans were performed annually worldwide and the sale of the radiopharmaceuticals used exceeded US$3B. Approximately 85% of SPECT procedures are for cardiology using ^{99m}Tc usually derived from a generator of ^{99}Mo produced in a reactor. However, there is increasing interest in making ^{99m}Tc directly with a particle accelerator, in large part due to concerns over recurring shortages of reactor-produced ^{99}Mo. The other SPECT procedures use mainly cyclotron-produced radionuclides such as ^{201}Tl, ^{123}I, ^{67}Ga, and ^{111}In. Each of these has particular medical uses. Thallium-201 is used for the diagnosis of coronary artery disease and other heart conditions such as heart muscle death, and for the location of low-grade lymphomas. Iodine-123 can be incorporated into a wide variety of compounds and is used for many procedures, including the diagnosis of thyroid function. Gallium-67 is used for tumor imaging and localization of inflammatory lesions (infections). Indium-111 is used for specialist diagnostic studies, e.g., brain studies, infection, and colon transit studies.

PET imaging is similar to SPECT in that detectors are used to determine the distribution of the radiotracer that was injected or inhaled into the body. The difference is that PET relies on the detection of the two coincident, back-to-back 511 keV gamma rays that are created when the positrons emitted by the PET radiotracer quickly annihilate with electrons

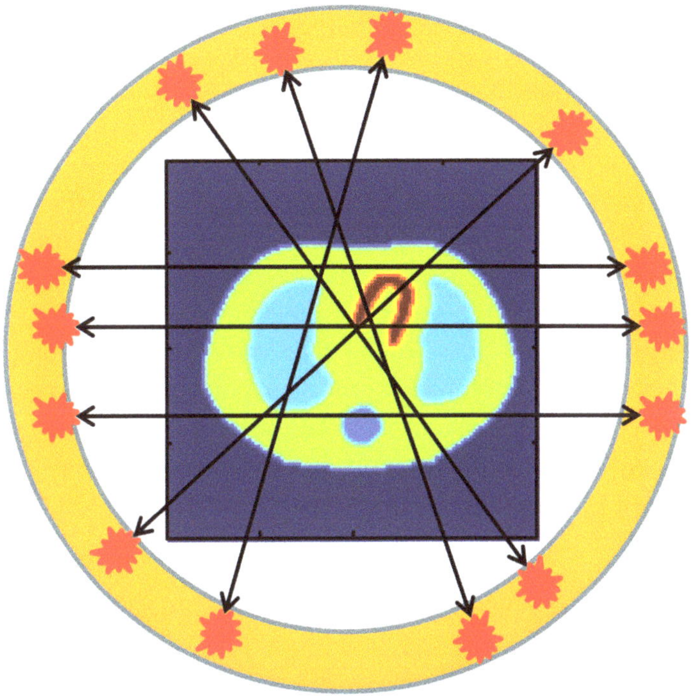

Fig. 2. Schematic diagram of PET imaging illustrating the colinear detection of the back-to-back gamma rays from the local annihilation of the positron emitted by the radiation source in the body.

in the body. Hence, electronic collimation can be employed in place of the heavy metal collimators used in SPECT. If two gamma rays arrive at detectors on opposite sides of the body in coincidence (i.e., within a few ns) then the location of the original positron decay must have taken place somewhere on a straight line between them. This principle is illustrated in Fig. 2. As in SPECT, a 3D image of the location of the radiation in the whole body can be reconstructed by moving the patient through rings of detectors. However, using the detection of only the 511 keV colinear coincident gammas, modern PET clinical scanners have a resolution of ~4 mm while the resolution of a typical SPECT image is ~9 mm.

PET was initially utilized by medical researchers to identify tumors using a variety of positron emitting radionuclides without a great deal of success. A key breakthrough came with the development of FDG (^{18}F-tagged glucose analog). Initially, FDG was used to study how the brain functions by injecting this tracer into the blood stream and imaging the radioactivity under various conditions of brain activation. FDG is a good tracer for glucose metabolism because it is transported into the cells just like normal glucose but then remains and decays in the cells because it cannot be completely metabolized like normal glucose. PET has since been recognized as an extremely powerful tool for the detection of many localized tumors and metastatic cancer, due mainly to the use of the four most common PET radionuclides: ^{11}C, ^{13}N, ^{15}O, and ^{18}F. The stable isotopes of these elements are among the most basic building blocks of life, and therefore these positron-emitters can easily be incorporated into organic molecules that can act as true tracers for an incredible variety of biological processes.

The use of this powerful medical diagnostic imaging tool has experienced rapid growth in recent years as many of the procedures for use in oncology and cardiology have received reimbursement approval from medical payment entities around the world. In 2010, more than two million PET procedures were performed at more than 2000 facilities in the US (currently the largest market segment for PET imaging). The sale of PET equipment and radiopharmaceuticals now exceeds US$2B per year worldwide according to a number of market surveys, and this business is growing at a compounded annual growth rate exceeding 8%.

2.3. *Therapeutic and other medical applications*

The idea behind the therapeutic application of radionuclides is based on the desire to link a radioactive atom having a high linear energy transfer associated with its decay products (Auger electrons, β-particles, or α-particles) to a biologically active molecule that can be directed to a tumor site. Since the β^--emitting radionuclides are neutron rich, they have in general been produced in reactors. There are, however, a few therapeutic radionuclides that are routinely produced with an accelerator and a few more that show promise for future radiotherapeutic applications.

2.3.1. *Brachytherapy*

In brachytherapy, radioactive seeds or sources are placed in or near a tumor to deliver a high radiation dose to the tumor while reducing the radiation exposure in the surrounding healthy tissues. The term "brachy" is Greek for "short distance"; hence, brachytherapy is radiation therapy given at a short distance resulting in a localized and precise dose to the tumor. Because the radiation sources can be precisely positioned at the tumor treatment site, brachytherapy enables a high dose of radiation to be applied to a small area.

Brachytherapy dates back to 1901 when Pierre Curie suggested that a small tube of radium could be inserted into a tumor (a similar but independent suggestion was made by Alexander Graham Bell in 1903), and early trials showed that the radiation did cause tumors to shrink.[15] Brachytherapy is now commonly used to treat cancers of the cervix, prostate, breast, and skin, but it can also be used in the treatment of a wide variety of other types of tumors. Because the radiation sources are placed in or next to the target tumor, the sources maintain their position in relation to the tumor when the patient moves or if there is any movement of the tumor within the body. Therefore, the radiation sources remain accurately targeted.

The suitability of a given radionuclide for brachytherapy is determined by its half-life and the energy, type, and abundance (number per decay) of its emission. Presently, a wide range of radionuclides including ^{125}I, ^{103}Pd, ^{90}Y, ^{90}Sr, ^{144}Ce, and ^{106}Ru are being used.[16] The most prominent recent developments in sealed sources include ^{192}Ir sources for high-dose-rate brachytherapy and ^{125}I and ^{103}Pd seeds for the treatment of prostate and brain cancers. Palladium-103 with a 17 day half-life is a very-low-energy photon emitter (21 keV) used for permanent interstitial implantation. It has better energy and safety characteristics when compared to ^{125}I. Its initial peripheral dose rate is approximately three times greater than ^{125}I. This potentially provides improved control of rapidly proliferating tumors. It has been used in the treatment of various cancers such as eye, brain, neck, uterus, and colon, but it is almost exclusively used now for prostate cancer — one of the most common cancers in men and the second leading cause of cancer death in men in the US alone.

Historically, ^{103}Pd was generated in reactors via the ^{102}Pd(n,γ)^{103}Pd reaction that relied on the availability of 1% naturally abundant ^{102}Pd in an enriched form and its moderately high neutron capture cross section. However, for the last 20 or so years, ^{103}Pd has been produced via the reaction ^{103}Rh(p,n)^{103}Pd by irradiating the metal rhodium with relatively low energy protons from accelerators. With more than 220 000 new incidences of prostate cancer reported in the US in 2007 (the most recent statistics available)[17] and estimates by the US National Cancer Institute that there will be in excess of 240 000 new cases in 2012, there is a substantial market for the patented ^{103}Pd seeds used to treat it. To keep up with the demand, the only US manufacturer of these seeds (Theragenics Corporation) currently has as many as 14 large cyclotrons dedicated to their production, and recently a manufacturer in Europe has also brought the US-patented type of ^{103}Pd seed implants to the world market.

2.3.2. *Targeted radiotherapy*

Although most patients with localized cancers can be cured by surgery, radiotherapy, chemotherapy, and combinations thereof, those with distant metastases generally need systemic therapy. One type of systemic radiotherapy is referred to as targeted radionuclide therapy. This involves the use of radiolabeled tumor-seeking molecules to deliver a cytotoxic dose of radiation to tumor cells. Although still in the early stages, targeted radionuclide therapy is beginning to be useful in the treatment of some tumor types. During the last decade, a considerable effort has been made to obtain stable binding of many radionuclides to monoclonal antibodies using a variety of bifunctional chelating agents (BFCA). As a result, many antibody-based agents have now been cleared for human diagnostic use.[18] The main theoretical advantage of targeted radionuclide therapy is that radiation can be delivered selectively to subclinical tumors and metastases that are too small to be imaged and thus cannot be treated by surgical excision or local external beam radiotherapy. In addition, the absorbed doses to tumors achieved by targeting may be higher than can be delivered by other systemic methods.

For any radiation-based therapy, the likelihood of tumor cure depends on three factors:[19] (1) the radiation dose absorbed in the tumor and its pattern of delivery (e.g., dose rate and fractionation); (2) the number of clonogenic tumor cells (i.e., the cells are clones of one another) present — these all have to be sterilized to cure the tumor; and (3) the response of the tumor cells to radiation (e.g., radiosensitivity, repair capacity, and proliferation rate).

Radioimmunotherapy (RIT) is a targeted radionuclide therapy approach that uses monoclonal antibodies, or parts thereof, to deliver the radiation from radionuclides to a specific type of cell.[20] Based on the experience gained with diagnostic agents, it can be expected that monoclonal antibody-based therapeutic radiopharmaceuticals in clinical application will grow substantially. Moreover, the introduction of small peptides (molecules consisting of 2 or more amino acids) as carriers for radionuclides has opened up new possibilities for development of diagnostic agents as evidenced by the use of ^{111}In-octreotide. Efforts are underway to extend the use of peptides as carriers for other therapeutic radionuclides such as ^{188}Re, ^{90}Y, and ^{153}Sm. The initial successful attempts to use Auger electrons from ^{111}In for therapeutic effect by high dose administration of ^{111}In-octreotide has provided a strong impetus for the search, development, and use of more accessible and perhaps more economical radionuclides like ^{153}Sm, ^{90}Y, and ^{177}Lu coupled with peptides and monoclonal antibody carriers. The incorporation of beta emitters into a variety of bifunctional chelates of bioactive peptides, including octreotide and VIP (vasoactive intestinal peptide) analogues, is an active area of research at several clinical research centers.[21]

Most of the radionuclides used for radiotherapy are produced in a reactor since one of the desirable characteristics of these radionuclides is that they will emit beta particles and therefore tend to be neutron rich. Indium-111 is an exception since it is routinely produced with a cyclotron.

Another class of radiotherapeutic agents is the alpha emitting radionuclides. These have the advantage of delivering a very high dose of radiation in a very small volume and are therefore particularly suited for micrometastases and blood-born cancer cells. The main radionuclides being developed in this class are ^{211}At, ^{213}Bi, and ^{225}Ac. These can all be produced with an accelerator and are being investigated as therapeutic agents. Actinium-225 and its decay product ^{213}Bi are also available as a generator system usually produced from ^{229}Th, which can be derived from ^{233}U. Unfortunately, ^{233}U is a "special nuclear material" and as such is under very strict control.

2.3.3. *Other medical applications*

Accelerator-produced radionuclides are also used as calibration sources for detectors and to determine the attenuation correction for PET cameras. The most common radionuclide for this purpose is ^{68}Ge with a 275 day half-life. It decays to ^{68}Ga, which is a positron emitter that can be used to directly measure the attenuation correction in PET tomographs. However, since the development of PET/CT (a combination of a PET imaging camera and a conventional X-ray CT scanner), the attenuation correction is usually done with the CT. Sodium-22, a long-lived positron emitter, is also used to create test-objects and point-sources for calibration of PET cameras and ion chambers.

2.4. *Industrial*

As mentioned earlier, there are few non-medical industrial applications of radionuclides produced with accelerators — the vast majority of radionuclides used in industry and consumer products are produced with reactors. However, there are some that are worth noting.

Accelerator-produced ^{22}Na and ^{68}Ge are used to calibrate PET cameras and ion chambers. The agricultural industry uses accelerator-produced radioactive carbon and oxygen isotopes to look at synthesis and water uptake from the soil in the study of plant biokinetics. Finally, thin layer activation (TLA) is a specialized industrial application of radiotracers that has been used for over two decades to study wear and corrosion under real operating conditions of critical machine parts, such as bearings, camshafts, vehicle brake disks, piston rings, and internal combustion engine cylinder housings, as well as to monitor corrosion and erosion of surfaces such as in pipelines, turbine blades, off-shore platforms, and nuclear power plants.[22] However, unlike the other applications described in this chapter, the radiotracer is not made in a stand-alone form. Rather, an accelerator is used to directly create a thin radioactive layer under the surface of a part or component of interest by charged particle or neutron activation. This is also considered to be an ion beam analysis technique and is presented in more detail in Chapter 5 under Charged Particle Activation Analysis (CPAA).

3. Accelerators for Radionuclide Production

The two main requirements for producing radionuclides via nuclear reactions using a charged particle beam are that the beam energy must be higher than the reaction threshold and the beam current must be sufficient to produce a usable quantity. Cyclotrons were the first accelerators used for this purpose and they continue to dominate the radionuclide production equipment market. As the use of radionuclides in nuclear medicine has progressed, several other types of accelerators have become available. Recent innovations include superconducting magnet cyclotrons, compact low energy RF ion linacs, tandem cascade accelerators and helium-particle-only linacs. While a few RF linacs have been sold in recent years for PET radionuclide production, none of these other accelerator types has gained wide acceptance for this application.

Almost all cyclotrons and linacs used to make radionuclides are produced by industry. The fastest growing segment of the equipment market is for systems being provided to commercial radionuclide production and distribution businesses. These accelerators range in energy from 7 MeV for PET radionuclides to 30 MeV for the large scale production of longer-lived gamma-emitters such as ^{201}Tl, ^{111}In, and ^{67}Ga. A few new cyclotron designs have emerged very recently (2010) in an attempt to capture anticipated markets where single dose PET is seen as the new paradigm for "personalized medicine". ABT (ABT Molecular Imaging Inc., USA) has designed and sold a 7.5 MeV mini-cyclotron that can deliver a single dose of ^{18}F-FDG every 20 min or so (after the first run). The ACSI (Advanced Cyclotron Systems Inc., Canada) 24-MeV TR24 cyclotron has two potential markets in sight — one to produce ^{123}I from the ^{124}Xe(p,X) reaction in smaller quantities for local use and the other for the direct production and distribution of ^{99m}Tc.

(The $^{124}Xe(p,X)^{123}I$ reaction is already widely used by large producers of ^{123}I for distribution using much more powerful and expensive 30 MeV cyclotrons.) The direct production of ^{99m}Tc has become increasingly interesting for two reasons: it is the most widely used radionuclide for nuclear medicine, and there is growing concern over past and future disruptions in the supply of ^{99}Mo generators (at present the only source of ^{99m}Tc) due to the planned or unplanned shutdown of more and more of the nuclear reactors used to produce it. It remains to be seen if either or both of these new systems prove to meet the demands of routine production and operation.

Based on numbers obtained from the available manufacturers' data and press releases from various market surveys, at the time of this writing there are an estimated 700 cyclotrons in operation worldwide for radionuclide production, with a regional distribution as shown in Fig. 3. The number of systems has increased by more than 10% annually in the last decade and is continuing to grow at more than 50–60 machines per year due primarily to the demand for PET radiopharmaceuticals, ^{18}F-FDG in particular.

The major cyclotron equipment manufacturers are listed in Table 2. A typical cyclotron used for the production of PET radionuclides is shown in

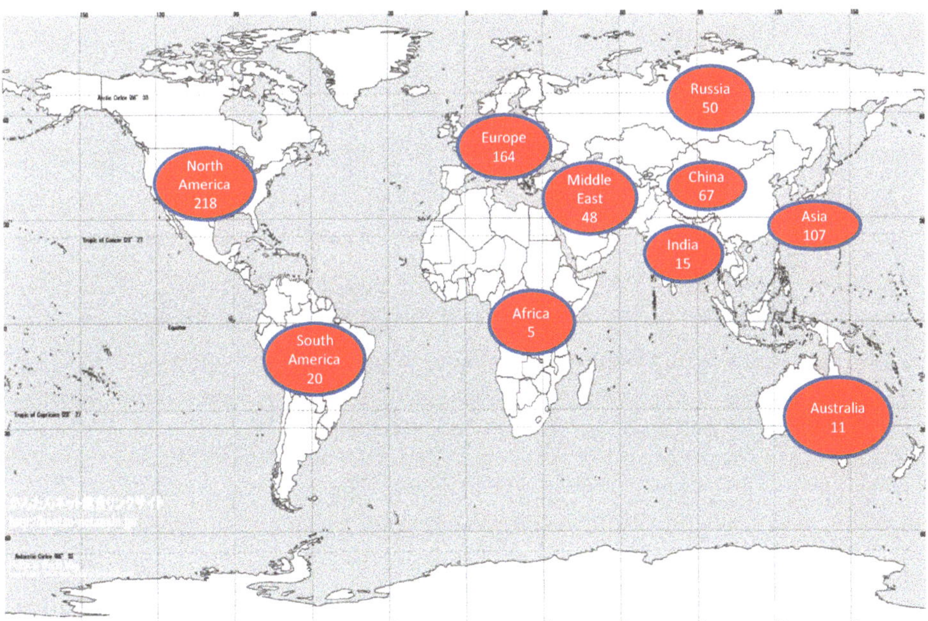

Fig. 3. Estimated worldwide distribution of commercial cyclotrons based on 2010 data collected from major manufacturer sales figures and various market surveys.

Table 2. Major cyclotron manufacturers worldwide.

Manufacturer	Location
GE Healthcare	Sweden
Siemens Healthcare	USA
Ion Beam Applications SA	Belgium
Advanced Cyclotron Systems Inc.	Canada
Best Cyclotron Systems, Inc.	Canada
Sumitomo Heavy Industries, Ltd.	Japan
Samyoung Unitech Co., Ltd.	Korea
NPKLUTS	Russia

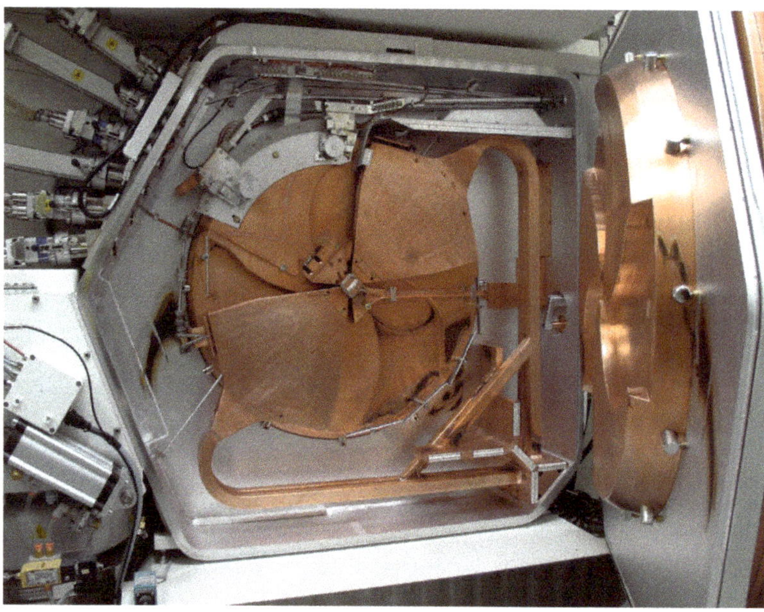

Fig. 4. GE PETtrace cyclotron with the vacuum tank open. Visible in the upper left are the targets attached to the external wall of the vacuum tank. The accelerating "dees" are the pie shaped objects in the foreground. The magnet poles lie beneath and are copper coated. The extraction foils are housed in the two small carousels in the left corner of the vacuum tank. (Courtesy of Dr. Michael Kovacs, Lawson Health Science Centre, London, Ontario, Canada.)

Fig. 4. The annual worldwide market for radionuclide production cyclotrons is estimated to now exceed US$100M.

Of the other major types of accelerators capable of producing radionuclides, the number of electrostatic accelerators in use for that purpose

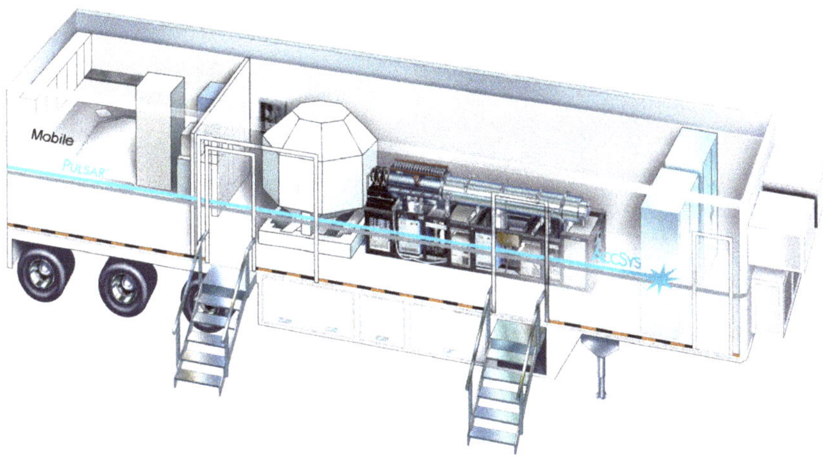

Fig. 5. Artist's rendition of a mobile PULSAR® 7 MeV proton linac for the production of PET radionuclides, showing the shielded radionuclide production target area (octagon shaped enclosure) and area for a chemistry lab (back of trailer). (Courtesy of AccSys Technology, Inc.)

is very limited and these will not be described further. As for RF linacs, several compact RF proton linacs are in current or planned operation for radionuclide production, mainly for PET radionuclides. These systems are based on the RFQ (Radio Frequency Quadrupole) and DTL (Drift Tube Linac) accelerator structures. Proton linacs for radionuclide production range in energy from 7–11 MeV, although higher energies can be provided by adding modular accelerating sections. The 7 MeV linacs can produce proton beam currents of more than 100 μA on target. At least two of these linacs have been installed in trailers (see Fig. 5), demonstrating the versatility of this compact, light-weight technology. The major equipment vendor for these RF ion linacs is AccSys Technology, Inc. (USA).

As already pointed out, the most common accelerator used for the production of radionuclides by far is the cyclotron. Cyclotrons have evolved from the first early machines that required a team of physicists to keep them operating to the present-day stable and reliable systems that can be operated by almost anyone after a few lessons. However, the principle components and theory of operation have remained essentially the same. The following sections give a basic description of the operating principles of a cyclotron along with an overview of the main components of a full radionuclide production system. A brief overview of modern RF ion linacs used for radionuclide production is also given.

3.1. *Cyclotrons*

There are three basic physics principles at work in a cyclotron: (1) a charged particle (ion) can be given energy (i.e., accelerated) by an electric field; (2) a charged particle moving in a magnetic field perpendicular to its direction of motion will travel in a circular path; and (3) according to the theory of electrodynamics, the rotational frequency of that charged particle is independent of the radius of its orbit. Through the application of these principles, cyclotrons can produce moderately energetic beams (10–70 MeV) of charged particles in a relatively confined space. For instance, a typical 18 MeV cyclotron for the production of PET radionuclides has a radius of only 1 m.

The general principles of cyclotron operation are relatively simple. A "packet" of positive or negative ions is released near the center of the cyclotron and moves gradually outward in a semicircular path through two or more hollow D-shaped electrodes (called "dees") placed between the poles of a strong magnet. A rapidly alternating RF voltage is applied to the dees to provide the accelerating electric field. The frequency of the voltage is chosen such that the ions always see an attractive force in the direction of travel as they enter the accelerating gap between the dees. That is, as the ions exit from one dee they are pushed by that dee and pulled by the other one, as illustrated in Fig. 6. The beam is shielded from the electric field while inside the dees, but the magnetic field keeps it moving in an outwardly spiral motion until it reaches the maximum energy. All of this occurs under vacuum so that the accelerating particles have minimal interaction with other materials such as gaseous molecules.

Each time a packet of particles passes through the gap between the dees, it is given a boost in energy closely related to value of the voltage

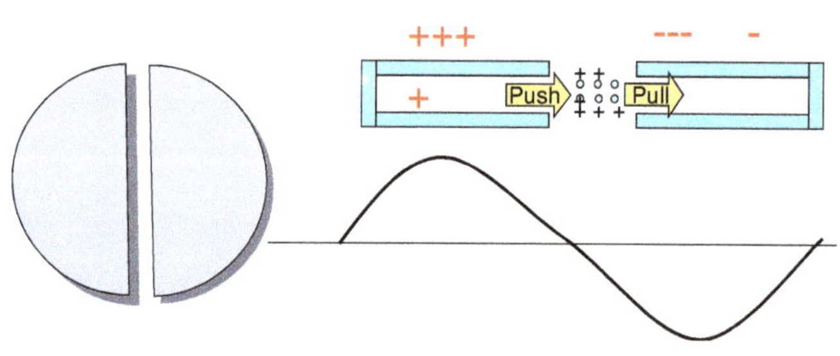

Fig. 6. Schematic representation of a simple RF acceleration field applied to two dees.

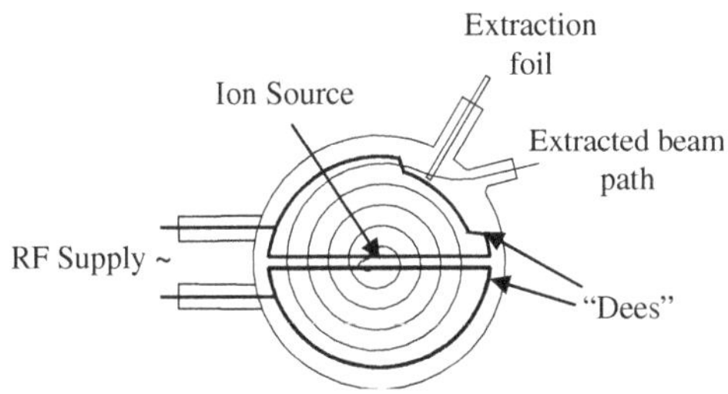

Fig. 7. Schematic diagram of a cyclotron showing the main accelerator components and beam path (magnet not shown).

applied. In a perfect case, these two conducting surfaces with a large voltage difference V_0 would result in a particle with charge q gaining a kinetic energy of qV_0. Usually the timing of the peak RF voltage is a little behind the packet of ions as they circulate. This allows the trailing particles to get a little extra boost in energy while the leading particles get less of a push. The result is that the packet stays together and is in better position for the next transition, but the energy gain is slightly less than qV_0.

The energy of the beam should increase smoothly from the center to the outer edge. As the energy increases, the path becomes longer but the time required to make one loop remains the same because the velocity increases in exact proportion to the increase in distance. When the energy exceeds about 20 MeV, relativistic effects cause the mass of the particles to increase, requiring a stronger magnetic field to keep the beam synchronized with the RF voltage.

The major components of a cyclotron system are the magnet that supplies the magnetic field in the acceleration region, the ion source, the RF power source that supplies the alternating voltage to the dees, and the beam extraction system. The magnet is usually a large multiple-coil, iron-core electromagnet with its two poles arranged to provide a magnetic field perpendicular to the dees. Since the operating frequency of the system is usually in the few to 10s of MHz range, the RF power system is similar to that of a commercial broadcast unit. Such systems are usually based on a high power electron tube (e.g., triode or tetrode), but there is now a move toward using solid-state units. A schematic diagram of the main accelerator components and beam path is shown in Fig. 7.[23] The methods used for ion production, ion injection, and beam extraction in cyclotrons are discussed in the following subsections along with a brief introduction to radionuclide production targets and shielding issues.

3.1.1. *Ion source*

The purpose of the ion source is to create either positive or negative ions to be accelerated. In a cyclotron, the ions are usually generated in a plasma discharge in the center. The ion source can be either external or internal depending on the type of ion and beam current required. In most cyclotron ion sources, a gas of neutral atoms or molecules is "heated" into a plasma state where ions and electrons are dissociated and move independently as free particles. The heating mechanism can be of various kinds: thermal, electrical, or even using laser light. Once the ions are created in the source, they are then extracted from the plasma and accelerated.

The most common type of ion source uses a gaseous discharge to create the ions. In any gaseous discharge, both positive and negative ions exist in approximately equal numbers. The objective of ion source design is to optimize the desired ion yield and beam quality. Most modern cyclotrons accelerate negative ions, and the ion sources are usually designed to maximize the production of negative hydrogen ions. Electron bombardment of the neutral atoms in the plasma is the most usual method for creating ions. Increasing energy is required to remove increasing numbers of electrons from the atom, so multiple-charged ions, such as helium, require much higher ionization energies.

The two main versions of gaseous discharge ion sources used are the hot cathode and cold cathode types. In the hot cathode type, a heated filament is used to create a sea of electrons and thereby maintain the arc. In the cold cathode type, once the discharge is initiated with a high voltage, a hot filament is not needed to maintain the plasma during normal operation of the ion source. An example of a simple hot cathode ion source is shown in Fig. 8.

In both types, the electrons are confined radially by the magnetic field and axially by the electrostatic potential well. In cyclotrons, it is possible to use the magnetic field of the accelerator to contain the plasma. The spatial distribution and output of an internal source is not well defined and the ions usually have a broad distribution of energies. These problems are usually overcome by using slits on the ion source that decrease the beam intensity but produce a well defined beam profile. Once the ions exit the ion source, the path of the beam is further defined with a "puller". This is a small slit in a metal plate that accepts only those ions with the proper energy and position to continue past the first orbit. The position of the ion source is critical to an efficient beam extraction. The relationship between the slit on the ion source, the puller, and the magnetic field in the central region must

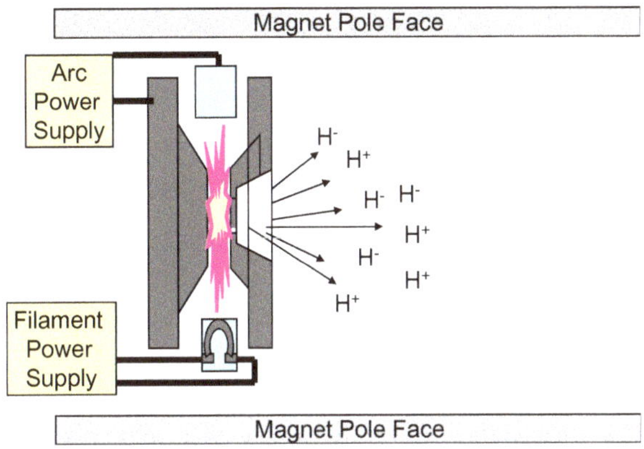

Fig. 8. Typical cyclotron hot cathode ion source producing both positive and negative ions.

be in the proper alignment in order to achieve efficient extraction of the beam into the first orbit for acceleration.

3.1.2. *Ion injection*

As the ions exit the ion source, the path they take is determined by the environment of the central region. This includes the magnetic and electric fields in the region that pull the ions from the plasma out of the ion source and into the first orbit. One of the critical parameters in ion sources is shaping the electric and magnetic fields so that the ions are drawn out of the plasma and into the dee region where they can be accelerated. Even with modern modeling of fields, this is often a trial and error sequence in order to maximize the efficiency.

3.1.3. *Beam extraction*

Once the beam of ions has been accelerated to the desired energy, it usually must be extracted from the cyclotron in order to bombard a target and create the radionuclide. There are two major methods for doing this depending on the charge of the particle being accelerated. Positive ions (e.g., H^+) are extracted by electrostatic deflection from their final orbits. Negative ions (e.g., H^-) are usually extracted by stripping the electrons and creating positive ions that are then magnetically deflected from their final orbits out of the machine.

For positive ion extraction, the beam is deflected out of the cyclotron by using an electric field applied to an electrode placed at the outer edge of the beam orbits. The advantages of this method are that the vacuum requirements are not as stringent and positive ion sources have in the past been capable of higher beam currents. The disadvantage is it is not as efficient as negative ion extraction. This results in loss of beam and activation inside the cyclotron, which can result in a higher radiation dose to the operators during internal machine maintenance.

In negative ion extraction, the beam is passed through stripper foils (the beam can be extracted at multiple exit points), and the resulting positive ions are deflected out of the cyclotron by its magnetic field. The advantages of this system are that the extraction efficiency is very high and there is very little internal activation of the cyclotron. The disadvantages are that the vacuum requirements are slightly higher than for a positive ion machine and the stripper foils need to be changed regularly.

Since most modern cyclotrons accelerate negative ions, a better vacuum is needed to minimize beam losses compared to positive ion machines. The vacuum systems have improved with the use of turbomolecular pumps and cryopumps that do not need oil, such that vacuums of 10^{-7} Torr are now common. The added advantage of very little activation of the interior of the cyclotron makes the radiation dose received by the operators during machine maintenance much lower. Improvements in ion source technology have produced negative ion sources that are capable of producing beam currents of the same order as positive ion sources, so the limitations on beam current no longer exist.

3.1.4. *Beam transport*

The ideal situation is to place the radionuclide production targets far enough away from the accelerator so that maintenance of the two systems can be performed independently. This is especially the situation when using very high beam power accelerators. The residual radiation fields around target stations can prevent access for many hours to days, while most accelerators can be accessed fairly quickly once the beam is turned off. To minimize beam loss, a high-vacuum beamline must be used to transport the beam to the target station.

On negative ion cyclotrons, there can be two or more beamlines from different ports on the machine that can be operated simultaneously. Each beamline may be further split to several target stations. Positive ion cyclotrons can only support one exit and therefore one beamline. This beamline however can also be split into several target stations using a switching magnet.

Quadrupole magnets are usually required to keep the beam focused for delivery to each target station.

3.1.5. *Targets*

The radionuclide production target material may be in gas, liquid, or solid form. Each of these has unique problems that must be addressed in the design and construction of the targets, a topic that will be discussed in Section 6.

Regardless of the target type, thermal issues are the dominant concern as they are the major cause of target failures. While solid targets are in general easier to cool with high water flow placed on the target backing, the actual target material will invariably have low thermal conductivity, making the operating temperature still a large concern.

Typically, gas and water targets are also water-cooled but they employ a double entrance foil system where one foil isolates the target from the vacuum of the accelerator or beamline and the second foil contains the fluid or gas target material. High pressure helium gas flows between the foils to remove the deposited heat. These foils must be thin in order to minimize energy loss as the particle beam passes through. The characteristics of the foil materials with respect to strength and thermal conductivity are usually not coupled in a positive way: the strongest foils tend to have lower thermal conductivity and thus are prone to heat-related failures.

3.1.6. *Radiation shielding and facilities requirements*

Radiation shielding is a major concern for all nuclear facilities, including accelerators. Regulators usually classify accelerator systems according to particle beam energy and current. Shielding requirements are aimed at safe operating conditions but also safe maintenance of the accelerator and target systems. Invariably, the targets represent a major shielding issue since they become a huge source of neutrons that are difficult to shield. Thermal neutrons flow as if they were a gas and thus can go through cracks and around corners.

Most facilities place the accelerator in a concrete bunker made of low sodium content materials to minimize ^{24}Na production from neutron capture. These facilities use either a maze for access to the machine, or if space is limited, a sliding shielding door. The alternative approach is to use what is misnamed a "self-shielded" accelerator. So-called "self-shielded" cyclotrons take advantage of the fact that the iron in the cyclotron magnet provides some built-in neutron shielding; however, the majority of the shielding comes from concrete or other composite materials put in close contact with the accelerator in order to diminish its footprint. Although these systems can operate in fairly open areas with restricted access, they do have operational constraints and challenges with respect to target area access.

3.2. *Linear accelerators*

The basic principle of acceleration in a linear accelerator is the same as in a cyclotron: a charged particle can be accelerated by an electric field. The main difference is that the cyclotron uses a magnetic field to confine the particles to a spiral orbit. The concept of the RF linear accelerator actually predates the cyclotron by several years, the idea of using an alternating voltage to accelerate a particle in several small linear steps having been first proposed by Gustav Ising in 1924 and taken up by Wideröe in 1928.[6] In fact, as already pointed out, it was Wideröe's work that sparked Lawrence's invention of the cyclotron.

The energy increase in an RF linac is supplied by an alternating (radio frequency) electric field that must be applied in exactly the proper sequence to keep accelerating the particles in a straight line. In its simplest form, this is achieved with the use of hollow electrodes (drift tubes), as illustrated in Fig. 9, that allow the particles to drift at constant velocity within the tube and then be accelerated in the gaps between the tubes. Particles are accelerated into the tube by an electric field that is opposite in sign to their charge. As they pass through the hollow tube, the phase of the electric field is changed, and at the exit of the tube, they are accelerated with a push from the field, which now has the same sign as the particles.

Just as in the cyclotron, the charged particles in the linac can be kept together by applying the maximum potential just after the particles have

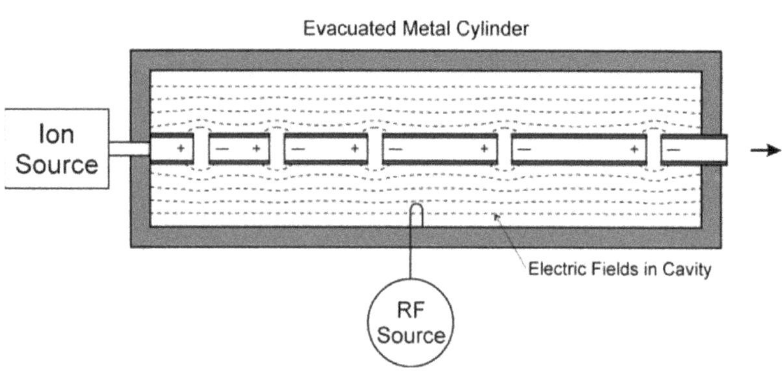

Fig. 9. Schematic representation of a drift tube linac.

passed the mid-point in the accelerating gap. If a particle arrives too early, the potential applied will be slightly less than optimal. It will traverse the next section more slowly and thus be in phase for the next accelerating potential.

Modern ion linacs built for radionuclide production (mainly for PET radionuclides) consist of an RFQ accelerator usually injecting a relatively low energy proton or deuteron beam into a drift tube linac. The RFQ linac is a relatively new ion linac structure that uses an RF electric quadrupole geometry with longitudinal perturbations added to continuously focus and accelerate a low energy ion beam.[24] Since the particle path is linear instead of circular and the acceleration occurs in the gaps during every RF cycle, in order to build a linac with reasonable energies and of reasonable size, a high frequency is essential. The development of the RFQ, which allowed the efficient acceleration of low energy ions to a reasonable (few MeV) output energy, led to the development of the modern high frequency ion linac structures that could be powered by the RF tubes that had been previously developed for long range radar applications. Although modern RF ion linacs are very reliable and stable, they are not commonly used for radionuclide production. This is due in part to a series of dramatic failures of the more conventional linacs in the past that caused some concern with potential customers. To a great extent, these problems have been overcome, but linacs have not caught up to cyclotrons in widespread acceptance.

3.3. *Choice of accelerator*

In choosing an accelerator for radionuclide production one must be clear regarding its purpose. This includes what radionuclides are to be produced and the quantities required. Potential growth should be kept in mind especially when the accelerator is meant for commercial production or is associated with a research program where the direction of the program may change.

Once the list of radionuclides has been defined, the particle–energy combination can be determined keeping in mind the required target material in terms of natural abundance or enriched materials. If an enriched target material is required, the availability as well as the costs should be kept in mind. All of these factors must be considered in order to choose the right accelerator system needed to meet the market demand or research project requirements.

In the case of a cyclotron versus a linac, there are differences in their beam characteristics that have consequences in the design and operation of the radionuclide production targets. The most obvious is the beam structure. Cyclotron beams are nearly uniform in intensity, and the frequency of the beam pulses (usually on the order of 40–70 MHz) makes the beam appear to be continuous. Linacs, on the other hand, have uniform intensity but operate at a much lower pulse frequency, typically 10–300 Hz. This results in different characteristics for beam heating of the target. In the linac, the target foil can heat up and then cool down during one cycle for the same time-averaged current as a cyclotron. This can put mechanical stress on the foil that in turn can cause metal fatigue, resulting in a weaker foil and one more likely to fail. The solution to this problem is to use metals that are more resistant to metal fatigue, thinner foils, and larger beam spots, or to change foils more frequently. On the other hand, the linac output beam has more uniformity and does not have the "hot spots" seen in some cyclotron output beams that are caused by the extraction of multiple orbits in the cyclotron.

4. General Principles of Radionuclide Production

Radionuclide production is indeed true alchemy — the atoms of one element are transformed into those of another. This transformation involves altering the number of protons and/or neutrons in the nucleus of the target element. If the target nucleus is bombarded by a charged particle, i.e., a proton, deuteron, or α-particle, the product nucleus will usually be that of a different element.

There are several models that can be used to predict the probabilities (cross sections) for nuclear interactions. One of the most comprehensive is the EMPIRE computer code.[25] EMPIRE is a modular system of nuclear reaction codes based on various nuclear models that is designed for calculations over a broad range of energies and incident particles. A projectile can be a neutron, proton, photon, or any ion (including heavy ions). The code accounts for the major nuclear reaction mechanisms, including direct, pre-equilibrium, and compound nucleus channels.

4.1. *Nuclear reactions*

Nuclear interactions between an energetic charged particle (projectile) and the target nucleus can take several forms. The projectile may simply be scattered (direct or indirect scattering) away, or if it has enough energy, it can combine with the target to form a composite (or "compound") nucleus that may then decompose along one of many possible exit channels. It is this latter process that is of most interest in radionuclide production.

In order for a nuclear reaction to occur, the projectile must have enough energy to overcome two barriers. First it must overcome the Coulomb barrier that is the result of the natural electrostatic repulsion between the protons in the projectile and the protons in the target; otherwise the two will not get close enough for the nucleons to experience the attractive nuclear force. The second barrier depends on whether the reaction releases energy (exoergic) or requires energy to be supplied by the projectile (endoergic). This energy excess or deficit is referred to as the Q value of the reaction, which is in turn used to determine the threshold energy — the amount of energy the projectile needs for the reaction to be energetically possible according to the physics laws of conservation of total energy and momentum.

4.1.1. *Coulomb barrier*

In a simple head-on collision model, the minimum energy a projectile of mass number (total number of nucleons) A_p and atomic number (number of protons) Z_p must have to overcome the Coulomb barrier of a target nucleus of mass number A_t and atomic number Z_t can be approximated by:

$$E_c \cong \left(\frac{Z_p Z_t}{A_p^{1/3} + A_t^{1/3}}\right)\left(\frac{A_t + A_p}{A_t}\right) \text{MeV}. \tag{1}$$

The first term is the approximate Coulomb barrier energy alone while the second term is due to the requirement to conserve momentum. If the projectile energy is below this minimum, it is unlikely a nuclear reaction will take place (reactions can take place well below the Coulomb barrier due to the effects of quantum tunneling, but the probability is exceedingly small). This energy is plotted for protons, one of the main projectiles used for radionuclide production, in Fig. 10 as a function of the atomic number of the target.

4.1.2. *Reaction models*

Among the many possible reaction paths (e.g., direct scattering, indirect scattering, particle transfer or knockout, absorption or compound nucleus formation, etc.), the ones that result in the reaction products being different from the reactants are of interest for radionuclide production. Two such reaction paths are discussed here.

- **Direct reactions.** As a result of the interaction, a particle is transferred from the projectile to the target or from the target to the projectile. The projectile changes its direction and generally transfers part of its kinetic

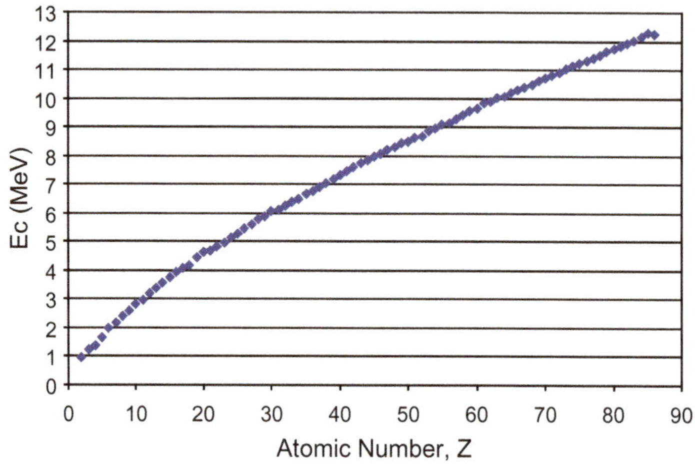

Fig. 10. Minimum energy needed for protons to overcome the Coulomb barrier as a function of target atomic number.

energy to the target. The respective mass numbers A (total number of nucleons) and atomic numbers Z (number of protons) after interaction differ from those before (in the example shown, a neutron is transferred).

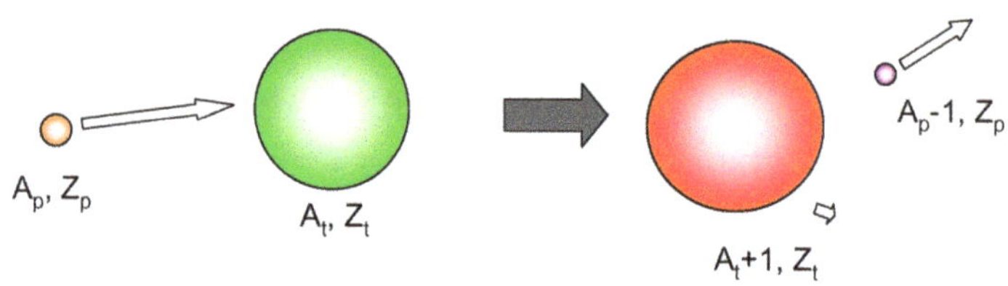

- **Absorption**. The projectile is captured by the target and a highly excited compound nucleus is formed. This nucleus then distributes the excitation energy over its internal degrees of freedom until equilibrium is reached. The compound nucleus may emit particles while still redistributing the energy, or it may emit particles after it has reached equilibrium.

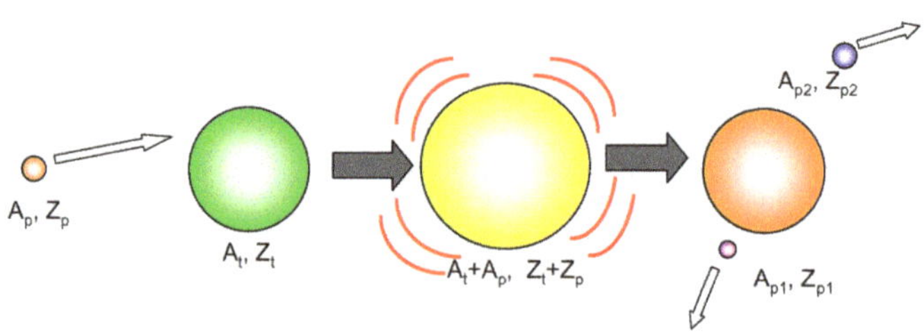

An important corollary of this model is that the highly excited intermediate nucleus "forgets" how it was formed except that all the fundamental conservation laws (total energy, total momentum, charge, etc.) remain in effect. As the incident particle combines with the target nucleus, it loses its identity, and the total energy of the compound nucleus is shared among all of the nucleons before it decomposes. This decomposition can take place by numerous exit (reaction) channels that are independent of the way in which the compound nucleus was formed. Some of the initial kinetic energy may be converted to mass in the decomposition or some mass may be converted to kinetic energy, meaning that the total kinetic energy of the products may be greater or less than the total kinetic energy of the reactants. In exoergic reactions, the kinetic energy of the products is greater and energy is released. In endoergic reactions, the kinetic energy of the products is less and energy must be supplied.

4.1.3. *Q value and threshold energy*

The kinetic energy difference described above is known as the *reaction energy*, or more commonly the *Q value*. Since the total energy of the reactants must equal the total energy of the products, any change in kinetic energy must be balanced by a change in mass. Therefore, Q can be calculated as the difference between the total rest mass of the reactants and the total rest mass of the products. Exoergic reactions have positive Q values, while endoergic reactions have negative Q values. For the two-body reaction $M_p + M_t \rightarrow M_R + M_r$:

$$Q = 931.4\{(M_p + M_t) - (M_R + M_r)\}\,\text{MeV}, \tag{2}$$

where M_p and M_t are the atomic masses of the projectile and target, respectively, and M_R and M_r are the atomic masses of the reaction products R and r, respectively. All masses are in atomic mass units (u), with 931.4 being the MeV energy equivalent of 1 u. The Q value is more conveniently determined using tables of mass excess (the difference between the atomic mass and the mass number) for the reactants and products. These are readily available from a number of sources.

The *threshold energy* E_t for an endoergic reaction is the minimum kinetic energy the projectile must have for the reaction to be energetically possible. This will always be slightly larger than the Q value because a fraction of the projectile's energy must be used to impart momentum to the products in order to also satisfy the conservation of momentum. It can be readily calculated from the Q value and the mass numbers of the projectile (A_p)

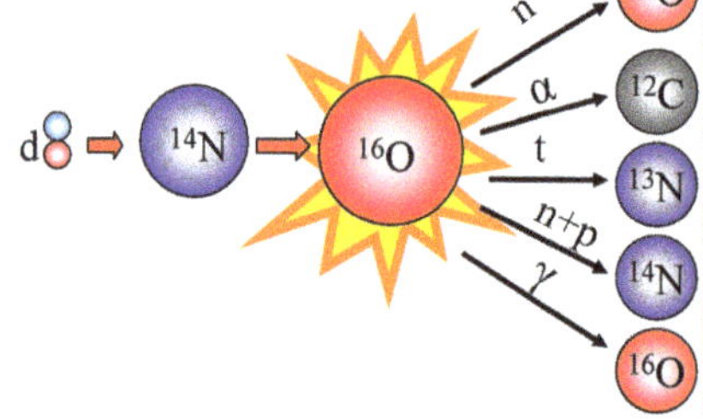

Q-Value MeV	Threshold MeV
5.1	0
13.6	0
-4.3	4.9
-2.2	2.5
20.7	0

Fig. 11. Q values and threshold energies for various channels for nuclear decomposition of the compound nucleus ^{16}O formed by the collision of a deuteron and a ^{14}N nucleus.

and target (A_t) by:

$$E_t = -Q\left(\frac{A_p + A_t}{A_t}\right). \tag{3}$$

Note that the threshold energy for exoergic reactions is always zero.

As an example, Fig. 11 shows the wide variation of Q values and threshold energies for a particular projectile–target combination with many possible reaction channels. In reality, the minimum projectile energy is the greater of the threshold energy or the energy needed to overcome the Coulomb barrier plus satisfy the conservation of momentum (Eq. (1)). Thus even for zero threshold energy, the projectile still needs energy for the reaction to proceed.

4.1.4. *Cross section*

While the laws of physics determine whether a particular reaction is possible, the reaction cross section represents the probability that the reaction will actually occur. Cross sections are usually denoted as σ and are expressed in units of an area (e.g., cm^2). Cross section values, both predicted and measured, exist for a vast number of nuclear reactions. The commonly used units are based on the "barn" (b), which is approximately equal to the physical cross section of a typical nucleus ($1\,b = 10^{-24}\,cm^2$).

When the cross section for a reaction is plotted as a function of projectile energy, this is often called the excitation function. An example is shown in Fig. 12. Excitation functions are useful in determining the amount of a radionuclide that may be made with a given accelerator. If several reactions are possible, the excitation functions can also indicate the levels of

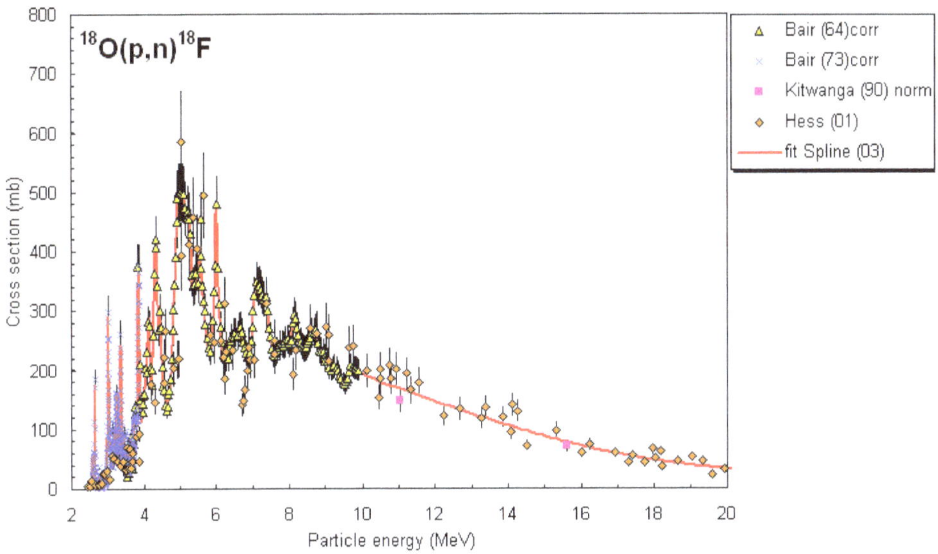

Fig. 12. Excitation function for the $^{18}O(p,n)^{18}F$ reaction illustrating the energy dependence of the cross section.

contamination of other radionuclides that may be produced in the target material.

4.2. *Optimizing production*

For a given combination of projectile, target, and beam energy, kinematics can determine which radionuclide will predominate where several reaction channels are possible. The irradiation time will depend on the half-life of the desired product and the half-lives of the potential contaminants. All of these factors must be taken into account when designing a production process.

A detailed treatment is beyond the scope of this chapter, but some useful generalizations can be made about the reaction cross section:[26]

- It is small but finite for bombarding energies that are far below the Coulomb barrier.
- It increases very rapidly with incident particle kinetic energy when the kinetic energy is less than the barrier height.
- It does not reach its maximum at a point where the kinetic energy is equal to the barrier height.
- It asymptotically approaches a maximum value that is simply the geometrical area of the target nucleus — πR^2, where R is the nuclear radius — when the kinetic energy is much greater than the barrier height.

It is unlikely that reactions between protons and heavier target nuclei will have a very significant cross section below the Coulomb barrier. If the projectile is an α-particle, the barrier height increases to 25 MeV so, the reaction cross section is usually small below this energy. The Q value of the various reactions often predict how far above this barrier the reaction will occur. These rules of thumb can often give insight to the probabilities of the desired and interfering nuclear reactions in a given situation. In a similar vein, it is possible to estimate the energy required to release one neutron, two neutrons, etc. This is important because neutron emission is much more probable than proton emission due to the exiting Coulomb barrier.

4.2.1. *Production rate*

Even though there are several models for nuclear reactions that give increasingly accurate results, the best source for a cross section is still an experimental measurement. This is because the theoretical estimation of these parameters is not as accurate as one would desire, especially for the light nuclei.

Experimentally, the cross section can be determined from the mathematical relationship between the various parameters affecting the production:[27]

$$R = Inx\sigma, \tag{4}$$

where R is the activity of the product radionuclide in disintegrations/s, I is the number of incident particles/s, n is the number of target nuclei/cm^3, x is the target thickness in cm, and σ is the cross section in cm^2/nucleus. This relationship assumes that (1) the beam current is constant over the course of the irradiation, (2) the target nuclei are uniformly distributed in the target material, and (3) the cross section is independent of energy over the energy range of the incident particles in the target, which is a good approximation for a thin target.

The rate of production is, of course, affected by the fact that the product nuclei are radioactive and are thus undergoing radioactive decay. For short-lived nuclides, the competition between formation and decay will come to equilibrium at sufficiently long bombardment times. This point is called saturation, meaning that no matter how much longer the irradiation is carried out, the production rate is equal to the rate of decay and the amount of radioactivity stays constant. Taking this into account, the amount of radionuclide produced at the end of an irradiation time t becomes:

$$R = Inx\sigma(1 - e^{-\lambda t}), \tag{5}$$

where t is the irradiation time in seconds and λ is the radionuclide's decay constant, which is related to its half-life by $\lambda = 0.693/t_{1/2}$. The term $(1 - e^{-\lambda t})$ is called the saturation factor and accounts for the competition of the production of nuclei due to the particle reaction and the radioactive decay of the nuclei that have already been produced. This term shows that the saturation factor approaches unity when the irradiation time is sufficiently longer than the half-life of the radionuclide being produced.

4.2.2. *Saturation factors*

As shown above, the competing rates of production and decay will affect the amount of radionuclide present at any point during the irradiation. At short irradiation times, the fraction of radionuclide produced is related to the saturation factor $(1 - e^{-\lambda t})$. From this, it is evident that an irradiation equivalent to one half-life would result in 50% of the maximum amount that can be produced. For practical reasons, irradiations rarely exceed three half-lives (90% of saturation) except for the shortest-lived radionuclides. It is relatively easy to come close to saturation for the production of ^{15}O (~2 min half-life), but it is not reasonable to irradiate a target for the production of ^{18}F (~110 min half-life) to the point of saturation because of the excessively long times involved. When the saturation factor is plotted as a function of the ratio of irradiation time to half-life, as shown in Fig. 13, it can be clearly seen that for longer half-lives, a compromise must be made with respect to expected yield versus irradiation time.

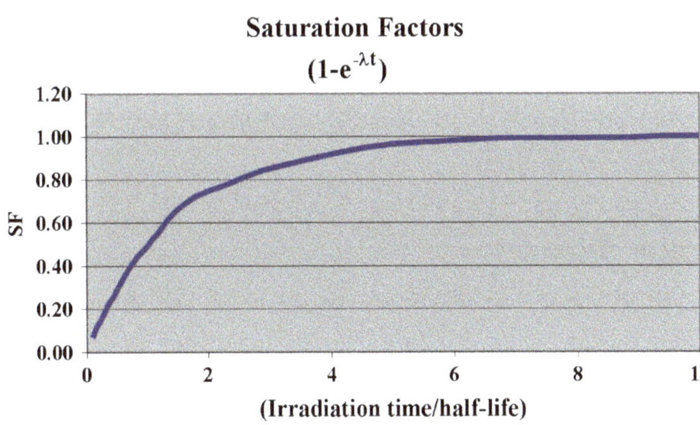

Fig. 13. Saturation curve as a function of the ratio of irradiation time to the half-life of the radionuclide. 50% of the saturation yield is reached in one half-life, 75% in two half-lives, etc.

For long-lived species, the production rates are usually expressed in terms of integrated dose or total beam flux (μA·h). For example, with a long-lived radionuclide like ^{82}Sr ($t_{1/2} = 25$ d), the amount of radionuclide at the end of bombardment will be essentially the same whether it is produced from 100 μA in 1 hour or 2 μA in 50 hours (both represent 100 μA·h of beam). For a shorter-lived radionuclide like ^{18}F, the same two irradiation conditions will produce very different amounts.

4.2.3. *Specific activity*

Another important factor to be considered is the specific activity of the radionuclide produced. Specific activity is a measure of the number of radioactive atoms or molecules as compared to the total number of those atoms or molecules present in a sample. The specific activity is usually expressed in terms of radiation units per mass unit. The traditional units have been Ci/mole (or Ci/g) or fraction thereof (nowadays expressed as GBq/mole). If the only atoms present in the sample are those of the radionuclide, then the sample is said to be carrier free. For example, a compound labeled with ^{211}At will be carrier free since there are no stable isotopes of astatine. In most systems, some of the stable isotope is present with the radionuclide. High specific activities are essential in nuclear medicine where the biological effectiveness of the radiopharmaceutical used to diagnose or treat diseases is greatly dependent on having the highest concentration of radionuclide with respect to carrier.

The principle advantage of accelerator-produced radionuclides is that high specific activities can be obtained through the (p,xn), (p,α), and other reactions involving charged particles that result in the product being a different element than the target. Reactors, on the other hand, often use an (n,γ) reaction that results in a different isotope of the same element and therefore low specific activity.

5. Accelerator Targetry

The routine production of radionuclides with an accelerator requires specially designed targets that are tailored to the target material, the desired end product, and the characteristics of the accelerator to be used (e.g., particle, beam energy, and beam current). Important considerations are (i) the chemical and physical form of the target nucleus, (ii) the chemical and physical form of the desired product, and (iii) the ease of separation of the product from the target.

The purpose of the target is to get the material to be irradiated into the beam, keep it intact during the irradiation, and facilitate the removal of the product radionuclide from the target material quickly and efficiently. There are a myriad of physics and chemistry phenomena that must be considered in the design and engineering of a target.

The scientific and engineering principles involved can be complicated and are beyond the scope of this chapter. However, because targets are the most critical part of a radionuclide production system, we will briefly describe how some of these phenomena can affect the yield and purity of the desired radionuclide in the expected chemical form. More detailed treatments can be found for instance in the IAEA Technical Report Series 465 and references given therein.[28]

5.1. *Stopping power and energy loss*

The slowing down (i.e., loss of kinetic energy) of a charged particle in a material by any and all means is called its "stopping power", which depends on the type and energy of the particle and on the properties of the material it passes through. Stopping power is defined as the average energy lost by the particle in the material per unit path length and is written as $S(E) = -dE/dx$, where E is the particle energy in MeV and x is the distance traveled in cm through the material.

Tables of stopping powers and dE/dx for protons and other particles exist in the literature for many materials. If the stopping power of the proton is known, the stopping power of heavier particles can be determined from the relations given in Table 3.[29] In a similar fashion, the stopping power of a particle in a chemical compound, alloy, or composite material is the stopping power of the particle in each element of the compound times the atom fraction of that element in the compound. Although this method entails making some simplifications and assumptions, the results are accurate enough for most target calculations.

5.2. *Energy straggling*

Due to the statistical nature of atomic collisions, an initially monoenergetic source of particles travelling through a material will have a distribution of ranges centered about a mean value. This phenomenon is called energy straggling. Energy straggling has significant implications for determining the expected yield from a target that does not fully stop the incident beam of charged particles.

Table 3. Stopping power of various particles relative to the stopping power of the proton.

Particle	Stopping power relative to protons
Deuteron	$S_d(E) = S_p(E/2)$
Triton	$S_t(E) = S_p(E/3)$
^{3}He	$S_{He\text{-}3}(E) = 4S_p(E/3)$
^{4}He	$S_{He\text{-}4}(E) = 4S_p(E/4)$

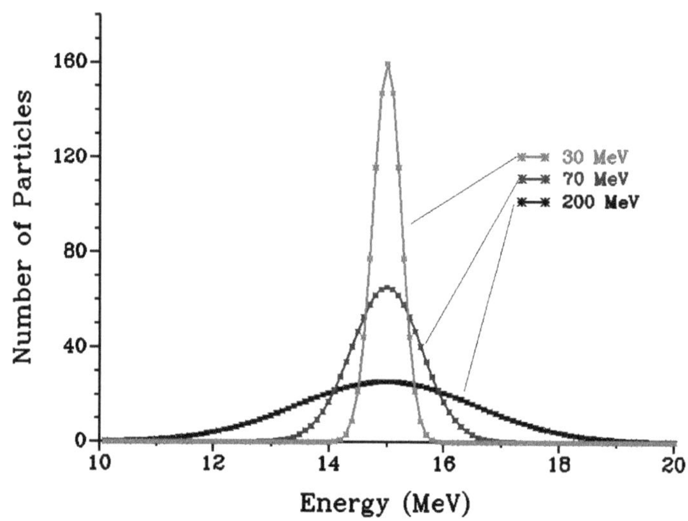

Fig. 14. Beam energy distributions when protons are degraded from an initial energy of 200, 70, or 30 MeV to a final energy of 15 MeV.

Using a foil to degrade the beam energy can have a particularly significant effect on the target yield even if the calculated beam energy after passing through the foil would be the same as that of a beam entering the target with no degrader foil. As an example, Fig. 14 shows the distribution of energies resulting from degrading beams of protons from 200, 70, and 30 MeV to 15 MeV, all with the same number of particles. It is clear from this figure that if the initial energy is 30 MeV, the energy spread will be ~2 MeV, whereas if the initial energy is 200 MeV, the energy spread is nearly 10 MeV. The width of this energy distribution at a given point in the target can be fairly accurately determined using statistical models of the beam interactions. A more complete discussion of this subject can be found in the IAEA Technical Report Series 465.[28]

5.3. *Small angle multiple scattering*

As charged particles travel through any material, they also undergo small angle multiple scattering. This phenomenon has an important implication for the design of accelerator targets because any foils or target material in the beam path will cause an angular spreading of the beam. The shape of the target must then account for this increase in beam diameter in order to maximize the yield. This is especially important in the design of gas targets in which the target medium is usually at a lower density than in a liquid or solid target, meaning that the gas target will be thicker (longer beam path) and therefore the effects of the small angle multiple scattering will be greater. A more complete description of the phenomenon can be found in the literature.[30]

5.4. *Beam heating and density reduction*

The energy lost when charged particles pass through the target medium is ultimately dissipated in the form of heat. One of the most challenging problems in the design of targets is finding methods to remove this heat from the target during irradiation. The heat generated in the target can often have several detrimental effects. These include target density reduction, chemical reactions occurring in the target material or products, and damage to the target foil or body.

For this reason, the choice of materials for the target body and entrance foil depends not only on their strength and chemical stability, but also on their thermal properties. It is important in the design of a target to understand how heat is transferred in order to efficiently remove it. Any mode of heat transfer involves a temperature differential as the driving force, and as one might expect, the greater the temperature gradient, the faster the heat transfer will be. Three modes of heat transfer are involved to one extent or another in accelerator targets: radiation, conduction, and convection.

The heat loss by radiation is usually minor, except for target foils or solid targets with low thermal conductivity and high beam current. Conductive heat transfer is especially important in the construction of the target body since this is where most of the heat from the beam will be deposited. The final mode of heat transfer is convection, which is the most difficult to estimate accurately. There are two modes of convective heat transfer: free and forced. In free convection, the flow patterns are determined mainly by the buoyant effects of the fluid; in forced convection, the flow pattern is determined by other forces such as fans or jets. In most cases, the efficiency of a particular system must be experimentally determined in order to obtain an accurate value. Some empirical relationships have been tabulated in reference books on heat transfer and may be used to estimate this quantity for a wide range of situations, but these are not as accurate as the experimentally determined value.

5.5. *Ionization of target materials*

As the beam passes through the target material, the charged particles lose energy and slow down through various mechanisms. Collisional interactions may result in excitation or ionization of the target material depending on the physical state. Excitation raises an electron to a higher energy shell, whereas ionization completely removes the electron from the atom. Ionization creates an electron–ion pair consisting of the electron ejected from the atomic shell and the positively charged atom from which the electron was removed. The freed electron may possess sufficient kinetic energy to cause further ionization events. A complete description of these calculations is beyond the scope of this chapter but may be found in the literature.[31] In practical terms, these ionization events will affect the chemical form of the radionuclide and will result in an increase in temperature in the target material.

5.6. *Radiation damage and activation*

The degradation of certain materials in the target due to radiation exposure can also be a problem. Gamma and beta radiation have little effect on metals, but they can have a dramatic effect on the structure of organic compounds. One important aspect of the beam passing through target foils is that they will become activated. Typical target foils such as Havar contain a number of metals. Each of the components can undergo nuclear reactions and create radionuclides. Usually these radionuclides are contained in the foils, but in certain circumstances such as in a water target, they can leach out of the foil and be transferred along with the product. For this reason, it is always important to check the radionuclidic impurity levels of any product made in these types of targets.

5.7. *Chemical reactions*

The chemical reactions of the highly excited nucleogenic atoms (i.e., those created by the nuclear reactions) with the surrounding target material during their de-excitation will often determine the molecular form of the radionuclide being produced. The state of the matter inside the target depends on the state of the matter being bombarded.

Fig. 15. View inside a gas target during irradiation where the highly excited and ionized gas molecules emit light and heat (Picture courtesy of Sven-Johan Heselius).

In a gas target, the gas is thermally very hot and highly ionized. This can result in a large number of possible chemical reactions, including ion–molecule reactions. A typical view inside a gas target during irradiation is shown in Fig. 15. A large number of chemical species can be formed in this gaseous plasma or "ionic soup". When other gases are present, either as contaminants or as additives, the number of possible reactions increases dramatically. In most cases, the final product distribution will be determined by the thermodynamics of the situation since there is more than enough energy to overcome the kinetic activation barriers that would place constraints on the product distribution at lower temperatures.

In a liquid target, the material may exist as a liquid or as a vapor, depending on the beam density and temperature, and the water molecules are highly excited and ionized. The target material may boil intermittently as has been seen in water targets for the production of ^{18}F.[32,33] An example of the water boiling and the excited water molecules giving off light is shown in Fig. 16. In this target as in the gas target, there are excited molecules and ions reacting to form new species such as hydroxyl radicals and peroxide radicals that can react with contaminants in the water. In a solid target, the solid may liquefy or sublime and there will be excited species created, although the types of reactions that occur in liquid and gas targets are somewhat suppressed.

5.8. *Pressure increases*

As the temperature rises in the target, the pressure will also increase proportionately. This pressure rise is critical in the choice of the foil that

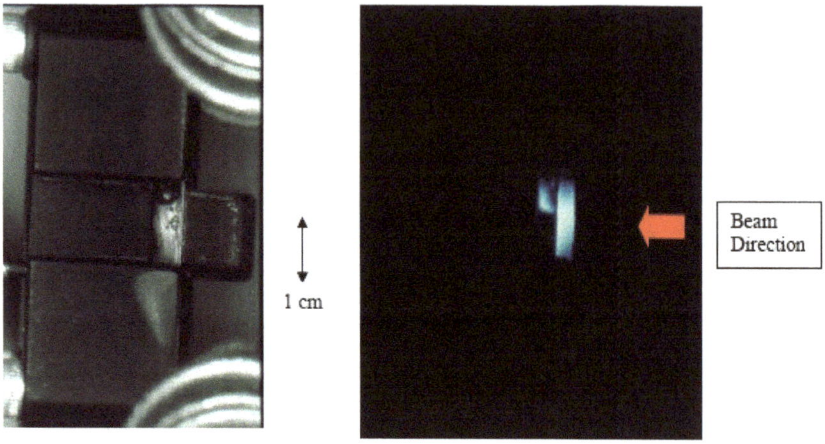

Fig. 16. Water boiling (left panel) and emitting light during irradiation (right panel).

separates the target material from the vacuum of the accelerator or beamline. If the stress on the foil from the pressure increase exceeds its tensile strength, the foil will burst. When the foil bursts, the target material may be lost and the accelerator will probably be shut down.

5.9. *Beam focusing*

The extracted beam is often transported from the accelerator to the target by a beam transport system. This can be a series of dipole and quadrupole magnets, magnetic lenses, collimators and beam monitors, or as in the case of many small cyclotrons where the targets are attached directly to the machine, it is just a beam collimator.

The beam exiting an accelerator can have unusual shapes depending on the dynamics of the acceleration and extraction as well as the fringing field in cyclotrons, although linacs don't have this latter issue. It is sometimes desirable to focus the beam with quadrupole magnets. This also allows some smoothing of the beam in those accelerators where the beam may have "hot spots" or spikes in the intensity profile, or where the beam spot is intrinsically very small.

6. Conclusions and Future Directions

The production of radionuclides with accelerators seems to have a very bright future. The number of cyclotrons has been steadily increasing throughout the world, with more than 700 now in use for this purpose. Many of the newer cyclotrons are lower energy (<20 MeV) and are used primarily for the production of PET radiotracers such as FDG for clinical use. Compact RF linacs are also being offered commercially for PET radiotracer production, including mobile systems installed in trailers. There does not seem to be an end in sight as the power of PET imaging is used more extensively for medical diagnostics and treatment monitoring. The development of new tracers and new radiopharmaceuticals used to diagnose or treat specific diseases is close at hand. When this occurs, there may be an even greater demand for accelerators for radiopharmaceuticals.

One example of a future application of PET lies in the diagnosis of neurodegenerative disorders, the most common of which are associated with intracellular or extracellular deposits of abnormal proteins. These disorders, which may result in dementia, include Alzheimer's disease (AD), frontotemporal dementia (FTD), dementia with Lewy bodies (DLB), and dementia in Parkinson disease (PDD). New tracers are under development that may be able to diagnose these conditions much earlier and therefore allow treatment to start before the disease progresses to the point where the symptoms become noticeable.[34] PET can also be used to diagnose non-neoplastic conditions such as infection and inflammation, which appear to have increased utilization of sugar and are therefore readily visualized by FDG PET imaging.[35]

There are a number of new developments and new ideas about radiopharmaceutical delivery. One of the more novel concepts is that of an accelerator and chemical synthesis module that produces a single dose of a radiopharmaceutical on demand. This moves away from the idea of a centralized distribution system and aims to put one of these devices in every hospital acting like a generator for PET radiopharmaceuticals.

Another concept is that of a true generator such as the $^{68}Ge \rightarrow ^{68}Ga$ generator system. This is becoming very popular in Europe and the number of radiopharmaceuticals incorporating ^{68}Ga is increasing. If radiopharmaceuticals are developed that have attractive characteristics for clinical diagnosis, these generators will definitely have a place in the future of nuclear medicine.

References

1. W. G. Myers, *J. Nucl. Med.* **20**(6), 590–594 (1979).
2. E. Rutherford, *London, Edinburgh and Dublin Phil. Mag. J. Sci.* **37**, 581 (1919).
3. *The First Cyclotrons — Ernest Lawrence and the Cyclotron: AIP History Center Web Exhibit*, http://www.aip.org/history/lawrence/first.htm, accessed July 2011.

4. E. O. Lawrence and M. S. Livingston, *Phys. Rev.* **37**, 1706 (1931).
5. E. O. Lawrence and N. E. Edlefsen, *Science* **72**, 376 (1930).
6. A. Sessler and E. Wilson, *Engines of Discovery: A Century of Particle Accelerators* (World Scientific Publishing, Singapore, 2007).
7. F. Joliot and I. Curie, *Nature* **133**, 201 (1934).
8. M. D. Kamen, *Science* **140** (3567), 584 (1963), doi:10.1126/science.140.3567.584.
9. E. Lebowitz *et. al.*, *J. Nucl. Med.* **6**(2), 151 (1975).
10. J. S. Fowler and T. Ido, *Seminars Nucl. Med.* **32**(1), 6 (2002).
11. *Directory of Cyclotrons Used for Radionuclide Production in Member States*, IAEA-DCRP/CD (International Atomic Energy Commission, Vienna, 2002), ISBN 92–0–133302–1.
12. *Cyclotron Produced Radionuclides: Physical Characteristics and Production Methods*, IAEA TRS **468** (International Atomic Energy Agency, Vienna, 2009).
13. *Cyclotron Produced Radionuclides: Guidelines for Setting Up a Facility*, IAEA TRS 471 (International Atomic Energy Agency, Vienna, 2009), ISBN:978-92-0-103109-9.
14. J. S. Keppler and P. S. Conti, *Am. J. Roentgenology*, **177**(1), 31 (2001).
15. V. K. Gupta, *J. Med. Phys.* **20**, 31 (1995).
16. A. Flynn, in *Radiotherapy in Practice — Brachytherapy*, Eds. P. Hoskin and C. Coyle (Oxford University Press, New York, 2005), pp. 1–20.
17. U.S. Cancer Statistics Working Group, *United States Cancer Statistics: 1999–2007 Incidence and Mortality Web-based Report*, (US Dept. Health and Human Services, Centers for Disease Control and Prevention, and National Cancer Institute, Atlanta, 2010), available at http://www.cdc.gov/uscs.
18. R. M. Sharkey and D. M. Goldenberg, *J. Nucl. Med.* **46**, Suppl. 1, 115S (2005).
19. J. A. O'Donoghue, M. Bardiès and T. E. Wheldon, *J. Nucl. Med.* **36**(10), 1902 (1995).
20. D. E. Milenic, E. D. Brady and M. W. Brechbiel, *Nature Rev. Drug Discovery* **3**, 488 (2004).
21. R. Schibli and P. A. Schubiger, *Eur. J. Nucl. Med. Mol. Imaging* **29**(11), 1529 (2009).
22. IAEA Bulletin 1/1994, accessed 30 Oct 2011 at http://www.iaea.org/Publications/Magazines/Bulletin/Bull361/36101081618.pdf.
23. T. Ruth, *Reviews of Accelerator Science and Technology* **2**, 17 (2009).
24. T. Wangler, *Principles of RF Linear Accelerators* (Wiley & Sons, New York, 1998), p. 225.
25. M. Herman *et al.*, *EMPIRE: Nuclear Reaction Model Code System for Data Evaluation, Nuclear Data Sheets* **108**(12), 2655 (2007).
26. R. D. Evans, *The Atomic Nucleus* (McGraw-Hill, New York, 1955).
27. G. Friedlander, J. W. Kennedy, E. S. Macias and J. M. Miller, *Nuclear and Radiochemistry*, 3rd Ed. (Wiley & Sons, New York, 1981).

28. *Cyclotron Produced Radionuclides: Principles and Practice*, IAEA TRS **465** (International Atomic Energy Agency, Vienna, 2008).
29. J. Janni, *Stopping Power and Ranges*, Air Force Weapons Laboratory Report AFWL-TR-65-150 (1965).
30. D. J. Schlyer and P. S. Plascjak, *Nucl. Instr. Meth. Phys. Res.* **B56/57**, 464 (1991).
31. W-M Yao *et al.*, *J. Phys. G: Nucl. Part. Phys.* **33**, 1 (2006).
32. S. J. Heselius, D. J. Schlyer and A. P. Wolf, *Int. J. Radiat. Appl. Instr. A, Appl. Radiat. Isot.* **40**, 663 (1989).
33. B. W. Wieland, D. J. Schlyer and A. P. Wolf, *Int. J. Appl. Radiat. Isot.* **35**, 387 (1984).
34. A. Kadir and A. Nordberg, *J. Nucl. Med.* **51**(9), 1418 (2010).
35. M. Hashefi and R. Curiel, *Ann. NY Acad. Sci.* **1228**, 167 (2011).

Chapter 5

INDUSTRIAL ASPECTS OF ION BEAM ANALYSIS

Ragnar Hellborg*
Department of Physics
Lund University, SE-221 00 Lund, Sweden
ragnar.hellborg@nuclear.lu.se

Harry J. Whitlow
Institut des Microtechnologies Appliquées Arc
Haute Ecole Arc Ingénierie Arc, CH-2300 La Chaux-de-Fonds, Switzerland
harry.whitlow@he-arc.ch

Ion beam analysis techniques are finding more and more industrial applications both for product- and process-development. This chapter presents an overview of these techniques. The methods are broadly divided into content-measuring and depth distribution-measuring techniques. Moreover, guidelines for introduction of accelerators for industrial materials analysis into an industrial environment and the instrumentation are discussed.

1. Ion Beam Analysis in Industry

Today's high technology society is placing more and more demands on new high performance materials. This is especially true in the emerging nanoscience and technology field where the developments are currently moving out of the laboratory and pilot-plant stage into industrial production. This means that there is an increasing demand for materials characterisation techniques for Quality Control (QC) capable of working on a nanometre scale. In the early days, high-tech was often considered synonymous with microelectronic technology where advanced materials and associated characterisation methods were focussed on process technology for future generations of central processing units and memory. Today's high technology industry has undergone a dramatic evolution. It is now much more diverse, and while a considerable focus remains on the microelectronic industry, new industries such as the pharmaceutical industry, the renewable energy industry, and the environmental monitoring industry are ranked as high

*Corresponding author.

technological industries. Examples are the pharmaceutical industry where there is an increasing demand for methods to detect the metabolism of pharmaceutical preparations in the body, follow biological pathways, and determine the efficiency of new drug-delivery systems such as functionalised nanoparticles. In the renewable energy field, the advent use of compound solar cells capable of working with a solar energy concentrator where the combined heat and current flow implies reliability and lifetime is directly related to the choice of semiconductor contact and packaging technology. As is discussed below, conventional standard "industrial" surface characterisation becomes less and less applicable as the dimensions shrink down to the nanometre level. This has created new demands for characterisation and modification of advanced materials, e.g., for QC and environmental stewardship where extremely low-levels of toxic matter must be detected in industrial feedstock and waste.

In this chapter, we present an overview of the different methods for ion beam analysis that are based on high energy ions. Most of the text is general and guides the reader through the different ion beam analytical methods that have either been used for industrial QC applications or as a process development tool. To emphasise the quantitative nature of methods such as RBS, a step-by-step guide to quantitative analysis of RBS spectra is given.

The use of ion beams for analysis of materials is a relatively new industrial application of accelerators. Accelerators were originally developed and used for nuclear physics research. Starting in the early 1970s, applications in material analysis led to the development of a field commonly referred to as Ion Beam Analysis (IBA). A chronology of this development is given in the introduction of a recent book edited by the authors and Y. Zhang.[1] The unique energy resolution of the beam and the possibility to easily and continuously change the beam energy of accelerators that use the direct voltage technique (see Sec. 2) make them ideal for analytical analysis techniques. They are today used in many fields ranging from materials science and environmental science to the study of cultural heritage and biological samples. Direct voltage accelerators are today used by the semiconductor industry for QC and by environmental sciences for pollution monitoring.

There is a need to detect very low concentrations of atoms in a sample for a number of research fields and technical applications. Several analytical techniques have evolved from nuclear physics methods, some of them with a quantitativeness, depth resolution, and sensitivity that can not be achieved with any other technique, either physical or chemical. In a particle accelerator, beams of fast, ionized atoms are produced. The type of ion, ion energy, intensity, and geometrical dimensions of the beam can often be chosen rather freely. This has made it possible to develop a number of analytical techniques with extremely high resolution and sensitivity. Some of these techniques and their qualities are listed in Table 1. For a detailed description of most of these techniques, see, for example, the book by Brune *et al.*[2] The use of 100s of keV to MeV ions as a probing beam implies that the energies involved in the primary scattering and X-ray excitation processes, on which these IBA techniques are based, are orders of magnitude greater than the chemical binding energies. The advantage of this is that, unlike most competing techniques (e.g., Secondary Ion Mass Spectrometry (SIMS), sputter-Auger and sputter-XPS[2]), the total amount of the element is measured without having to take chemical effects into account. The success and impact of these IBA methods can be judged from the observation that many more new accelerators have been delivered for materials characterisation and modification than were delivered for the original task of basic nuclear physics studies.

Materials analysis techniques all rely on the vast database of knowledge that exists for most of the nuclei in the periodic table. The following physics techniques are described in this chapter:

- Rutherford backscattering spectrometry — RBS
- Elastic recoil detection analysis — ERDA
- Particle induced X-ray emission — PIXE
- Particle induced gamma ray emission — PIGE
- Nuclear reaction analysis — NRA
- Charged particle activation analysis — CPAA
- Accelerator mass spectrometry — AMS

A technical development that has played an important role for the field of IBA is the nuclear microprobe. The possibility to collimate and/or focus an ion beam in the MeV range from the normal mm diameter to μm dimensions has opened quite a few possibilities for different IBA techniques.

The basic idea behind the nuclear microprobe seen in Fig. 1 is to produce μm or sub-μm ion beams of high enough intensity to allow the use of a number of IBA techniques in combination with precise knowledge about the beam position. In most laboratories, the nuclear microprobe uses a direct voltage accelerator (see Sec. 2 for a discussion of different types of accelerators), and the experimental measurements are carried out on a beamline that contains magnetic focussing devices, such as quadrupoles, that have been designed to minimize aberrations, as seen in Fig. 1(a). The quality of the quadrupole magnets used for focusing the ion beam to μm dimensions is crucial for achieving a small beam spot size. Misalignment and parasitic

Table 1. Different analytical techniques using beams of fast ions.

Technique[a]	Information depth (μm)	Resolution depth/Lateral	Sensitivity[b]	Sample	Facilities worldwide/ Cost per sample (€)	Typical beam	Measurement complexity
RBS	2	25 nm/0.5 mm	10^{-4}–10^{-7}	Solid/0.25 mm^2	500/70	2 MeV $^{4}He^{2+}$	Simple
ERDA	1	20 nm/1 mm	10^{-4}–10^{-5}	Solid/5 $mm^{2,c}$	300/100	2 MeV $^{4}He^{2+}$	Complex ToF detector often needed
PIXE	50	10 μm/1 μm	10^{-6}–10^{-7}	Solid/powder/1 mm	300/50	2 MeV $^{1}H^{+}$	Simple
PIGE	50	10 μm/1 μm	10^{-5}–10^{-6}	Solid/powder/1 mm	200/70	2 MeV $^{1}H^{+}$	Simple
NRA	0.1–1	5–10 nm/1 mm	10^{-4}–10^{-7}	Solid/3 mm	300/200		
CPAA	100	–/1 mm	10^{-7}–10^{-8}	Solid/1 mm^2	100/1000	20 MeV $^{2}H^{+}$	Simple
AMS	—	—	10^{-14}–10^{-15}	—	100/400	12 MeV $^{14}C^{3+/4+}$	Complex

[a]The definition of the acronyms used is given in the text.
[b]Number of atoms per number of host atoms.
[c]Glancing incidense beam and glancing exit beam.

(a)

(b)

Fig. 1. The nuclear microprobe system in Uppsala, Sweden. (a) General overview showing the magnetic scanning system and sample chamber. The entire lens and sample chamber assembly is mounted on a solid concrete plinth to achieve stability. (b) Inside of the sample chamber showing detectors for ERDA, PIXE and RBS.

aberration due to mechanical misfit together with the aberrations inherent to each optical design will limit the highest lateral resolution obtainable for an ideal incoming ion beam. It is most important to minimize spherical and chromatic aberrations. In the experiment chamber, an array of different detectors are used for measurements using different IBA methods. Often, optical microscopes are also incorporated into the target chamber to precisely position the sample for analysis.

Those techniques mostly used in a nuclear microprobe setup include particle induced X-ray emission (see Sec. 3.2), backscattering spectrometry (see Sec. 4.4), elastic recoil detection analysis (see Sec. 4.5), scanning transmission ion microscopy, particle induced gamma ray emission (see Sec. 3.3), the broader field of nuclear reaction analysis (see Sec. 4.6), and others. The detector systems in a typical sample chamber are shown in Fig. 1(b). A comprehensive overview on the microbeam method can be found in a newly published handbook.[3]

A good example of a non-academic laboratory using the IBA technique, among others, is the Louvre museum in Paris.[4] This laboratory, Centre de Recherche et de Restauration des Musées de France, is in charge of studying collections at public museums in France. Multidisplinary investigations are performed by art historians, chemists, physicists, geologists, conservators, etc. The laboratory is using specific analytical techniques,[5] many of them involving nuclear physics such as IBA. A 2 MV tandem accelerator is installed in the basement of the Louvre museum. External beams are used, allowing the analysis to be performed at atmospheric pressure (see Sec. 3.2 for further details about external beams). All different types of IBA techniques are used. PIXE is the most common technique, but also PIGE and NRA are regularly used. RBS with a helium beam as well as ERDA analysis with helium to obtain hydrogen profiles are performed.

An international conference on IBA is held every second year. The most recent one[6] was held in April 2011 in Itapema, Brazil.

2. Accelerators for IBA

Accelerators used for IBA are nearly exclusively of the direct voltage type. In accelerators utilizing this principle (other names are "potential drop" or "high voltage dc accelerators" because the current is dc, unlike all other accelerator types), the particle (after ionization) is accelerated through an accelerator tube in one step. The tube is constructed as a long rectilinear drift tube with a number of electrodes along the axis with a controlled voltage for each electrode, partly to aid in focusing the beam and partly to distribute the voltage gradient uniformly along the insulation surfaces. The positive ions (or electrons) to be accelerated are generated in an ion source located at high voltage (except for tandem accelerators — see below — in which negative ions are generated at ground potential). Direct voltage accelerators are often identified by the type of high voltage generator used. The high voltage can be generated by rectifying ac-voltage (often called cascade generator) or using electrostatic charging in which a mechanical system carries the charge to the high voltage terminal (called electrostatic generator). The open-air accelerator fails above a few MV mainly because electrical breakdown of the air results in sparking. If the accelerator is enclosed in a tank with a suitable gas of high pressure, the voltage available today is up to a few tens of MV. Normally for industrial applications, much more compact accelerators with voltages of 2 MV or less are used.

The direct voltage principle has been used in thousands of accelerators around the world. The reasons why this type became so popular are:

- All types of ions can be accelerated.
- The ion energy can be changed continuously.
- The high-voltage stability is extremely good, and therefore
- The ion energy has a very low energy spread.

An overview of different types of accelerators, including the direct voltage type, can be found in the introduction to a book on electrostatic accelerators edited by Hellborg.[7] We describe them briefly below.

Cascade accelerators. The high voltage unit of a cascade accelerator consists of a multiplying rectifier-condenser system. The original design of a cascade generator (first used by Cockcroft and Walton[8]) can be seen in Fig. 2. The circuit is asymmetrical. In practice, the asymmetrical principle was soon replaced by a symmetrical one to reduce the ripple. A third principle, introduced much later for obtaining the high voltage, is a parallel-driven circuit. Accelerators of this type are enclosed in a pressure tank and are today designed for use up to 6 MV. The main advantage is that a cascade accelerator has a large output current of up to several hundred mA. This type of accelerator has for many years been used as the injector to higher voltage accelerators due to the high beam current that such an accelerator

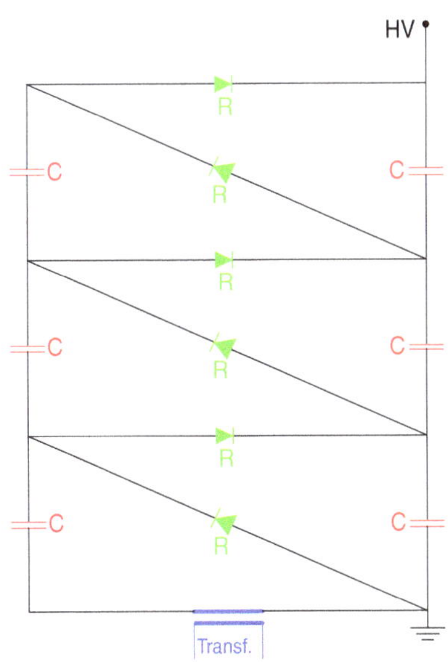

Fig. 2. Asymmetrical circuit cascade generator consisting of capacitors C and rectifiers R.

Fig. 3. The 5 MV parallel-driven cascade tandem accelerator, a Tandetron, for the Universidad Autonoma de Madrid under test at the HVEE plant in Amersfoort, the Netherlands. For further details see the text. (Courtesy of HVEE.)

can handle. A detailed description of different types of cascade generators can be found in the recent book on electrostatic accelerators.[9]

High Voltage Engineering Europe (HVEE) and Kobe Steel Ltd, Japan, are the main producers of commercial cascade accelerators today. Both single stage and tandems (see below) are available. The HVEE accelerators are sold under the trademark Tandetrons. In Fig. 3 we can see a photo of a 5 MV tandem cascade accelerator with the generator part designed along the parallel driven principle.[10] The solid-state power supply is wrapped around the high energy accelerator tube (to the left in the photo). The RF field is supplied by a pair of large electrodes named dynodes. The end of the upper one of these dynodes can be seen at the top of the accelerator tank opening. The RF voltage present on the dynodes is capacitively coupled to the power supply diode stack via the corona rings, also seen in the photo.

Electrostatic accelerators. In 1929, Robert Van de Graaff demonstrated the first generator model of this type.[11] An electrostatic charging belt is used to produce the high voltage. Two rollers are provided, one at ground

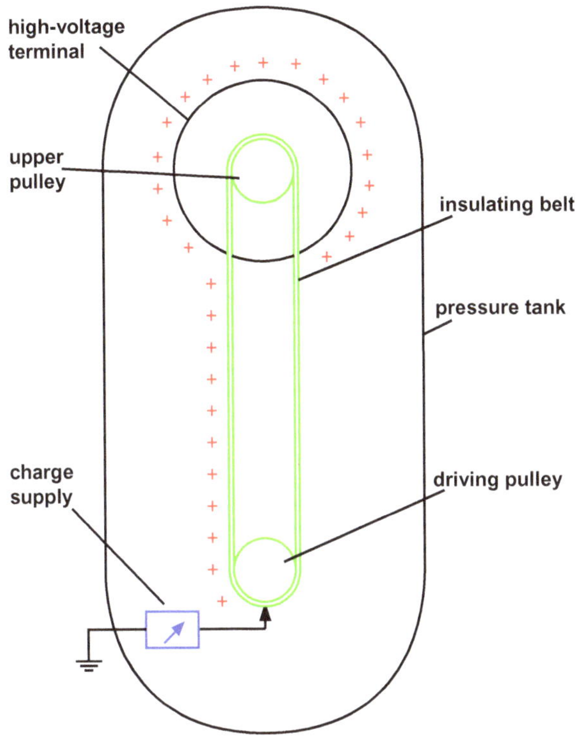

Fig. 4. Principle of an electrostatic generator. By including an ion source and an accelerator tube, the generator is converted to an accelerator.

potential driven by a motor and the other located in the high-voltage terminal, well insulated from ground. An endless belt of insulating material passes over the rollers. Charge is sprayed from sharp corona points onto the moving belt. The belt conveys the charge to the insulated high voltage terminal within which the charge is removed by collector points and allowed to flow to the surface of the electrode. The principle of an electrostatic generator is illustrated in Fig. 4.

A disadvantage of an electrostatic accelerator is the low current output compared to the cascade accelerator. All modern accelerators of the direct voltage type are enclosed in a tank with gas of high pressure to reduce the size and to be independent of moisture in the air.

The insulating belt for transport of the charge has been replaced almost entirely in many modern electrostatic accelerators by a chain of metal cylinders. The individual cylinders are connected to each other by insulating links. Each metal cylinder is charged inductively. In this way, a more robust transport system with much more well-defined charge transport and hence a better voltage stability is obtained compared to using an insulating belt.

Fig. 5. Picture of the 1.7 MV Pelletron tandem accelerator facility in Jyväskylä, Finland used for industrial consulting and materials research and development. (Left photo.) The low-energy end of the accelerator with two different ion sources. One is for production of ions of gaseous elements (H, He, O) while the other is for production of ions from solid materials (C, Cl, Cu, Br, I etc.). (Middle photo.) The accelerator tank. (Right photo.) The different materials measurement and modification stations. The RBS measurement station (Fig. 24) used for industrial consulting work is on the far right. To the left of this is a ToF–E ERDA system. On the far left is a MeV ion beam lithography system used for materials modification.

More details about different types of accelerators and especially electrostatic accelerators can be found in an extensive textbook.[12]

NEC is the main producer of commercial electrostatic accelerators today. Their accelerators are sold under the trademark Pelletrons. Several photographs showing different sections of a Pelletron facility are shown in Fig. 5.

Tandem (cascade and electrostatic) accelerators. During the 1950s, negative ion sources were developed, i.e., beams of ions with an extra electron added to the neutral atom became available. This development made it possible to build two-stage (or tandem) accelerators. In a tandem, the high voltage is utilized twice. Negative ions are formed at ground potential and injected into the first stage where acceleration to the positive high voltage terminal takes place. The energy gain is $e\,U_T$ eV, where e is the elementary charge and U_T is the terminal voltage. In a stripper system located in the high voltage terminal, the negative ions lose a few electrons and become positive ions. In the second stage, the positive ions once more gain energy. Now the energy gain is $q\,e\,U_T$ eV, where q is the charge state of the positive ion. Thus a total energy gain of $(q+1)\,e\,U_T$ eV is obtained. For heavy ions and high voltages, i.e., high speed ions undergoing stripping, q can be quite

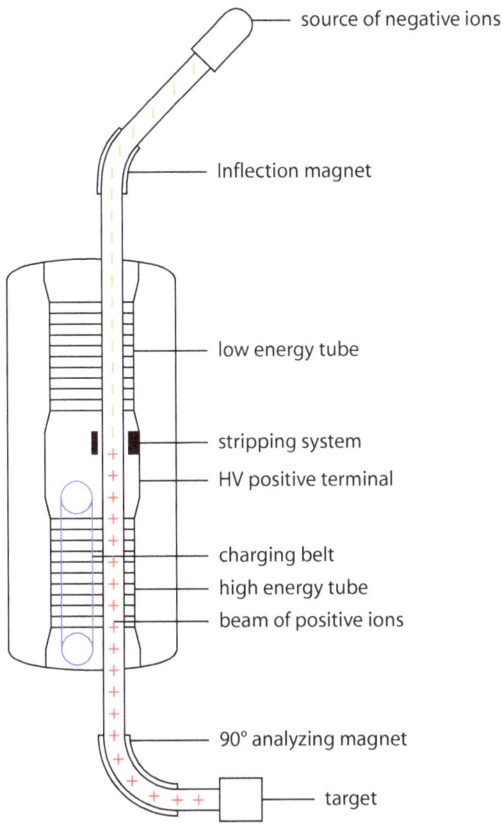

Fig. 6. Principle of a two-stage (tandem) accelerator.

high and therefore the final energy of the ions can be hundreds of MeV. A drawing illustrating the principle of a tandem accelerator is shown in Fig. 6.

An international conference on heavy ion accelerators is organized every third year. Most types of accelerators used in IBA techniques are dealt with in these conferences. The most recent conference was held in Venice, Italy in June 2009.[13] The next one will be held in Chicago, USA, in June 2012.

3. Quantity Analysis

3.1. *Introductory remarks*

Among the different techniques used for Ion Beam Analysis, the following are mainly used for quantity analysis: PIXE, PIGE, CPAA, and AMS. These methods are normally not useful as depth dispersive methods. *Quantity analysis* refers to the determination of the amount of different elements in a sample. As such it does not give information directly as to how the elements are distributed in the sample. Techniques where the depth distribution of elements in a sample are characterised are dealt with in the next section.

3.2. *PIXE technique*

Particle induced X-ray emission (PIXE) is a well-known analytical method applied in a variety of fields.[14] Ions of an energy greater than a few MeV per nucleon impinging on a sample have a high production cross section for inner-shell vacancies in the target atom. In other words, they readily knock-out the innermost and most tightly bound electrons. The subsequent de-excitation results in an emission of X-rays with an energy characteristic for each type of sample atoms, as illustrated in Fig. 7. Alternatively to X-ray emission, an Auger electron can be emitted. However, particle induced Auger electrons have not been significantly used for industrial characterisation and will therefore not be considered further. The PIXE technique was conceived by Sven A.E. Johansson in 1969.[15] It can be used for measuring all elements heavier than about Na. For lighter elements, characteristic Auger electrons dominate over X-rays and their characteristic X-rays have too low energy to be easily detected. The characteristic X-rays are detected by a high resolution semiconductor detector. This type of detector, as well as a similar type for gamma ray detection, became available in the mid 1960s. The characteristic X-ray peaks in a PIXE spectrum are superimposed on a continuous background, as can be seen in Fig. 8.

A similar and older technique compared to PIXE is X-ray fluorescence (XRF). In this technique, an X-ray beam is used to produce the characteristic

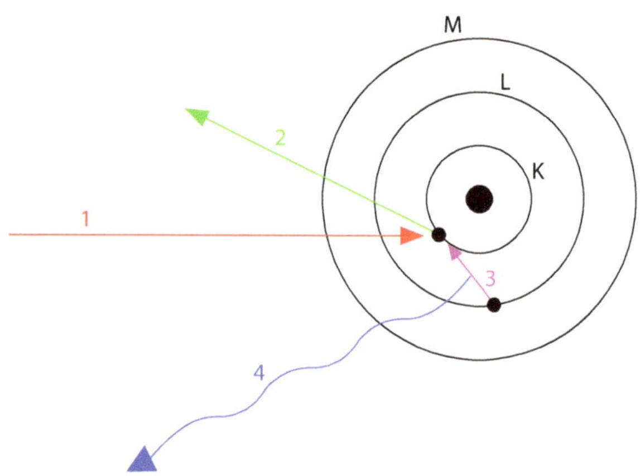

Fig. 7. Illustration of characteristic X-ray production. Legends: 1. Incoming ion beam, 2. Inner shell vacancy production, 3. De-excitation, 4. Emission of an X-ray photon.

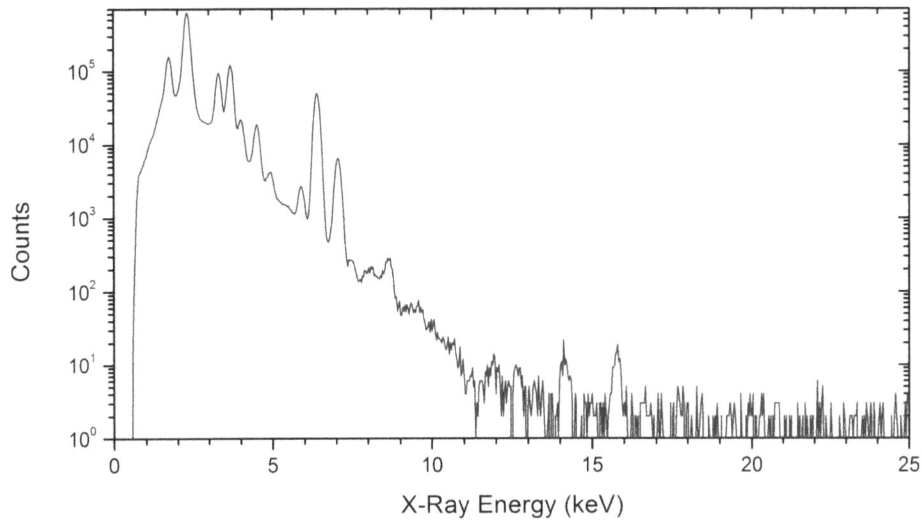

Fig. 8. PIXE spectrum of an aerosol sample taken on May 16, 2010 from the CARIBIC platform (Civil Aircraft for Regular Investigation of the Atmosphere Based on an Instrument Container) in a volcanic cloud from Eyjafjallajökull. (Courtesy of Bengt Martinsson, Lund University.)

X-rays. Charged particles have a clear advantage compared to X-rays for producing the characteristic X-rays as they give rise to a dramatically lower continuous background intensity in the X-ray spectrum. As a result, the relative detection limits (for example, in μg per g) are typically two orders of magnitude lower using the PIXE technique compared to using the XRF technique. Furthermore, an ion beam can be easily focused in contrast to an X-ray beam, therefore the incident beam density can normally be much higher in PIXE experiments compared to XRF experiments. Moreover, focusing of an ion beam down to μm dimensions is possible, as was described above in Sec. 1.

The interaction when a beam passes through a material occurs mainly by inelastic encounters with the electrons in the target. This not only results in excitation and ionisation of the atoms but also in other phenomena like electron bremsstrahlung that contribute to the background. This background will determine the ultimate detection limit for the PIXE technique.

When a beam of ions penetrates material, the ions will gradually lose energy until they finally stop within a very well-defined range. The cross sections for PIXE vary slowly with ion energy. In the proton energy range of 1–4 MeV, which is a typical range for PIXE experiments, the ionisation and X-ray production cross sections increase with increasing proton energy and decrease with increasing atomic number of the target atom.

An experimental setup for PIXE analysis includes a system for producing an ion beam (i.e., an accelerator). The beam should irradiate the target homogeneously. The homogenized beam is defined by a pair of collimators, normally circular, with diameters between 1 and 10 mm. The energy dispersive X-ray detectors used for PIXE have an effective sensitive area of 10 to 100 mm^2.

Most PIXE measurements are carried out in vacuum. But external-beam approaches are also used, either under atmospheric pressure of He[a] or in moderate vacuum. The vacuum requirements are less stringent, thus making it easier to design a low-cost chamber. Using an external beam, the heat conductivity from the target is increased and therefore the target temperature can be kept at a lower level.

Because the beam through the accelerator system needs a low pressure so as to minimise divergence, the external ion beam must be extracted into the region of moderate vacuum or atmospheric pressure either through a thin exit foil or through a narrow orifice combined with differential pumping. The exit foil will be strongly irradiated and has therefore a limited lifetime. Its material must be carefully selected. Thin polymer or metal foils have been used. A good example is commercially available Kapton foils, which only contain low-Z elements and can withstand a high intensity and a high radiation dose before breakdown. A recent improvement is the use of ultrathin membranes (down to 100 nm) of Si_3N_4. The beam size is usually defined by collimation in vacuum a few cm before the exit window at a typical value of 1 mm × 2 mm. The beam current used varies from a few nA to a few tens of nA. The sample is, if possible, placed a few cm from the exit window. The strong argon K X-ray line produced when the beam travels in air can alternatively be used to monitor the beam charge. This is otherwise difficult to do in non-vacuum PIXE experiments. Delicate and/or large objects can be examined by using an external beam. Examples are objects of interest in archeology or the history of art.[16] Wet samples or liquid samples are difficult to use in vacuum, therefore complex sample preparation procedures are needed. Such problems are overcome by using an external beam. Having the target in atmosphere also greatly facilitates and speeds up all kinds of operations, especially when large numbers of samples have to be analysed. A detailed description of an experimental setup for an external beam facility used for environmental studies can be found in Calzolai *et al.*[17]

---

[a]He is often used instead of air because air contains 1% Ar which can give rise to interfering X-ray lines.

In a PIXE nuclear microprobe, the beam current for very narrow beams ($\sim \mu$m) may be very low, e.g., tens of pA. New multicrystal detectors are therefore being developed. Up to about ten detectors, each about 100 mm^2 in area, are arranged in an annular geometry around the beam entrance hole. In the case of an external beam, the sample is placed a few mm after the exit foil which can be as thin as 100 nm with an area of 1 mm $\times$ 1 mm. The spectral shape can be advantageously modulated by placing an X-ray absorber between the target and the detector. The aim of the absorber is to reduce or eliminate unwanted continuum background and/or intense X-ray peaks and their associated pileup peaks, while at the same time allowing bombardment at higher beam intensities so that the elements of interest can be measured in shorter bombardment times with fewer spectral interferences. The electronics (amplifiers and pulse processors) require long time constants for optimal energy resolution. This implies that pulse pileup can become a serious problem at relatively low count rates. An electronic pileup rejector and its associated dead-time correction circuitry is therefore often incorporated in the pulse processing chain. When analysing thin targets, the ions pass through the target and are stopped in a Faraday cup for current integration. When using thick targets, the beam will stop completely in the target and the beam current must be measured either indirectly or by insulating the whole chamber from ground. If possible, the target should be mounted electrically-insulated from the chamber. The PIXE target setup, an outline for which is shown in Fig. 9, often also includes other detectors like a γ-ray detector for PIGE analysis (see Sec. 3.3).

In PIXE experiments, elements with Z $\lesssim$ 50, i.e., Sn, are normally detected by their emitted K X-rays. For heavier elements, the emitted L X-rays are used for detection because they lie in the energy range (1–10 keV) where the X-ray detector is most sensitive. The basis for a quantitative analysis is the relationship between the number of detected X-rays and the amount of element present in the sample. Mostly, the PIXE setup is calibrated experimentally using thin film standards. If this is not possible or suitable, an absolute measurement has to be done. For a proton beam and an infinitely thin sample (by this is meant a target that is sufficiently thin so that matrix effects become negligible), the relation between the number of X-ray quanta $Y_p(Z_2)$ obtained in a certain peak p in the X-ray spectrum and the concentration C_2 of the particular element in the sample is given by:

$$Y_p(Z_2) = C_2 \frac{N_0 \rho t}{MW} \frac{d\sigma}{d\Omega} \delta\omega\eta_p, \tag{1}$$

where Z_2 is the atomic number of the particular element in the target related to peak p in the spectrum, $\mathrm{d}\sigma/\mathrm{d}\Omega$ is the differential cross section for X-ray

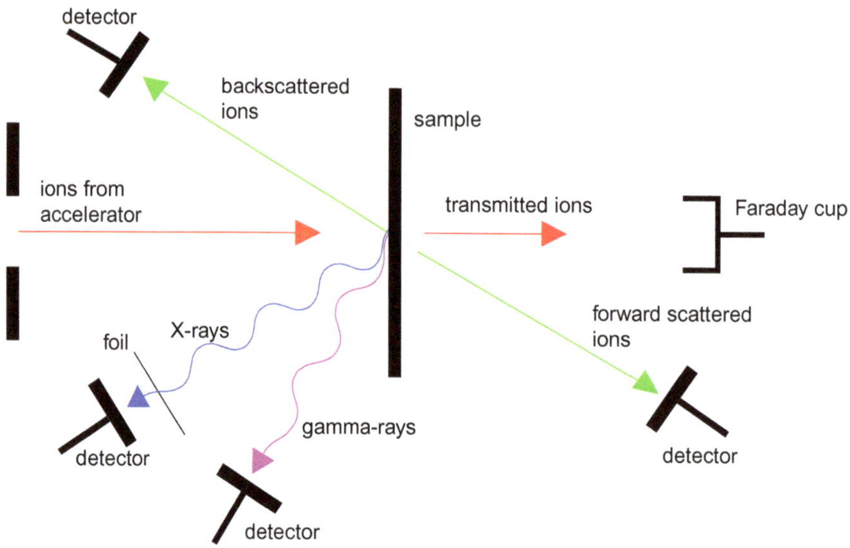

Fig. 9. Experimental setup showing some of the different detectors that can be used in an IBA experiment.

production for peak p in a small angle $\delta\omega$, η_p is the detection efficiency for X-ray peak p, N_0 is the number of incident ions, ρ is the sample density, t is the sample thickness, and MW is the molecular weight of the particular element related to peak p. Here, the subscripts follow the normal notation (used in the following section) where subscript 1 and 2 represent the projectile ion and target atom, respectively. In formulating this equation, it is implicitly assumed that the target is uniform and that the beam size is smaller than the sample area.

In practice, samples are rarely thin enough for matrix effects to be entirely negligible. For samples of intermediate thickness and for infinitely thick samples (the latter are samples that are thicker than the particle range), the quantification of Eq. (1) has to include a correction for the slowing-down of the ion and attenuation of the characteristic X-rays.

The minimum detection limits for PIXE depend both on matrix composition and on the element under investigation. By using both K and L X-rays, a good sensitivity across the periodic table can be obtained. In routine analysis, the detection limits are of the order of 0.1–1 μg/g. With an optimization of the experimental parameters, an improvement by a factor of 10 can in best cases be achieved. In some cases depth information can also be obtained, however depth resolution is rather poor.

The PIXE technique has been sucessfully applied in a number of fields. Materials science is perhaps the most important application, others include environmental science, art and archeometry, geophysics, and biology-medicine. Samples used can be of paper, papyrus, paintings, ceramics, glass artifacts, semiprecious stones, gold, bronze and other metals, blood, aerosol particles, tissue sections, etc. A comprehensive presentation of the PIXE method can be found in a newly published handbook.[18]

3.3. *PIGE technique*

The abbreviation PIGE stands for Proton Induced Gamma-ray Emission. As was mentioned above in Sec. 3.2 dealing with the PIXE technique, the availability of semiconductor detectors in the mid 1960s significantly improved the resolution for gamma ray detection, making it possible to use PIGE for quantity analysis.

The cross section for a nuclear reaction depends in a complicated manner on the incident ion energy and the internal properties of the beam nucleus and the target nucleus involved. The projectile will not normally undergo a nuclear reaction unless the incident energy is greater than the Coulomb barrier. The Coulomb barrier is approximately:

$$E = \frac{Z_1 Z_2}{(M_1^{1/3} + M_2^{1/3})} \text{ MeV}, \tag{2}$$

for an incident ion of atomic number Z_1 and mass M_1 and a target nucleus of mass M_2 and atomic number Z_2. For moderate ion energies, nuclear reactions will therefore only occur with the lightest target element. Hence, a light ion of MeV energy can overcome the nuclear charge of low-Z matrix elements, and inelastic reactions can occur. The existence of particular nuclear reactions with relatively high cross sections represents a potential for accurate trace element determination of specific light isotopes in high-Z matrices.

The nuclear reaction is either a direct reaction or compound nucleus reaction. In the latter case, an intermediate-state excited nucleus forms that will eventually decay by gamma rays or charged particle emissions. By detecting this radiation, analysis of light elements can be performed. There are many combinations of ions and elements that facilitate analysis by specific nuclear reactions. The possibility of measuring lighter elements, such as boron, beryllium, and nitrogen, is a major advantage in combination with techniques such as RBS (discussed in Sec. 4.2) and PIXE, which are more suited for measuring medium-heavy and heavy elements.

PIGE is normally based on detection of prompt gamma rays produced during MeV ion bombardment of a sample that is homogeneous in depth. As was mentioned above, the cross sections are lower than for PIXE, and PIGE typically represents a less sensitive analytical technique than PIXE. However, the gamma ray peaks are generally well isolated, and the energy of the emitted gamma ray photons is high enough so that correction for attenuation will not be necessary. PIGE may hence in some situations be favorable to PIXE for elements like Na, Mg, Al, Ca, and K. The self-absorption of the X-ray photon inside an aerosol particle is in some cases non-negligible. PIGE has no problems of this kind because of the higher gamma ray energies. PIGE is also useful in the analysis of paintings because it can detect the presence of low-Z elements in the paint layers below the protective varnish.[20]

The high penetrability of gamma rays simplifies the experimental arrangements, and the nuclear interaction also facilitates isotopic information. External beams are a standard method in the case of PIGE[20] and offer the same advantages as discussed above in the case of the PIXE technique. Some examples of detection limits of PIGE are given in Tables 2 and 3. They are given only as indications of attainable sensitivities since the specific experimental conditions will have a strong impact on the detection limits.

Table 2. Detection limits (atomic fraction) for PIGE using protons ($E_p < 9\,\mathrm{MeV}$).[19]

>1%	0.1–1%	<0.1%
Pd, Sm, Gd, Hf, W, Au, Pb	S, K, Sc, Ti, Co, Cu, Ge, Y, Zr, Mo, Ru, Ag, Sn, I, Ta, Pt	Li, Be, B, C, N, Na, Mg, Al, Si, Cl, Ca, V, Mn, Fe, Ni, Zn, Nb, Cd, In, Sb

Table 3. Detection limits (atomic fraction) for PIGE using α-particles (5 MeV).[21]

> 1%	0.1–1%	<0.1%
Sc, Cr, Cu, Zn, Rb, Zr, Er, Hf, Hg	O, Mg, Si, Cl, K, Ti, Fe, Br, Mo, Ru, Pd, Ag, Cd, Ta, W, Re, Ir, Pt, Au	Li, B, N, F, Na, Al, P, V, Mn, Rh

As the gamma rays are a result of a nuclear reaction, the PIGE technique is part of a broader field called "nuclear reaction analysis" (NRA — see Sec. 4.6). Any ion with appropriate energy can be used for PIGE. However, mostly light ions like protons, deuterons, and helium are used.

Homogeneous concentrations of elements in thick samples can be obtained by comparison with standards. However, this method only gives good results when the standard is very similar to the sample to be analysed. Some attempts have been made to perform a quantitative analysis without using a standard. It is done by integrating the nuclear reaction cross section.[22]

PIXE and PIGE techniques are often used simultaneously. The setups are designed to accommodate the relevant detectors and other equipment required for this combination, as can be seen from the experimental setup shown in Fig. 9.

The use of a nuclear technique (as in PIGE) is facilitated by a low Coulomb barrier, i.e., a light ion in combination with a light target nucleus. On the other hand, as was mentioned in Sec. 3.2, the soft X-rays from low-Z elements make the PIXE technique suitable for analysis mainly of elements heavier than Na. If both are employed in a common experimental setup, it is thus possible to measure both light and heavy elements and to compare independent results from both techniques for quality assurance. Both techniques or a combination of the two have been used extensively in many fields of applications, for instance in geology, biomedicine, environmental studies, and archeology. A more detailed discussion of the PIGE method is presented in a newly published handbook.[23]

In summary, PIXE and PIGE techniques are often applied to the same fields and can be considered complementary for measurement of heavy and light elements, respectively. PIGE does have one important difference from PIXE in that it is isotope specific, rather than element specific. Otherwise, the instrumentation and formalism are very similar.

3.4. *Accelerator Mass Spectrometry (AMS)*

One of the applications of nuclear physics techniques listed in Table 1 that has been of great benefit to other fields of scientific endeavor is Accelerator Mass Spectrometry (AMS). The capability of AMS in extremely sensitive isotope measurements has been extensively demonstrated over the past 35 years. For example, AMS has allowed refinements in the technique of ^{14}C dating in the fields of archaeology and quaternary geology. The most important improvement compared to traditional radiometric methods (i.e., measurement of the emitted radiation) is the possibility to date small samples with AMS. Using AMS, the radiocarbon age of a 10 000-year-old sample of 1 mg or less can be determined with a precision of 40 years in less than 30 minutes. To achieve the same precision with radiometric methods, the sample must contain more than 1 g of carbon and counting must proceed for over 24 hours.

We will illustrate this consideration with the following example for ^{14}C: One gram of modern carbon contains 6×10^{10} atoms of ^{14}C (the number of ^{12}C atoms is 1.2×10^{12} times more). Only around 13 of these ^{14}C atoms will decay per minute. To obtain a statistical precision of 0.5% (which is often required in radiocarbon dating) using 1 g of carbon, it is necessary to count the decays for more than 48 hours. When using mass spectrometer (MS) techniques, it is not necessary to wait for the atoms to decay, and this alternative is therefore, in principle, much more efficient. However, conventional MS cannot be used because the ^{14}C ions are masked by an intense flux of the atomic isobar ^{14}N and by molecular isobars with the same mass, such as ^{13}CH, $^{12}CH_2$, ^{12}CD, and $^{7}Li_2$. Also, the "tails" of the stable isotopes ^{12}C and ^{13}C contribute to the background in MS. These factors mean that MS can only be used for isotope ratios of $^{14}C/^{12}C$ of about 10^{-7} or higher. By employing AMS instead, the $^{14}C/^{12}C$ detection limit can be lowered to about 10^{-15}.

The two standard methods that have been used for a long time to determine the isotopic composition of an element are MS and decay counting. While MS can be used for all isotopes, decay counting is restricted to radioisotopes. For long-lived radioisotopes, decay counting is inefficient because only a small fraction of the nuclides in a sample decays during a reasonable measurement time. Conventional MS has a high efficiency, but it is limited to isotope ratios greater than about 10^{-7}. In AMS, the efficiency of MS is combined with extremely good discrimination against isobaric, isotopic, and molecular interference. In Fig. 10, the efficiency of decay counting and AMS are shown as a function of the half-life ($T_{1/2}$) of the measured isotope. The efficiency is defined as the number of atoms detected compared to the number in the sample. The counting time for AMS is in Fig. 10 assumed to be 1 h, while the efficiencies for decay counting periods of 24 h and 14 days are shown in the figure. The high discrimination of AMS is obtained by accelerating the ions to a high energy, usually employing a tandem accelerator (see Sec. 2). In a few cases, other types of accelerators are used. The most significant technical development during the past 10 years is the trend towards smaller accelerators and also towards small

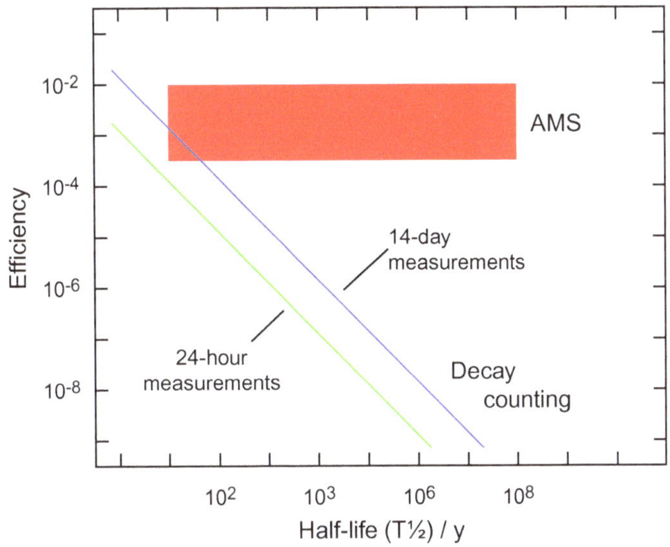

Fig. 10. The efficiencies of AMS and decay counting as a function of half-life. For details, see the text.

single-stage AMS systems.[24] The reduction in floor space and overall cost are the most attractive features of these new systems, as well as the limited technical resources and staff required to run the system. Single stage machines are smaller and easier to use. A single stage accelerator installed in Lund, Sweden is shown in Fig. 11.

A typical AMS system includes the following steps and components:

(i) Production of negative ions in a multi-sample, negative-ion source. The sample material containing the rare isotopes to be counted is placed in the ion source. Typical beam currents, selected ion of the stable isotope, and material to be put into the ion source are given for some AMS isotopes in Table 4.
(ii) Acceleration of negative ions from ground potential to a high positive voltage.
(iii) Recharging of all ions to positive by stripping off electrons and at the same time dissociation of all molecular ions.
(iv) Acceleration of the now positive ions back to ground potential.
(v) The removal of unwanted ions using electric and magnetic fields.
(vi) Identification and counting of the individual rare isotopes with nuclear detection techniques.
(vii) Computer control of the accelerator system to allow for unattended operation and to provide control of the AMS system parameters.

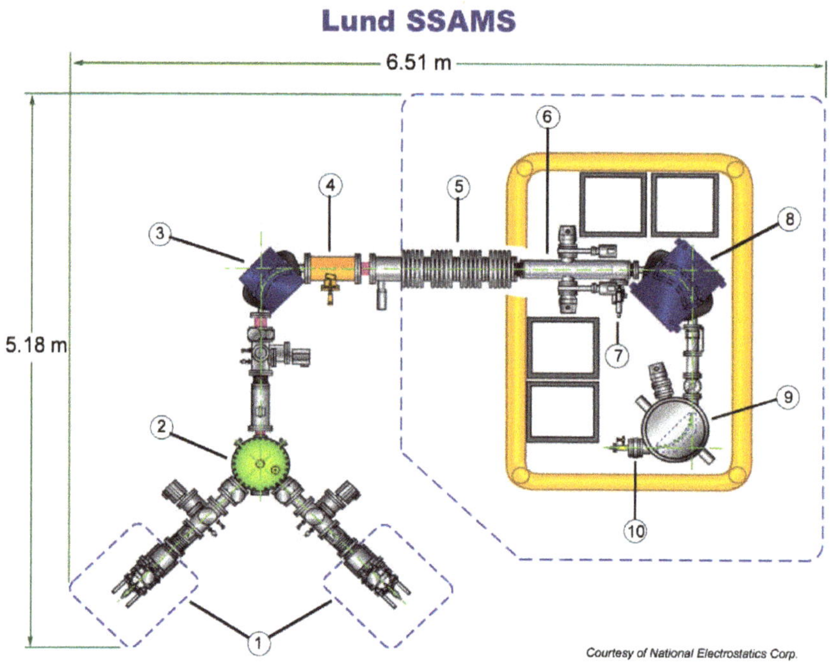

Fig. 11. The single stage AMS (SSAMS) system at Lund University. Legends: 1. Ion sources; 2. Low energy 45° spherical electrostatic analyser; 3. Low energy bending magnet; 4. Einzel lens; 5. Accelerator tube; 6. Argon gas stripper; 7. Argon valve; 8. High-energy bending magnet; 9. High-energy electrostatic spherical analyser; 10. Sequential post-accelerator deflector.

Table 4. Typical beam currents from high-intensity cesium sputter sources used for AMS.

AMS isotope	Material	Selected ion of stable isotope	Negative-ion currents (μA)
^{10}Be	BeO	^{9}Be^{16}O$^-$	1–20
^{14}C	Graphite	^{12}C$^-$	20–100
^{26}Al	Al_2O_3	^{27}Al$^-$	0.1–1
^{36}Cl	AgCl	^{35}Cl$^-$	5–25
^{129}I	AgI	^{127}I$^-$	2–10

AMS can be viewed as an extension of MS to include an accelerator. In Fig. 12, an MS system is compared with a simple AMS system. The introduction of the tandem accelerator, followed by several ion-filtering devices, reduces the background by a factor of the order of 10^8. Three especially

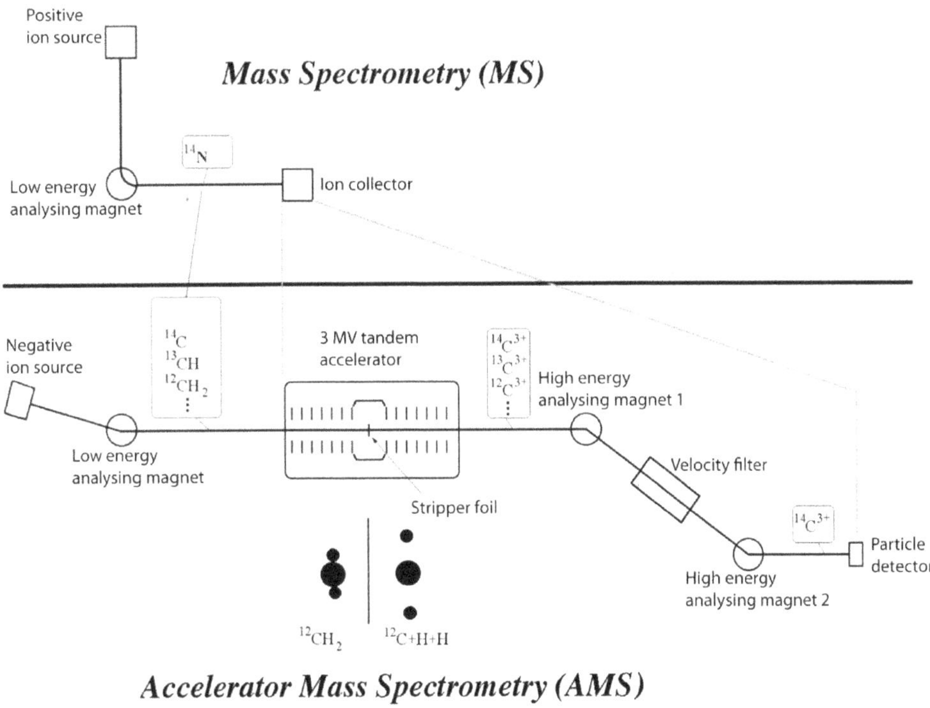

Fig. 12. Comparison between MS and a typical AMS system.

important characteristics of AMS allow the measurement of low isotope ratios (for example, for $^{14}C/^{12}C$ down to 10^{-15}):

(i) Interference by some isobars is avoided by using a negative ion source (for example, the interfering isobars $^{14}N^-$, $^{26}Mg^-$, $^{55}Mn^-$, and $^{129}Xe^-$, all with negative electron affinity, are supressed when detecting ^{14}C, ^{26}Al, ^{55}Fe, and ^{129}I, respectively).

(ii) Interference by molecules is avoided by using a stripper system in the high-voltage terminal of the accelerator (for example, $^{12}CH_2$ and ^{13}CH when detecting ^{14}C). In the stripping process, the negative ions become positive, and at the same time, nearly all the molecular ions will be dissociated. A few molecular ions with a maximum charge state of 2+ will survive. Molecular ions can thus be avoided by selecting a charge state higher than 2+ with the high-energy analyzing system after the accelerator.

(iii) The counting of individual ions in a silicon or gas-ionisation detector is possible due to their high final energy. The ions that reach the detector are easily separated by their energy difference.

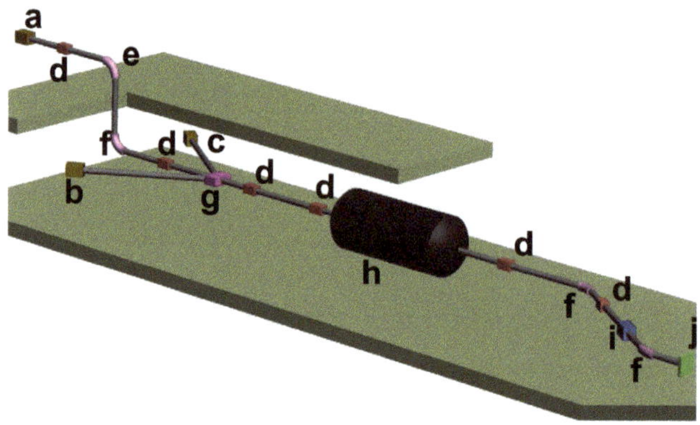

Fig. 13. An outline of the AMS system at the Pelletron in Lund (in use for AMS from 1988 to 2005). Legends: a. Second-generation AMS source; b. Off-axis duoplasmatron source for purposes other than AMS; c. First-generation AMS source (of Cs-gun type); d. Einzel/quadrupole lens; e. Spherical electrostatic analyzer; f. Magnetic analyzer; g. Magnetic analyzer with three entrance ports; h. Electrostatic tandem accelerator; i. Velocity selector; j. Particle detector. Total length between ion source (a) and particle detector (j) is about 35 m. The accelerator tank is just greater than 5 m long and nearly 2 m in diameter.

To illustrate some of the technical details of an AMS system, we consider the system at the Pelletron in Lund, shown in Fig. 13. A beam of negative ions is formed in the ion source (a), and the beam is analyzed by an electrostatic (e) and a magnetic (f) analyzer in the injector. The electrostatic analyzer bends the beam to make it vertical while the magnetic analyzer bends the beam back to the horizontal. The ion source is thus elevated relative to the accelerator. Beam optics elements (einzel lenses/quadrupole lenses) are placed at several positions along the beamline to keep the beam envelope as small as possible (d). The beam of negative ions is transported to the accelerator and accelerated to the positive high-voltage terminal of the tandem accelerator (h). In the high-voltage terminal, the negative ions pass through a foil or a gas cell. The negative ions, which are now at a high energy, will lose electrons and become positively charged. Molecular ions are dissociated in the foil or gas and can therefore not interfere. The positive ions are accelerated to ground potential through the second stage of the accelerator and gain a further increase in energy. The beam of positive ions is analyzed by a magnetic analyzer (f) and the ions of interest are selected. The beam is further analyzed by a velocity selector (i) and once more by a magnetic analyzer (f).

The selected ions will have well-defined charge state, energy, and mass. They are identified and counted one by one using a detector (j). The ions are identified by their energy. This is necessary as ions other than those desired can reach the detector. The various components of the AMS system are described in more detail in an AMS overview article.[24]

As the number of AMS facilities has grown, the number of applications has increased. ^{14}C is still the most important AMS isotope, and new applications in addition to ^{14}C dating are directed towards studies of atmospheric processes and ocean circulation to gain information about past climates. ^{14}C-AMS has also found applications in biomedical studies. Other isotopes such as ^{10}Be ($T_{1/2}$ = 1.5 Ma) and ^{36}Cl ($T_{1/2}$ = 301 ka) have been used to gain hydrogeological information. These two isotopes are also trapped in the ice sheets of Greenland and Antarctica where they can be used as tracers of the solar and geomagnetic modulation of the cosmic radiation reaching the earth. Cosmic ray exposure also causes the build-up of ^{10}Be, ^{26}Al ($T_{1/2}$ = 720 ka), and ^{36}Cl in surface rocks. The accumulated amounts of these isotopes can be measured by AMS and used to date the rock (so-called exposure dating). This is often crucial information in prospecting for industrial minerals and fossil fuels. ^{26}Al-AMS has also been used to study metabolic processes in living systems. The production of ^{41}Ca ($T_{1/2}$ = 103 ka) from nuclear weapons tests has been measured by AMS, and ^{36}Cl and ^{129}I ($T_{1/2}$ = 15 Ma) have been used to trace the migration of nuclear waste from nuclear storage and reprocessing plants, and nuclear power plants.

Industrial applications of AMS include uses in the applied fields of geology, oceanography, and biomedical science. The use of AMS in drug development is the fastest-growing industrial application of this technique. The high sensitivity permits use of microdosing for the determination of the pharmacokinetics of new drugs. Drug development regulations permit drug doses labeled with ^{14}C to be administered to human subjects at 1% of the nominal dose for studies using AMS to determine the drug uptake in various organs. This sensitive analysis tool continues to be used by a number of research facilities, but industrial use is following the same evolution as the other IBA methods.

An AMS system dedicated to biomedical applications is installed at a commercial company in York, UK. The system is based on a 5 MV tandem accelerator produced by National Electrostatic Corporation (NEC) in Wisconsin, USA. Two SSAMS machines, similar to the one in Lund (see Fig. 11), are to be used for drug development and were delivered by NEC in 2005. Another four NEC AMS systems used solely for biomedical applications have been installed at pharmaceutical and other companies during the last few years.

The major producers of turnkey AMS systems are NEC and High Voltage Engineering Europe (HVEE) in the Netherlands. A number of AMS systems are sold for industrial use. Rough prices for a complete system range from US$1.2M for a basic ^{14}C-system to over US$4M for a versatile system capable of chlorine and other isotope analysis.

Also, the AMS group at the ETH (Swiss Federal Institute of Technology in Zürich, Switzerland) is offering AMS accelerators for biomedical application. Their construction is a very compact tandem with only 200 kV on the high voltage terminal.

An international conference on AMS is organized every third year. The most recent one was held in March 2011 in Wellington, New Zealand.[25]

4. Depth Profiling Methods

4.1. *Introductory remarks*

Industrial technological thin film processing often requires quantitative measurements of: (i) the thickness of thin surface films; (ii) the presence of an interfacial oxide; (iii) the reaction of a one overlying film with another; (iv) the quality of epitaxial growth of epitaxial layers; and (v) the distribution of implanted atoms. These can be characterised by measuring how the elemental composition changes with depth in the sample. This type of measurement is commonly referred to as *depth profiling*. A number of depth profiling methods have up until now been widely used in industry to measure thin films such as the sputter-based methods Sputter-Auger Electron Spectroscopy (AES), Sputter-X-ray Potential Spectroscopy (XPS), and Secondary Ion Mass Spectroscopy (SIMS). These methods are essentially based on digging a pit in the surface of the material by sputter erosion. As the pit is dug, either the material ejected during the digging process is analysed or the surface layer of the bottom of the pit is analysed to give a *depth profile* (concentration vs. depth curve). These conventional sputtering-based methods are poorly suited to the emerging needs of industry for characterising nanometre-scale structures because the sputtering process mixes the surface and underlying layers in a complex way. This implies that these techniques require sophisticated time consuming and expensive calibration in order to deliver quantitative results.

Elastic Recoil Detection Analysis (ERDA) and Rutherford Backscattering Spectrometry (RBS) are based on a completely different principle where MeV ions penetrate through the surface layers and bounce off surface and subsurface atoms. As a result, the ion either rebounds out of the material (RBS) or knocks the atom in the sample out through the surface

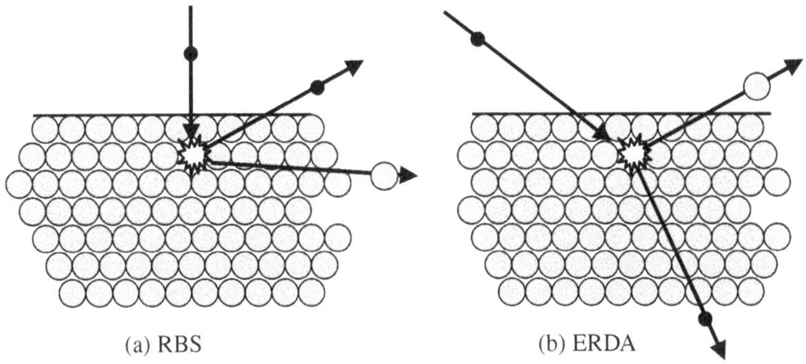

Fig. 14. Illustration of the basic scattering process in (a) Rutherford Backscattering Spectrometry (RBS) and (b) Elastic Recoil Detection Analysis (ERDA). The incident ion is indicated by a small black circle, while the recoiling target atoms are shown as larger grey circles.

in a single collision (ERDA). This is illustrated in Fig. 14. The essential difference between RBS and ERDA is that in the former the incident ion is detected after scattering while in ERDA the recoiling target atom is detected. Although the energies are much higher, these techniques are for many purposes non-destructive because material is hardly removed from the surface. Ion induced mixing and diffusion does take place, but the levels at which it occurs are much lower than for the sputter profile techniques. These methods provide true elemental analysis because the energies transferred are on the order of MeV and are hence insensitive to chemical bonding effects which take place on a scale of eV.

Two related methods for obtaining depth-distributions are Nuclear Reaction Analysis (NRA) where prompt nuclear reactions are used and Charged Particle Activation Analysis (CPAA) where the products from radioactive decay, emitted over long periods of time (half-lives ranging from minutes to years), are detected. These methods have been less widely used for industrial measurements and development than RBS and ERDA. They are included here for completeness. These two methods are isotope specific rather than element specific. For example, this enables NRA to be used to follow tracers, such as deuterium in normal hydrogen for such things as water ingression determination, and ^{15}N markers for determining wear erosion of hard nitride coatings containing ^{14}N on precision-machined metal components.

4.2. *Fundamentals of RBS and ERDA measurements*

Here we give a condensed account of the fundamental basis of depth profiling analysis. The description is of necessity technical. Therefore, some readers may prefer to jump to Sec. 4.3 where the main results are summarised. For brevity, we have considered the case for RBS. The ERDA case is very similar and can be simply obtained by changing the cross sections, kinematic factors, and stopping forces to take into account the fact that the outgoing particle is a recoiling target atom and thereby different from the incident projectile ion.

The fraction of the incident particle energy E_0 that is retained by the scattered ion (RBS case) or transferred to the recoiling target atom in ERDA is respectively:

$$K(\theta, M_1, M_2) = \left[\frac{\sqrt{M_2^2 - M_1^2 \sin^2\theta} + M_1 \cos\theta}{M_1/M_2} \right]^2 ; \quad M_1 \leq M_2, \tag{3}$$

$$E_2 = E_0 \frac{4M_1M_2}{(M_1 + M_2)^2} \cos^2\phi, \tag{4}$$

where E_2 is the energy of the recoil atom, M_1 and M_2 are the masses of the projectile ion and target atom, respectively, θ is the scattering angle of the projectile, and ϕ is the angle the recoil makes to the incident ion direction.

One of the simplest applications of RBS and ERDA is to measure the composition of a self-supporting foil. This is illustrated in Fig. 15. In the setup shown in this figure, $K(\theta, M_1, M_2)$ will only be a function of M_2, i.e., $K(M_2)$. This is because the angle θ is defined by the apertures and the mass M_1 is fixed by the beam selected from the accelerator. The number of counts in the peak corresponding to the element A is:

$$Y_A = \left(\frac{d\sigma_A}{d\Omega} \right) N_{\text{ion}} N_A \delta\Omega, \tag{5}$$

where $(d\sigma_A/d\Omega)$ is the differential cross section for scattering per atom into unit solid angle (this is a measure of the probability of scattering); $\delta\Omega$ is the solid angle subtended by the detector; N_{ion} is the number of incident ions; and N_A the number of A atoms per unit area. Writing a similar equation to Eq. (5) for element B, dividing Eq. (5) by the new equation, and rearranging gives the composition ratio N_A/N_B in terms of the ratio of the areas under the peak:

$$\frac{N_A}{N_B} = \frac{(\frac{d\sigma_B}{d\Omega})Y_A}{(\frac{d\sigma_A}{d\Omega})Y_B}. \tag{6}$$

The depth at which scattering takes place for each collision in both RBS and ERDA is determined from the energy-loss ΔE encountered by the incident ion on the inward path ΔE_{in} and the scattered ion, or recoil, on the outward path ΔE_{out}. This energy loss, $\Delta E = K\Delta E_{\text{in}} + \Delta E_{\text{out}}$, can be obtained

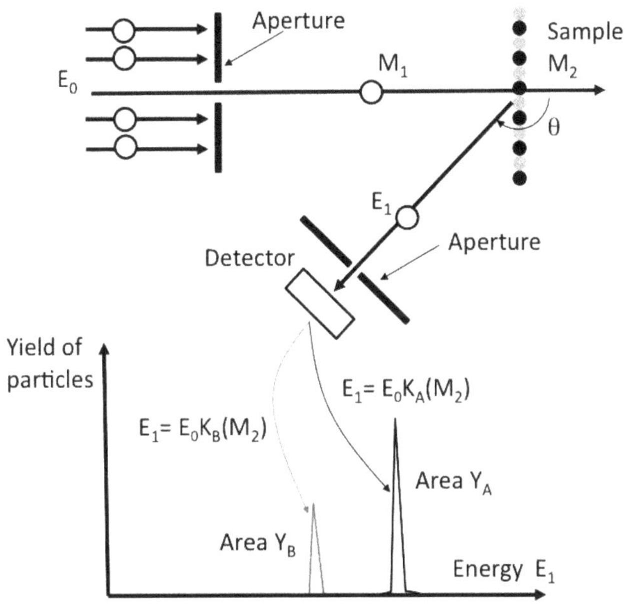

Fig. 15. Schematic illustration of a simple RBS measurement of the composition of a self-supporting bi-elemental film. Atoms of the two elements which have different masses M_2 are denoted by black and grey circles. The accelerator (not shown) produces ions with a fixed mass M_1 and energy E_0. The two apertures define the beam direction and the scattering angle θ, respectively. The detector functions as a spectrometer, and an electronics system sorts the pulses according to their amplitude to produce an energy spectrum. The different atomic masses give rise to peaks in the energy spectrum. The size of the peaks is proportional to the concentration of grey or black atoms, and their position corresponds to their mass M_2.

by measuring the energy difference between where the scattering takes place at the depth t and the surface, as illustrated in Fig. 16. For ions that have been backscattered from the surface with energy E_1 and at depth t with energy E_2 (see Fig. 16), we can write:

$$E_1 = E_0 K(M_2), \tag{7}$$

$$E_2 = (E_0 - \Delta E_{\text{in}})K(M_2) - \Delta E_{\text{out}}, \tag{8}$$

$$\Delta E = E_1 - E_2. \tag{9}$$

The kinematic factor $K(M_2)$ represents the fraction of energy that is retained by the projectile ion (RBS case) or the recoiling target atom (ERDA case) after the collision. On the inward and outward paths, the energetic ions lose energy mainly through collisions with electrons. This gives *stopping forces*

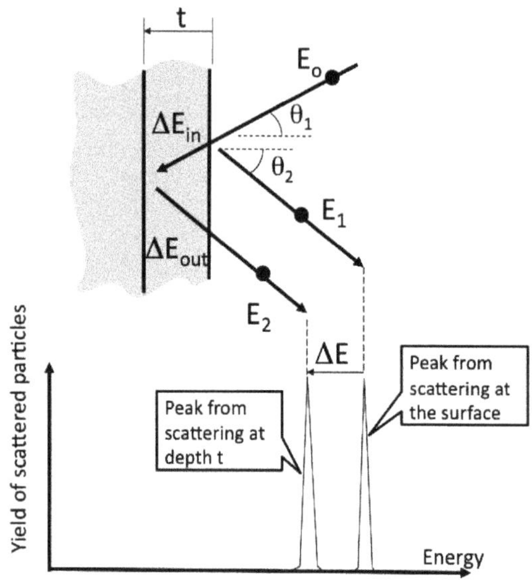

Fig. 16. Illustration of the basis of depth profiling with both ERDA and RBS. The energy loss along the inward and outward paths is a measure of the depth at which scattering took place.

dE/dx, so that for not too large t we have:

$$\Delta E_{\text{in}} = \left|\frac{dE}{dx}\right|_{\text{in}} \frac{t}{\cos\theta_1}, \tag{10}$$

$$\Delta E_{\text{out}} = \left|\frac{dE}{dx}\right|_{\text{out}} \frac{t}{\cos\theta_2}. \tag{11}$$

Substituting Eqs. (10) and (11) into Eqs. (7) and (8), we obtain for Eq. (9):

$$\Delta E = \left\{K(M_2)\left|\frac{dE}{dx}\right|_{\text{in}} \frac{1}{\cos\theta_1} - \left|\frac{dE}{dx}\right|_{\text{out}} \frac{1}{\cos\theta_2}\right\} t. \tag{12}$$

The stopping forces $|dE/dx|_{\text{in}}$ and $|dE/dx|_{\text{out}}$ depend on the sample material as well as the ion species and energy. For most ions and elements, these have been measured and tabulated with an uncertainty of 5–10%. Considering the stopping to be represented by a *stopping cross section* ϵ, which is the energy lost in traversing material with a thickness corresponding to one atom per unit area, we can write:

$$\Delta E_{\text{in}} = Nt\epsilon_{\text{in}} \tag{13}$$

$$\Delta E_{\text{out}} = Nt\epsilon_{\text{out}}. \tag{14}$$

Here, N is the number of atoms per unit volume. Nt is then the thickness of the layer in terms of atoms per unit area. The stopping cross sections are tabulated or can be conveniently obtained from predictor codes such as SRIM.[26,27] This approach has the advantage that we can easily determine the stopping in a chemical compound or mixture of elements by using *Bragg's rule*. Then for the compound A_mB_n, the stopping cross section is simply the weighted average of the stopping cross sections for the elements A and B:

$$\epsilon_{\text{in}}^{A_m B_n} = \frac{m\epsilon_{\text{in}}^{A} + n\epsilon_{\text{in}}^{A}}{m+n}. \tag{15}$$

A similar relation exists for $\epsilon_{\text{out}}^{A_m B_n}$, which is obtained by replacing the subscript "in" by "out" in Eq. (15). For ERDA, the corresponding Eqs. (5) through (15) are identical except that the kinematic factor K (Eq. (3)) must be replaced with recoil kinematic factor Λ (Eq. (4)) and the appropriate stopping cross sections used.

In the early days, the data from RBS and ERDA experiments were analysed by measuring the heights of steps, areas under peaks, and the position of edges to determine film thickness. A particular problem is that in the energy spectrum, the signals from different elements can be overlapped. For example, the signal from oxygen in an oxide layer is often superimposed on a signal from a heavier substrate like Si. In general where the signals are overlapping in a single energy spectrum, it is not possible to unambiguously solve for the depth profiles of different elements. With the advent of low-cost computing, a number of simulation programs have been developed, such as the early RUMP code,[28] SIMNRA,[29] and the IBA data furnace.[30] These allow the comparison of a simulated energy spectrum from a model structure with the measured spectrum. Then by adjusting the parameters that describe the depth profiles of different elements in the model structure, the simulated and experimental spectra are made to coincide. Simulation codes reduce the ambiguity in determining the elemental depth profiles from overlapping spectra by ruling-out intuitively physically unreasonable solutions. However, it should be born in mind that in some rather rare cases, the solutions for the depth distribution equations obtained from simulation may not be unique, implying the results have a large uncertainty. This is particularly troublesome where extremely thin films containing light elements (B, C, N, O, etc.) are measured when the substrate is a heavy element, e.g., Au. A further pitfall is that RBS using He ions is blind to hydrogen. Hydrogen is an important impurity and its effect may need to be incorporated in simulations of RBS spectra. However, because it does not contribute to the spectrum, the hydrogen profile has large uncertainties. Hydrogen, on the other hand,

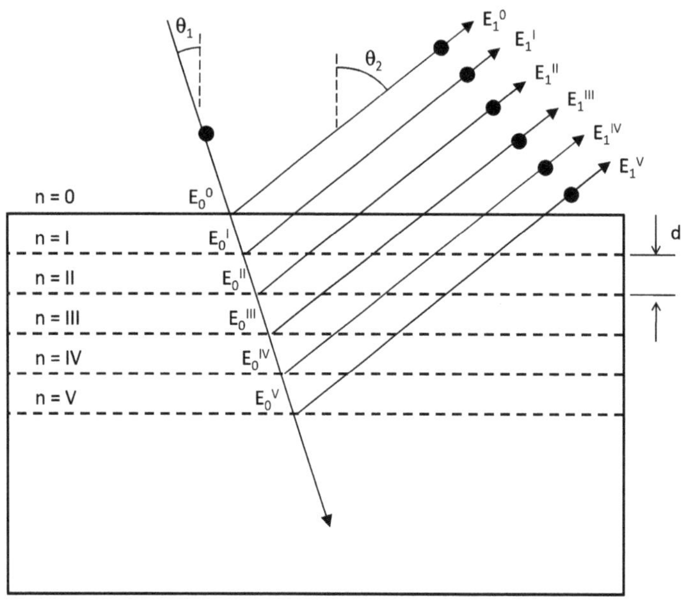

Fig. 17. Schematic illustration of the slab model used in data analysis by computer simulation. The sample is considered to be made up of slabs. The sample model is constructed by selecting the elemental composition and thickness of each slab. The energy E_0^i of ions arriving at the ith slab is then the incident energy less the energy loss in the preceding slices with $n < i$. This energy loss is calculated for each slice talking into account the compositional dependence through Eq. (15). Subsequently, for each element in the sample, the energy after scattering is calculated (Eqs. (3) (RBS), (4) (ERDA)). This is used to calculate the energy of outgoing particles E_1^n from the slab for each element in turn by considering the composition-dependent energy loss for each slab. Finally, the energy spectrum is obtained by superposing the energy spectra from each element.

is readily detected by ERDA, and its profile can be accurately determined with this method.

The simulation codes are based on a slab model, illustrated in Fig. 17. Modern RBS and ERDA analysis codes incorporate a number of additional features, for example, iterative parameter-refinement procedures that are useful when a large number of similar samples are analysed and energy straggling, which is helpful to diagnose interface roughening.

4.3. *Basis of the quantitative analysis in ERDA and RBS*

The quantitative nature of ERDA and RBS depth profiling measurements are based on two processes quantified through Eq. (12), which establishes the depth scale, and Eq. (5), which relates the yield of detected particles to the number of atoms of a given type in the sample via the scattering cross sections (probability of scattering per atom). The quantification in RBS and ERDA thus rests on the following two factors:

(1) The probability of the ion scattering from the nucleus of sample atom. For Rutherford scattering, which is generally the case, the nucleus of the ion penetrates the innermost electron shell of the sample atom so that no shielding from orbiting electrons takes place and the statistical differential cross sections can be exactly calculated.
(2) The energy loss of MeV ions. The stopping cross sections have been tabulated and stopping force predictor programs, such as SRIM,[26,27] are available which provide stopping cross sections to be estimated with an accuracy of ~10% or better. If greater accuracy is needed, relative measurements against a reference standard can be used.

The significance of this is that ERDA and RBS are inherently more quantitative than the sputter profiling methods that are widely used in industry, such as Secondary Ion Mass Spectroscopy (SIMS) and Sputter-Auger/XPS. The largest contribution comes from the uncertainty in the stopping cross section, which gives an absolute uncertainty of about 5–15% for RBS and 12–20% for ERDA with a coverage factor ($k = 2.0$). The relative uncertainties are dominated by counting statistics, and for typical counting rates, these can be several orders of magnitude smaller than the absolute uncertainty. For industrial QC applications, this gives both RBS and ERDA considerable advantages over the sputter profiling methods. This is because the sputter methods require an extensive, costly, and time-consuming calibration, and even then, they are subject to difficult-to-control effects such as beam current variations, memory effects, and matrix dependence of the secondary ionisation coefficients. These difficulties become more acute as the dimensions of interest for industrial QC shrink down to a nanometre scale.

4.4. *Rutherford Backscattering Spectrometry (RBS)*

The most straightforward MeV-ion beam analysis method for measurement of depth profiles is Rutherford Backscattering Spectrometry (RBS). The name is widely used even though under some circumstances the scattering is not strictly Rutherford scattering. This arises at energies that are so low that the electrostatic charge on the colliding nuclei is shielded by orbiting electrons and at high energies where the nuclei come so close that the nuclear forces start to interact.

The RBS method rose to prominence in the late 1960s and early 1970s driven by the needs of the emerging microelectronics industry where it was used for determining the depth distributions of implanted dopant ions, the thickness of gate oxide layers, and metallisation for interconnects in the emerging microelectronics industry. The great advantage this method had over alternative methods, such as Sputter-Auger and Secondary Ion Mass Spectroscopy (SIMS), is that the method is quantitative in that the elemental depth distributions could be given in terms of depths in numbers of atomic layers and concentrations in atomic %. An extensive review of these methods with an intercomparison is given elsewhere by the authors.[2] An important limitation of these sputtering methods is that they rely on the composition at the bottom of a crater formed by successive sputtering. Irradiation-induced segregation and diffusion effects can mean this layer differs significantly from the bulk composition a few 10s of nm below the surface. Commercial instruments for RBS and ERDA are produced that are capable of measuring thin films with a sub-nm resolution (see Sec. 5). These make use of the large cross sections for backscattering and stopping forces at low energies (100–700 keV). For this reason, the analytical method is often referred to as Medium Energy Ion Scattering (MEIS). In this energy region, the "Rutherford" is usually dropped because the force field responsible for scattering may not be Coulomb but is shielded by the orbiting electrons.

A number of different types of RBS measurements have been used. To optimize the depth resolution, glancing-incidence and -exit geometries are used, so that the path lengths in the inward and outward paths of the ion are maximized, respectively, in order to obtain a large energy difference ΔE (Eq. (16)). The depth resolution can also be optimized by using a high resolution magnetic spectrometer. RBS with heavy ions such as ^{12}C and ^{16}O can also be used to optimise the mass resolution. The different types of RBS and their applications are listed in Table 5. Comprehensive handbooks and overviews on the method have been published by Leavitt *et al.*,[31] Chu *et al.*,[32] Götz and Gärtner,[33] and Wang and Nastasi.[34] A closely related technique is *ion channelling*, which may be used to measure the depth distribution of defects in the near-surface layers of crystals as well as the location of the defects in the lattice. The method has been widely used for basic development work for the microelectronics industry, in particular for the measurement of the recovery of crystal damage resulting from ion implantation. The technique is well documented in standard texts on RBS[31–33] and also by Swanson.[35] The need to precisely align the crystal lattice in

Table 5. Rutherford Backscattering Spectrometry (RBS) methods and applications.

Method	Beam and detector	Depth resolution FWHM nm	Min. detectable concentration at. %	Applications	Comments
Normal incidence RBS	1–2.5 MeV ^{4}He Si charged particle detector	20–30	0.1	Ion implantation Sputtered film Atomic Layer Deposition (ALD) films Microcircuit metallisation Bio-interface layers	H not analysed
Glancing angle RBS	1–2.5 MeV ^{4}He Si charged particle detector	3–7	0.01	Gate dielectrics Microelectronics metallisation Optical films Barrier layers	H not analysed Sensitive to surface roughness
Magnetically-analysed RBS	0.5–2 MeV ^{4}He Magnetic spectrometer	0.2–1	0.01	Gate dielectrics Microelectronics metallisation low/high-k dielectrics Barrier layers	H not analysed
Heavy-ion RBS	10–20 MeV ^{12}C, ^{16}O Si charged particle detector Time of flight spectrometer	0.2–10	0.05	GaAs Microelectronics metallisation low/high-k dielectrics III-V semiconductors MOCVD	H not analysed

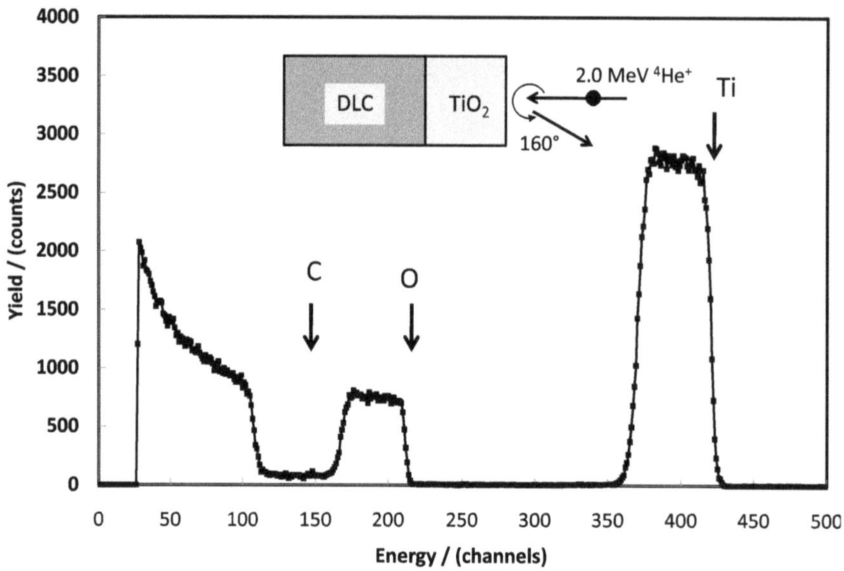

Fig. 18. RBS spectrum from a TiO_2 film on a diamond-like carbon (DLC) substrate. The inset shows the geometry. The energies associated with scattering from elements at the surface are denoted by vertical arrows.

each sample with the ion beam makes the method unsuitable for industrial QC. It is, however, useful in industrial development for modest numbers of samples.

In order to illustrate the quantitative nature of the RBS technique for analysis of thin films, a step-by-step guide is given below. This again is quite technical and readers may wish to move directly to Sec. 4.5. Consider the RBS spectrum in Fig. 18 that was measured from a sample consisting of a thin film of titanium oxide with a nominal composition TiO_2 on a substrate of diamond-like carbon (DLC). This example was chosen because of the industrial importance of TiO_2. The properties and some industrial applications of thin film TiO_2 are given in Table 6. Inspection of Fig. 18 shows three regions in the spectrum. The highest energy backscattered ions correspond to backscattered ions from Ti (the heaviest element in the sample) from the surface of the TiO_2 layer. As the ions penetrate deeper, they lose energy according to Eqs. (8), (9), and (12). This gives rise to the "plateau" in the channel interval 362–420. The low energy (rear) edge of the plateau arises because once the ions penetrate through the TiO_2 film and into the DLC layer underneath, they do not encounter any more Ti atoms and consequently they cannot be backscattered with high energy anymore. A similar plateau is seen between channels 167–215 that corresponds to oxygen in the

Table 6. Properties and applications of TiO_2 thin films.

Property	Applications
Superhydrophobicity	Antifouling, anti-fog coatings, self-cleaning glass
Biocompatibility	Osseointegration-promoting layer between bone and substrate Substrate for cell-based biosensors
Photocatalytic properties	Hydrolysis, deodorising films, dye-sensitised solar cells, enzyme-based electrochemical sensors for biomolecules
Refractive index ($n \cong 2.7$)	Dielectric mirrors, uv-blocking films

TiO_2. Oxygen is lighter than Ti; hence, the kinematic factor is smaller and thereby the signal from O is backscattered at lower energy. The signal in the channel interval 27–106 corresponds to backscattering from C in the DLC film. In the next two subsections, we go through the numerical analysis of the spectrum in Fig. 18 to obtain the thickness of the thin TiO_2 film and the stoichiometry. In order for the reader to follow in detail, the numerical results of each step are presented below in tabular form.

4.4.1. *Determination of the thickness of the TiO_2 film*

It can be seen from Fig. 18 that the signals for oxygen and titanium have a well defined low-energy edge. The position of this edge corresponds to scattering from ^{16}O and ^{48}Ti (the most abundant isotopes) at the DLC/TiO_2 interface. The energy width of the Ti and O signals, measured from the position of the half-heights on the front and rear edges, is related to the thickness of the film according to Eq. (12). For the measurement in Fig. 18, the energy calibration was determined to be 3.3713 keV/channel with a channel zero offset of 28.6 keV. Using this, the position of the edges and the widths of the O and Ti signals are specified in Table 7. The kinematic factor K in Table 7 can be calculated or taken from published tables.[31–33,36,37]

To transform the energy widths ΔE into a depth, it is necessary to calculate the stopping forces acting on the ion in the TiO_2. In order to demonstrate the method, we use a simple assumption called the *surface approximation*. This is done to overcome variation of the stopping force with energy. In this assumption, the energies of the ion along the inward

Table 7. Edge positions and energies for Ti and O from Fig. 18.

Element	E_0 (keV)	K	$E_1 = K(E_0)$ (keV)	Surface edge (chan #)	Rear edge (chan #)	ΔE (keV)
Ti	2000	0.7231	1446.2	420.5	375	153.4
O	2000	0.3708	741.6	211.5	167	150.0
C	2000	0.262	524	146.9		

Table 8. Stopping cross sections in Ti and O.

Element	E_0 (keV)	ϵ_{in} [eV/(10^{15} at. cm^{-2})]
Ti	2000	73.31
O	2000	35.84
Element	$E_1 = K_{\text{Ti}}(E_0)$ (keV)	$\epsilon_{\text{out}}^{Ti}$ [eV/(10^{15} at. cm^{-2})]
Ti	1446.2	84
O	1446.2	42.2
Element	$E_1 = K_{\text{O}}(E_0)$ (keV)	$\epsilon_{\text{out}}^{O}$ [eV/(10^{15} at. cm^{-2})]
Ti	741.6	95
O	741.6	47

path are taken to be E_0 and on the outward path $E_1 = K(E_0)$. That is, the energies are the same as if the scattering took place at the surface. The first step is to determine the stopping cross sections at the energy of the incoming ion and after scattering from each element in the film. The stopping cross section values can be taken from tables[31–33] or computer predictor codes such as the SRIM package.[26,27] Table 8 presents the stopping cross sections for He ions in Ti and O: ϵ_{in} at the incident energy (inward path), and $\epsilon_{\text{out}}^{Ti}$ and $\epsilon_{\text{out}}^{O}$ (outward path) after scattering from Ti and O nuclei, respectively. To obtain the stopping cross section in the compound film from the stopping cross sections for constituent elements of TiO_2, we use Bragg's rule (Eq. (15)) with $m = 1$ and $n = 2$ to obtain the results in Table 9. Using

Table 9. Stopping cross sections in TiO_2 from Bragg's rule (Eq. (15)).

Compound	E_0 (keV)	ϵ_{in} [eV/(10^{15} at. cm^{-2})]
TiO_2	2000	48.33
Compound	$E_1 = K_{Ti}(E_0)$ (keV)	ϵ_{out}^{Ti} [eV/(10^{15} at. cm^{-2})]
TiO_2	1446.2	56.13
Compound	$E_1 = K_O(E_0$ (keV)	ϵ_{out}^{O} [eV/(10^{15} at. cm^{-2})]
TiO_2	741.6	63

Eqs. (13) and (14), we can write Eq. (12) as:

$$\Delta E = \left\{ \frac{K(M_2)\epsilon_{in}}{\cos\theta_1} - \frac{\epsilon_{out}}{\cos\theta_2} \right\} Nt. \tag{16}$$

For the case of scattering of He ions from Ti, this can be written as:

$$\Delta E = Nt[\epsilon]_{TiO_2}^{Ti}, \tag{17}$$

$$[\epsilon]_{TiO_2}^{Ti} = \frac{K(Ti)\epsilon_{in}^{Ti}}{\cos\theta_1} - \frac{\epsilon_{out}^{Ti}}{\cos\theta_2}. \tag{18}$$

The quantity $[\epsilon]_{TiO_2}^{Ti}$ is termed the *stopping cross section factor*. A similar expression for He ions that have been scattered from oxygen atoms can be written by replacing the superscript "Ti" with "O". Then using the kinematic factor (Eq. (3)), energy width ΔE taken from Table 7, and the stopping cross sections for TiO_2 from Table 9, we can determine the thickness in terms of atoms per unit area given in Table 10. The intrinsic thickness unit in RBS and ERDA analysis is atoms per unit area, i.e., a coverage. This can be converted to a thickness with units of length if the density ρ and the molecular weight MW of the film is known or can be assumed, so that N can be calculated as:

$$N = \frac{\rho}{MWu}(m+n), \tag{19}$$

where $(m+n)$ is the number of atoms in one molecule of a compound A_mB_n. The thickness of the film determined from the energy widths of the O and Ti signals differs by a few percent (Table 10). This difference mainly arises because of uncertainties in the stopping cross sections (Table 8). Continuous

Table 10. Stopping cross section factors, energy width, and thickness (Nt) of TiO_2 film.

Element	$[\epsilon]^A_{\mathrm{TiO_2}}$ [eV/(10^{15} at. cm^{-2})]	Δ E (keV)	Nt (10^{15} at. cm^{-2})	t (nm)
Ti	108.7	153.4	1411	147
O	100.6	150	1491	156

efforts are made to improve the accuracy of stopping cross sections through experimental measurement and predictor computer codes based on improved semi-empirical fits to experimental data, as well as *ab initio* approaches. At the time of writing, the stopping cross sections have an accuracy of about 5% for He ions and somewhat better, 3%, for protons.

4.4.2. *Determination of the composition of the titanium oxide film*

So far we have assumed the composition of the film to be TiO_2. This entered into determination of the stopping cross sections (Table 9). From Eq. (5), we can write the ratio of yields from Ti and O as:

$$\frac{Y_{Ti}}{Y_O} = \frac{(\frac{d\sigma_{Ti}}{d\Omega})N_{\mathrm{ion}}N_{Ti}\delta\Omega}{(\frac{d\sigma_O}{d\Omega})N_{\mathrm{ion}}N_O\delta\Omega}, \tag{20}$$

where the yield Y is taken to be the counts H_A in the spectrum from the element A over a given energy interval ξ in the spectrum. The thickness spanned by ξ is different for He ions scattered from Ti and O because $[\epsilon]^O_{\mathrm{TiO_2}}$ and $[\epsilon]^{Ti}_{\mathrm{TiO_2}}$ are not exactly the same due to the different backscattering energies. Furthermore, for Rutherford scattering, the backscattering cross sections are approximately proportional to Z^2, where Z is the target atomic number. Then Eq. (20) can be rearranged to give:

$$\frac{N_{Ti}}{N_O} = \left(\frac{Z_O^2[\epsilon]^{Ti}_{\mathrm{TiO_2}}}{Z_{Ti}^2[\epsilon]^{O}_{\mathrm{TiO_2}}}\right)\frac{H_{Ti}}{H_O}. \tag{21}$$

Substituting the heights of the Ti and O signals over a 10-channel interval with the stopping cross section factors (Table 9) and the atomic numbers of O and Ti, the ratio of Ti to O in the film $m/n = 0.53$. This is within 8% of the stoichiometric value of $m/n = 0.5$ for TiO_2. The uncertainty can be attributed to the uncertainty in the stopping cross sections. The statistical counting uncertainty contributes ~0.1%.

4.4.3. *Analysis of complex RBS spectra*

Figure 19 shows the RBS spectrum from a multilayer thin film structure of $SnO_2/SiO_2/Si$ (substrate). Thin films of tin oxide have also been employed as resistivity sensors, e.g., for carbon monoxide. The main industrial use is in conductive optically transparent films of indium tin oxide. These are important in the emerging organic light emitting diode (OLED) display industry and for solar cell applications. The RBS spectrum in Fig. 19 is made up of overlapping signals from different elements and layers. The relation between the edges and the interfaces in the structure is indicated in this figure by dotted lines. While it is quite possible to obtain the thickness and composition of the SnO_2 and underlying SiO_2 film manually, as discussed above for Fig. 18, the best approach is to use a simulation program. This has the advantage that more accurate determinations of the different layers can be made because assumptions such as the surface approximation (see above) need not be invoked. Figure 20 shows the results of a simulation using

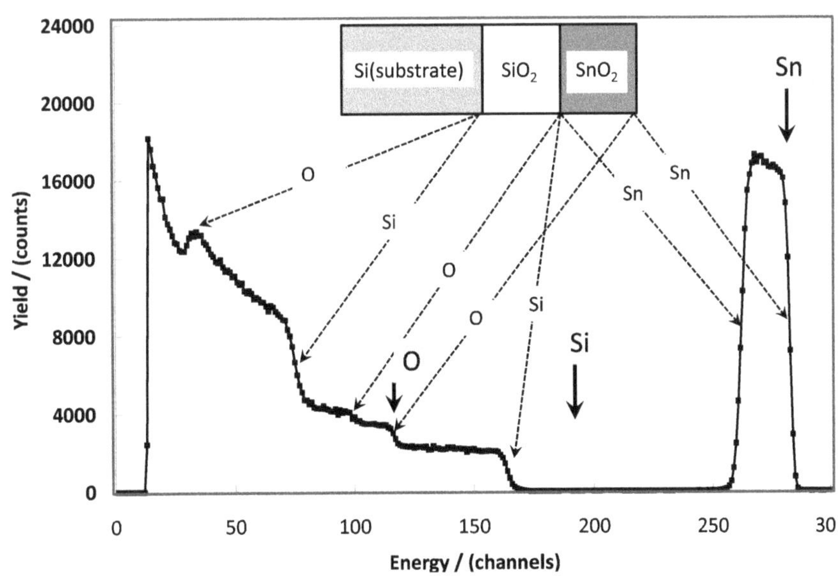

Fig. 19. RBS spectrum from a $SnO_2/SiO_2/Si$ (substrate) thin film. The energies associated with scattering from elements at the surface are denoted by vertical arrows. The dotted lines indicate the relationship between interfaces in the structure and edges in the spectrum.

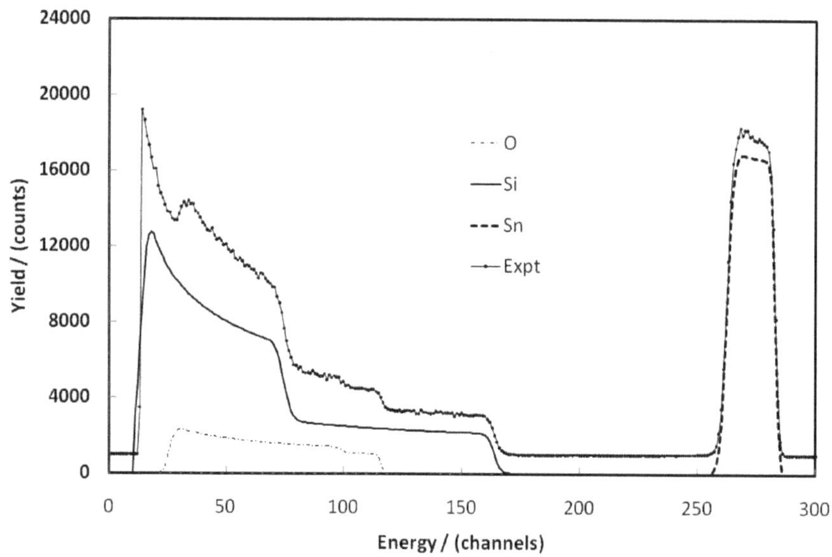

Fig. 20. Simulation of the RBS spectrum from the $SnO_2/SiO_2/Si$ (substrate) structure shown in Fig. 19. The experimental spectrum is displaced vertically to show the details of the simulated spectrum and the contributions from different elements. The sum curve (not shown) is in very good agreement with the experimental data. The simulations were made using SIMNRA.[29]

SIMNRA[29] for the spectrum shown in Fig. 19. Inspection of Fig. 20 shows that the contributions from the different elements to the RBS spectrum are calculated separately by adjusting the width, composition, and thickness of different slabs of material. To generate the composite RBS spectrum, the elemental signals from all elements in the sample are summed together. The composition and width of each slab are adjusted until the calculated and experimental spectra match. This may be done manually, or provided the changes needed are not too large, an iterative optimization procedure can be used. The power of this approach can easily be seen as it allows the features in a complex RBS spectrum to be identified and analysed in a simple and straightforward way.

4.5. *Elastic Recoil Detection Analysis (ERDA)*

Elastic Recoil Detection Analysis (ERDA) is very similar to RBS. The technique has the key advantage that the sample atom is itself detected, rather than the scattered particle. This means that in addition to the depth information that is carried by the energy of the recoil, the detected particle directly carries atomic number and mass information that can be measured using suitable detectors. A comprehensive presentation of the method is given in Barbour and Doyle[38] and the recent Handbook on Ion Beam Analysis.[34]

Three forms of ERDA are commonly used. The simplest uses a stopping foil to separate the recoils from the forward-scattered ions that, because they are more numerous, would swamp the signal from recoiling target atoms. The second uses a ΔE detector to determine the atomic number, while the third employs a time-of-flight detector to determine the atomic mass. These three forms are often referred to as stopping-foil ERDA, ΔE–E ERDA and ToF–E ERDA.

Stopping-foil ERDA is mainly used as a compliment to RBS for depth profiling hydrogen. This is because, as mentioned above, hydrogen cannot be detected in RBS using He ions. Stopping-foil ERDA may be carried out in the same system as for RBS and uses a 1–2 MeV ^{4}He ion beam. Hydrogen is important because it plays a key part in many processes of industrial importance, such as corrosion, and is often incorporated at buried interfaces as a result of hydrocarbon contamination. Hydrogen is also invisible in XPS and difficult to profile with SIMS because it is a common rest-gas in vacuum systems where it readily attaches to surfaces. Stopping-foil ERDA can also be used with deuterium as a tracer in diffusion and water incorporation studies by replacing H_2O with D_2O.

In *ToF–E ERDA*, a detector system consisting of two time detectors and an energy detector is used. This is often termed a detector telescope. The configuration is illustrated in Fig. 21. The mass M of each recoil can be obtained by measuring the flight time T between the two time detectors separated by a distance L and the kinetic energy E. Then $M = \kappa E(T - T_0)$, where κ is a calibration constant and T_0 is another constant associated with electronic delays. By recording the time and energy data for each recoil atom in coincidence, the different masses are separated on the 2-dimensional histogram of T vs. E as shown in Fig. 22(a). An overview of the time detectors is given in an early article by Whitlow.[39] Although ToF–E ERDA was originally developed by two groups in the 1970s, it was not until Whitlow *et al.* developed multi-dispersive measurements in 1987[40,41] that the full potential of the method became available. The multi-dispersive form of ToF–E ERDA is now used at laboratories worldwide.[b]

[b]Whitlow *et al.*[40] originally termed this Recoil Spectrometry. The generally accepted name that was adopted by consensus some years later is ToF–E Elastic Recoil Detection Analysis (ToF–E ERDA).

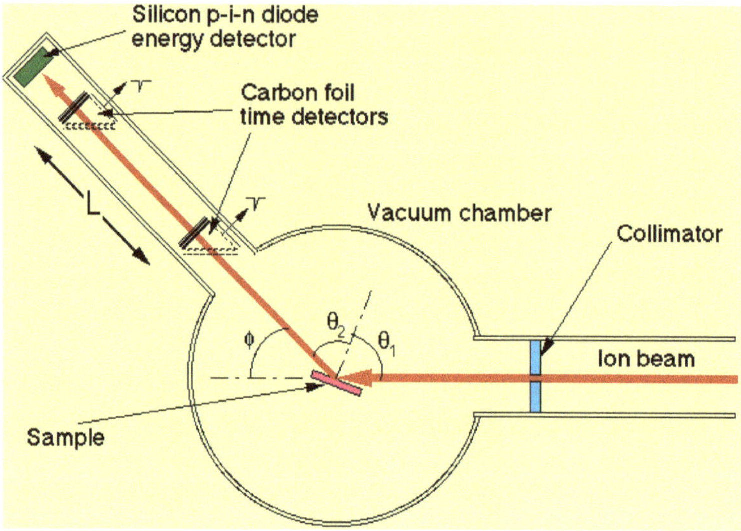

Fig. 21. Schematic illustration of a ToF–E ERDA measurement system. The ion beam is incident at a glancing angle on the sample and creates recoils that are ejected in the forward direction. The carbon foil time detectors generate a pulse when the recoils penetrate a thin 3–5 μg cm^{-2} carbon foil. These foils are so thin that they give rise to only a small energy loss as the recoils pass through them, which can usually be neglected. The time of flight is given by starting a clock with the first time detector and stopping the clock using the second time detector. The energy is then measured using a Si charged particle detector that can be a surface barrier or p-i-n detector.

A measurement end-station for ToF–E ERDA that was used for industrial consultancy work is depicted in Fig. 23. This system is fitted with a sample change mechanism for 10 large or 20 small samples, which facilitates unattended data collection. Figure 22 shows ToF–E ERDA measurement data for a thin hydroxyapatite-like film on borosilicate glass used as a substrate for culturing bone cells.[42] Figure 22(a) represents the 2-dimensional time of flight vs. energy histograms. The data is clustered into "bananas" with each banana corresponding to different mass isotopes in the sample material. It is seen that the bananas corresponding to the different mass isotopes can be clearly separated. Thus it is clearly seen from Fig. 22(a) that ToF–E ERDA allows the separation even of weak signals, such as from Na. This signal would be invisible in an RBS measurement because it would be completely buried in the background statistical noise on the spectrum. It is this power to separate the different elemental signals that makes ToF–E ERDA particularly useful because it completely overcomes the superposition problem encountered in RBS as discussed above. In Fig. 22(a),

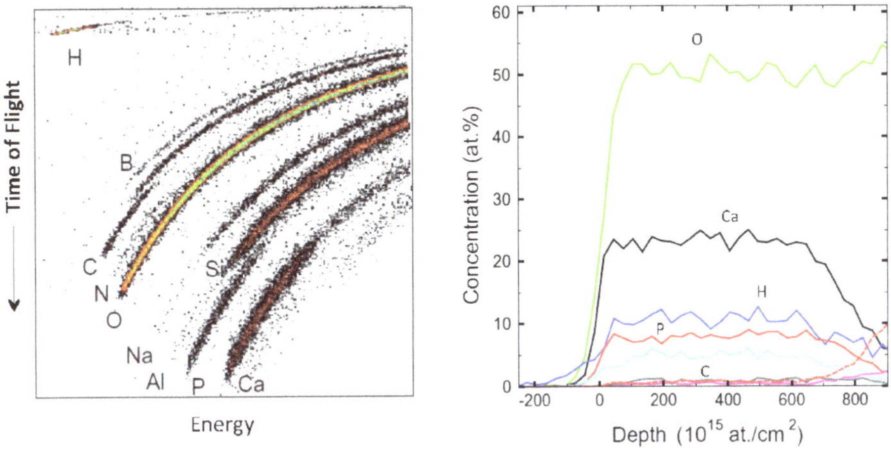

(a) (b)

Fig. 22. ToF–E ERDA data from a biomiometic ~100 nm thick Ca–P oxide film on a borosilicate glass substrate. The data were obtained using 16 MeV $^{63}Cu^{7+}$ ions at a facility located in an industrial research centre. (After A. Sagari, C. Lautaret, S. Gorelick, M. Laitinen, P. Rahkila, M. Putkonen, K. Arstila, T. Sajavaara, S. Cheng and H.J. Whitlow.[42])

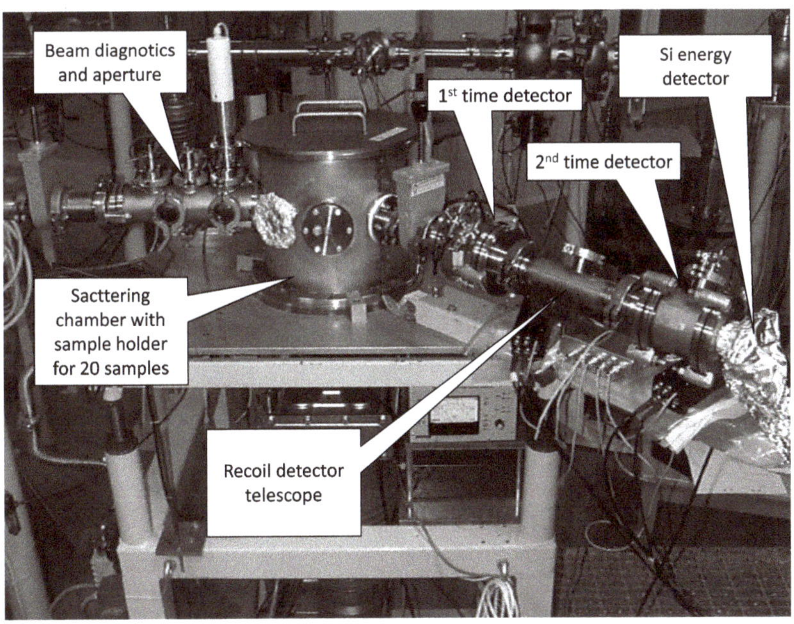

Fig. 23. ToF–E ERDA measurement end-station. The ion beam enters the beam diagnostic and aperture system from the left. The recoil detector telescope can have interchangeable sections to allow different flight lengths to be used.

a signal originating from hydrogen can also be clearly seen. This is particularly significant because hydrogen is an important contaminant in many industrial thin-film process technologies. Bearing in mind that there are no elements with significant natural abundances of isotopes that overlap with another element up to about $Z = 18$, this can be used to uniquely and quantitatively determine the elemental depth profiles seen in Fig. 22(b). The well separated depth distributions shown in this figure allow the Ca/P ratio to be determined with a high accuracy, which is important for the performance of the film as a cell-growth substrate.[42] It is clear that there is a considerable amount of hydrogen in the film. The depth scale in this figure is in terms of at. cm^{-2}, which, as noted above, denotes the thickness of the film in terms of quantity of material without invoking the uncertain density. (For hydroxyapatite with density 3.08×10^3 kg m^{-3} and composition $Ca_5(PO_4)_3(OH)$, 1 nm is equivalent to 8.13×10^{15} at. cm^{-2}.)

In *ΔE–E ERDA*, the atomic number of each recoil is determined by measuring the stopping force in the ΔE detector. In this case, the detector telescope is made up of a thin ΔE detector that is so thin that the recoils pass right through, followed by a thick detector that measures the residual energy E_r of the ions. The stopping force dE/dx is strongly dependant on the recoil atomic number Z_2 according to $dE/dx \,\tilde{\propto}\, Z_2^2/E$. The signal from the ΔE detector of thickness t_{det} is $\Delta E = (dE/dx)t$. Hence, the different recoil elements can be identified by plotting ΔE vs. E_r. Recoils with different Z_2 lie along different banana shaped regions in these plots. Usually gas ΔE detectors are used because, unlike Si ΔE detectors, these are insensitive to the radiation damage that arises from introduction of crystal damage centres by energetic heavy recoil bombardment. Generally, for ΔE–E ERDA, high energy probing ions such as 70 MeV ^{127}I are used so that the recoils have sufficient energy to traverse the gas ΔE detector. This means that the technique is only available at the relatively few centres in the world where such high energy beams are available and thereby only suitable for industry–university development consultancy work.

Table 11 compares the different types of ERDA analysis. The most widely adopted are stopping-foil ERDA and ToF–E ERDA. Stopping-foil ERDA can be used with the same accelerator as for RBS, and the data is collected as a 1-D spectrum that is easy to interpret. It is therefore suitable for QC applications. ToF–E ERDA requires different beams of high energy heavy ions like 20 MeV $^{63}Cu^{10+}$. These beams can be produced by small (1–3 MV) tandem accelerators, which are sometimes located in industrial research centres. The data from ToF–E ERDA measurements is usually in

Table 11. Elastic Recoil Detection Analysis (ERDA) methods and applications.

Method	Ion beams and detector	Depth resolution FWHM	Sensitivity at. %	Applications	Comments
RBS	1–2.5 MeV ^{4}He Si charged particle detector	20–30 nm (normal incident) 3–10 nm (glancing angle)	0.1 0.03	See Table 5	H not analysed
Stopping-foil ERDA	1–2.5 MeV ^{4}He Si charged particle detector	5–20 nm	0.1	Corrosion studies Amorphous-Si photovoltaic cells	Sensitive only to H and D, Surface roughness sensitive
ToF–E ERDA	10–120 MeV ^{35}He, ^{63}Cu, ^{81}Br, ^{127}I etc. Time of flight detector telescope	<1–50 nm	0.03	Corrosion studies Metal/III-V semiconductor contacts Optical and magnetic information storage films Biointerface films	Surface roughness sensitive, Sensitive to all elements
ΔE–E ERDA	50–120 MeV ^{35}He, ^{63}Cu, ^{81}Br, ^{127}I etc. ΔE-E detector telescope	<1–50 nm	0.1	MOCVD structures Metal/semiconductor contacts Optical and magnetic thin films	Surface roughness sensitive, Best for light elements (Z <8)

the form of 2-D spectra that require additional analysis steps. In analysis for industrial applications, the ability to analyse the data by simulation codes such as SIMNRA[29] considerably enhances the possibilities and applicability of this method. Although the ToF–E ERDA method is much more powerful than RBS, the need to analyse multi-dimensional data makes it best suited for industrial development work in industrial research centres. An example of such a centre is IMEC (Interuniversity Microelectronics Centre) in Belgium.

4.6. *Nuclear reaction analysis (NRA)*

Nuclear reaction analysis (NRA) covers ion beam analysis methods that are based on a specific nuclear reaction. PIGE is one such technique. However, it cannot be used to directly obtain the depth distribution of an isotope, and because it is often used in combination with PIXE, it was discussed in Sec. 3. The basis of all NRA methods is a characteristic nuclear reaction with a target nucleus X initiated by an incident ion a, which produces a reaction product b and a residuum Y that can be detected. The reaction $a \rightarrow X \Rightarrow Y + b$ can be written as $X(a, b)Y$.

Generally, the forward reaction $X(a, b)Y$ is used with a being a light ion, such as ^{1}H, ^{3}He, or ^{4}He. In a few cases for material studies, ^{2}H and ^{3}H have been used as a probe beam. However, this use has been very restricted because of radiation safety considerations associated with the significant neutron production with these ions. In the characteristic nuclear reaction, an amount of energy is absorbed or released (termed the Q-value). Unlike the case for scattering, energy (e.g., Eq. (4)) is not conserved, and as a consequence, the formalism describing the scattering and energy transfer is less straightforward. It is tabulated in detail in a number of texts (for example, see Mayer and Rimini[36] or the work by Kantele[43]). As for the case of elastic scattering, the probability of a reaction occurring can be described by a differential cross section. Unlike the case for Rutherford scattering where the scattering cross section has a smooth $1/E^2$ form, in NRA the reaction can have a smoothly varying, resonant, or sigmoid (threshold) energy dependence.[44]

In *charged particle reaction analysis*, the particle that is detected is either a charged particle(s) or charged residuum. The method is conceptually similar to RBS described in Sec. 4.4 because the energy is lost along the inward path of the ion and the outward path of the detected particle. The Q-value of the reaction implies the equations relating the incoming ion and outgoing particle energies are somewhat less straightforward than for the RBS and

ERDA cases. The equations have been tabulated in a handbook by Mayer and Rimini,[36] and useful descriptions of the formalism of scattering are given by Rickey[45] and Kantele.[43] The main use of the method has been measuring the depth distributions of light elements for atomic numbers below 20. The method has the advantage that it is isotope specific and thereby can be used with tracers. For example, ^{2}D atom tracers can be used to study hydrogen incorporated in a sample in the presence of a water or hydrocarbon layer containing ^{1}H atoms. Although the method is sensitive (at the ppm level in some cases), it has largely been superseded by ToF–E ERDA.

In *nuclear resonance profiling*, sometimes referred to as *Nuclear resonance broadening (NRB)*, a sharp resonance is used to sample the concentration from a depth region below the sample surface.[44] At each sampling step, the ions impinge at an energy exceeding the resonance energy. As the ions slow down, they reach the resonance energy and can undergo a nuclear reaction. The number of reactions taking place within the depth window spanned by the resonance width is governed by the differential cross section and the number of atoms of the isotope in question. Then by stepping the accelerator energy, the concentration of the isotope in different depth windows can be determined. The detected particle can be a charged particle or a γ-ray. Changing the accelerator energy in steps is generally a tedious process. Therefore, the method is poorly suited to industrial applications, except for development work where only a small number of samples are to be measured.[44]

This method is useful for measuring hydrogen using the 402 keV (centre of mass energy) resonance in the ^{1}H(^{15}N,$\alpha\gamma$)^{12}C reaction. In this case, the inverse reaction of ^{15}N(^{1}H,$\alpha\gamma$)^{12}C is used, so that the projectile is ^{15}N and the target ^{1}H. The resonant energy then corresponds to a ^{15}N energy of 6.385 MeV, which can be produced as a 2+ or 3+ charge-state ion using a small tandem accelerator.

Threshold reactions, e.g., for profiling ^{3}T for developing fundamental fusion reactor technology, have been rarely used for industrial purposes. They will therefore not be considered further.

4.7. *Charged particle activation analysis (CPAA)*

Charged particle activation analysis (CPAA) refers to methods where radioactive atoms are introduced into the surface layer of material. Normally, an ion beam with sufficiently high energy to activate the surface layer is used. A very similar approach is where a radioactive ion beam is implanted into the surface layer. In both methods, the radioactivity is progressively

Table 12. Charged particle activation analysis methods.

Activation method	Errosion method	Type of information
$^{1}H^{+}$, $^{1}D^{+}$ direct beam	Chemical attack	Errosion rate
$^{1}H^{+}$, $^{1}D^{+}$ direct beam	Mechanical abrasion	Wear rate
$^{1}H^{+}$, $^{1}D^{+}$ direct beam	Electrochemical erosion	Errosion resistance
Radiaoactive beam ion implantation	Sputter erosion	Depth profile Diffusion rate

measured as the sample is eroded away. The erosion process can be done in different ways that give different types of information (Table 12). The method generally has been used for measurement of C, N, and O. However, provided a suitable beam is available, other elements can be analysed as well. The matrices can be solid materials, such as semiconductors, ferrous metals, and glasses or environmental materials such as fly ash. Detailed overviews are given by Blondiaux, Debrun and Maggiore[46] as well as Strickmans.[47] A clear drawback of this method is that after irradiation, the sample is radioactive and thereby subject to legislation regarding its handling, storage, and disposal. In order to induce radioactivity, the beam must have enough energy to overcome the Coulomb barrier, which was discussed in Sec. 3.3. This means that for most elements, a proton or deuterium beam with an energy in the 5–40 MeV range must be used. This requires either a medium-sized (5–20 MV) electrostatic accelerator or a small cyclotron accelerator. A list of reactions has been tabulated by Blondiaux, Debrun and Maggiore.[48] Overall, although the method is well suited to industrial application, it has not been as widely used as RBS and ERDA due to the scarcity and high cost of suitable accelerators. The measurement can be based on monitoring the residual activity retained in the sample, or measuring the activity of the eroded material (differential measurement). For example, one application has been the study of the wear-rate of bearing surfaces. The advantage of using radioactive atoms is that the sensitivity can be very high. This is particularly so if the activity of the eroded material is measured and the induced activity is a short-lived isotope. This has been used with sputter-erosion to measure depth distributions where the peak concentration is as low as 10^{9} atoms cm^{-3} for determining the diffusion of ^{209}Pb in soda-lime glass.[49] The best competing technique is Dynamic-Secondary Ion Mass Spectrometry (dynamic-SIMS) where sensitivities of 10^{12} at. cm^{-3} can be achieved. This ^{209}Pb diffusion measurement was important for establishing a method for retrospective radon dose exposure in dwelling places. Another industrial application has been to measure the wear on engine bearing surfaces in testing oil additives. Normally, this measurement would require the engine bearing to be dismantled for each step. Instead, a sample of the oil could be taken sequentially, which is more realistic of normal running conditions. Historically, radioactive ion implantation has played an important role in establishing the ion energy–range distribution relationship in ion implantation, which is an important industrial materials modification technique.

5. Industrial Facilities for Ion Beam Analysis

As nanotechnology migrates to industrial production, the demands for quantitative methods for characterisation that are capable of quality assurance with traceability on a nanometre scale is increasing. The shrinking size means that existing conventional industrial methods[2] such as Glow Discharge Optical Emission Spectroscopy (GDOES) and SIMS become increasingly more difficult and expensive to reliably calibrate. Despite the need for the additional complication of an accelerator, MeV IBA methods are being increasingly used for industrial applications, particularly in high-tech companies in Japan and to a lesser degree in the USA.

The requirements for IBA in industry can be broadly divided into two categories: quality control (QC) and industrial development/consultancy work. The demands for industrial application are quite different from the academic environment where accelerators are developed mainly to serve fundamental research, at first in nuclear physics and later in materials development. Accelerators in an academic or research centre environment are usually highly flexible machines capable of providing ion beams of a wide range of elements and energies for many different applications, such as medical isotope production, ion implantation, ^{14}C AMS, and fundamental nuclear physics. This high degree of flexibility means the machines are expensive and complex, require specialist operators, and typically serve a wide range of different highly trained researchers. Usually the IBA facilities at these accelerators are geared towards sophisticated measurements with goniometer sample holders that can only handle a single sample or small numbers of samples per loading. This means the availability and reliability are often low because of time-sharing, and the frequent changes of operating parameters lead to failure associated with frequent start-stops and consequent stressing of components. However, many of these research centres serve industrial customers who are performing R&D of a new analytical technique of interest to their business or who are seeking help in solving a problem that they have encountered in a production process.

The needs of Quality Assurance (QA) and industrial research and development are quite different. QA is concerned with QC and also traceability for diagnostic purposes. IBA is well-suited for applications where the number of workpieces to be tested is small or for calibration of secondary measuring tools that are used for high-volume testing. Many high-tech production processes are batch rather than continuous production. Examples are thin-film metallisation of photovoltaic cells, microchips, sputter deposition of optical coatings, and electrochemical coating of catalytic films on sensors. Depending on the application, each individual component or sample may need to be measured or a test sample is included in the batch that is analysed.

Industrial IBA applications have quite different demands from academic research environments:

- *In industrial QA*, the samples will either be the manufactured components or special test samples that are processed at the same time and under identical conditions as the components. In both batch- and continuous-processing, large numbers of closely identical samples are involved that are analysed in exactly the same way. This places demands for sample handling systems that are able to handle large numbers of identical samples, measure them, and analyse in a standard way. These may also include test samples that are used for QC of the IBA measurement. Even in an industrial research and development environment, large numbers of samples are frequently required. An example is process parameter optimisation by the matrix approach. For example, optimisation of three parameters using ten different samples for each parameter inplies that 1000 different samples must be measured and analysed.
- *Reliability* is extremely important especially for QA applications because the accelerator is a part of the production process. Consequently, a breakdown means the entire production process must be halted, which will generally cause significant economic costs. Even if the accelerator is only used for process diagnostics, it is important that the analysis can be carried out quickly to minimise economic losses due to process downtime.
- *Operator skills.* In micro- and nano-technology fabrication, the actual QA work is usually carried out by operators who are not highly trained. It follows that start-stop, sample handling, and operation procedures for an IBA system must be straightforward and the data analysis sufficiently

straightforward to allow an unskilled operator to make a go/no-go judgement for a single defined system. The supervisory staff are generally trained engineers who require sufficient training to diagnose problems based on the IBA data and decide corrective action for out-of-tolerance measurements. In an industrial development facility, the staff are generally well-trained scientists and engineers. The work is generally less time critical than for production, and a specialist would normally be responsible for and operate the system as well as providing services to users.

- *Traceability* is an important issue for industrial analysis, particularly for QA applications. Consideration must be given to establishing reliable sample identification procedures so that the correct analysis result is connected to the correct sample. Generally, the raw data spectra and analysis report are generated as a computer file and these can be readily archived as an analysis protocol.
- *Turn-key installation* is important because often industrial organisations do not have the in-house competence required to construct an IBA facility from component parts. Furthermore, installation may require an expensive shut-down of a production facility.
- *Small footprint.* Floor-space is an expensive commodity in an industrial enviroment. This is particularly so if the IBA facility must be located in a so-called clean-room because the cost per unit area in such a facility is extremely high. Moreover, accelerators do require service from time to time and opening the pressure vessel may lead to release of dust particles formed on the walls and components inside the vessel by the inevitable gas discharges. It may therefore be advantageous to place the accelerator in an adjacent service area with a transfer line leading the ion beam to the analysis station inside the clean-room area. Generally, measurements will only need to be carried out in a clean-room in production QA applications. For industrial research and development and process diagnostic applications, it is generally sufficient to carry out the analysis in a semi-clean environment outside of the clean production area.

For both industrial development and QA applications, the number of samples to be handled is an important consideration. In both cases, the samples are usually closely identical in size and shape. The main limitation to measurement-throughput is the time taken to load the samples into the vacuum system and pump down to achieve a suitable low pressure for analysis ($<10^{-3}$ Pa). Sample identification is of central importance when large numbers of samples are to be measured. It may be desirable to laser-mark each sample with a bar-code and use machine-vision techniques to record

Fig. 24. Goniometer system developed for automated RBS and ERDA measurements at an industrial research centre. Up to 20 samples can be mounted on the sample wheel (Fig. 25). The paths of the analysing-beam and the backscattered ions are shown by arrows.

Fig. 25. Front (right) and rear (left) faces of the sample wheels for the goniometer (Fig. 24).

each sample so it can be associated with the correct measurement data set. Consideration should also be given to systematic measurement identification because a number of spectra may be measured from the same sample.

Figure 24 shows a sample handling system developed for work at an industrial development centre. This sample handling system (termed a goniometer) can be loaded with sample wheels each containing 20 samples (Fig. 25). The goniometer can be controlled by a LabView^{TM} program that allows unattended data collection from RBS and axial channelling measurements of damage distributions. The control of ion beam analysis systems is discussed elsewhere by one of us.[50]

Figures 26 and 27 show a modern system that is intended for thin film analysis. The system is based on a low energy 500 kV single-ended accelerator with a modern solid-state high voltage supply. The energy of scattered particles is measured using a double-focusing magnetic dipole spectrometer with a multi-channel plate detector. The low-energy ions in combination with the magnetic spectrometer give the capability of sub-nm resolution at the surface. The small footprint and availability as a turn-key system make it suitable for industrial research and development environments. This instrument and its predecessors have been installed at a number of Japanese industrial facilities. These systems are quite flexible and can be used to perform ERDA as well as channelling analysis of crystal lattice strain in thin-films. Lattice-strain measurements may be required to compliment and standardise conventional X-ray rocking curve measurements in QA applications (e.g., III-V epitaxial layers produced by Molecular Beam Epitaxy, MBE, because these require a skilful setup to align the sample). Although this procedure has been automated, using, for example, a LabView^{TM} environment, it is

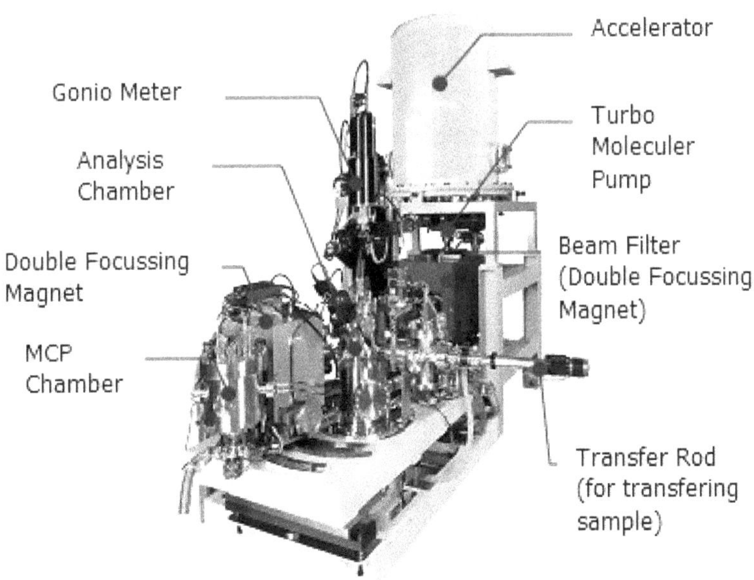

Fig. 26. Complete table-top RBS thin-film analysis system suitable for industrial development facilities. The system occupies 2 m × 1.5 m floorspace and uses a high voltage of 500 kV. The control electronics is contained in a separate double 19 inch rack. Image kindly supplied by Kobe Steel Ltd. Japan.

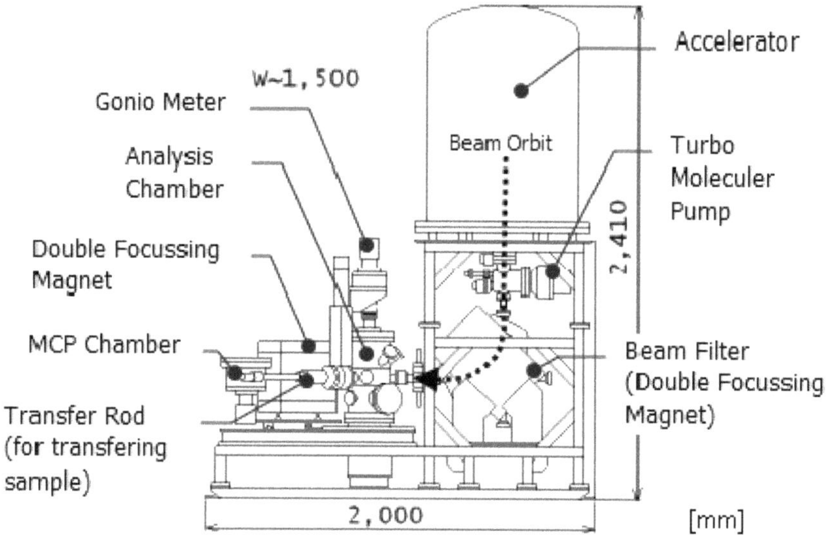

Fig. 27. Schematic illustration of the system in Fig. 26. Image kindly supplied by Kobe Steel Ltd. Japan.

probably too slow and difficult to interpret for QA applications. However, such measurements certainly lie within the scope of an industrial research and development facility. The footprint of $2\,m \times 1.5\,m$ of this analysis tool is quite comparable to or even smaller than that required for a SIMS instrument, and the price is similar. A similar automated RBS end-station capable of nanometre depth resolutions is marketed by NEC.

Hazardous material and environmental stewardship are becoming of increasing importance in industry. In addition to the samples themselves, which may present an environmental hazard, the accelerator often uses hazardous materials. These are, for example, alkali metals (Li, Na, Cs, Rb) and pressurised gases (He, H_2, O_2, etc.) that are used in the ion source, as well as the dielectric gas in the tank. SF_6 is widely used for the latter purpose. This is classified as a greenhouse gas alongside freons by the EU and is subject to special handling legislation. This however does not represent a significant problem because modern accelerator systems can be supplied with a re-cycling system that allows the gas to be reused to almost 100%.

Ionising radiation is produced by the accelerators. This arises from the back-streaming of electrons along the accelerator tubes as well as nuclear reactions caused by energetic ions in the beam hitting components such as defining apertures. Modern accelerator tubes have special electrode designs or incorporate magnetic suppression to mitigate ionising X-rays formed when the energetic back-streaming electrons impinge on acceleration electrodes or other accelerator components. Pb shielding may be mounted around the accelerator tank to attenuate the X-rays. Heavy ions, such as used in ERDA, have energies too low to induce significant nuclear reactions. For the moderate energies (0.5–2 MeV) of protons and ^{4}He ions used in IBA, the latter is not a significant problem. Enhanced radiation levels around apertures can be effectively reduced to sub-background levels using lead sheet. A more detailed discussion can be found in a newly published handbook.[51]

The total sales of IBA equipment to industry over the past 3 decades has been about US$23M. This represents 25–30% of the total sales, the remainder going to university laboratories (where they are often used for industry-based research). There has been some evidence of a maturing in the industrial market in the USA and Europe. It is worthy of note that the cost of a turnkey industry IBA system is about US$700 000, which is comparable with a good transmission electron microscope or an advanced scanning electron microscope/electron beam lithography system.

Acknowledgements

Timo Sajavaara, University of Jyväskylä, Finland, kindly supplied the RBS spectra (Figs. 18, 19 and 20) and performed the computer simulation. The samples for these figures were graciously supplied by Gomathi Natarajan and Prof. David Cameron of Lappeenranta University of Technology. Max Strandberg, Faculty of Engineering, Lund University, has kindly prepared Figs. 2, 4, 6, 7, 9, 10 and 13. I. Mure, G. Norton, and E. Yoshida have kindly helped the authors with details in different ways.

This work has been written while HJW was at the Department of Physics, University of Jyväskylä, Finland and supported by the Academy of Finland Centre of Excellence in nuclear and accelerator-based physics, Ref. 213503 and Grant no. 129999.

References

1. R. Hellborg, H. J. Whitlow and Y. Zhang, Eds., *Ion Beams in Nanoscience and Technology* (Springer, Heidelberg, 2009), doi: 10.1007/978-3-642-00623-4.
2. D. Brune, R. Hellborg, H. J. Whitlow and O. Hunderi, Eds., *Surface Characterisation* (Wiley-VCH Weinheim, 1997).
3. M. B. H. Breese, in *Handbook of Modern Ion Beam Analysis*, Eds. Y. Wang and M. Nastasi (Materials Research Society, Pittsburg, 2009), pp. 285–303.
4. J. Castaing and M. Menu, *Nucl. Phys. News* **16**(4), 4–10 (2006).
5. T. Calligaro *et al., Nucl. Instr. Meth.* **B161–163**, 328–333 (2000).
6. *Proc. 19th Int. Conf. on Ion Beam Analysis, Nucl. Instr. Meth. B*, to be published.

7. R. Hellborg, in *Electrostatic Accelerators: Fundamentals and Applications*, Ed. R. Hellborg (Springer Verlag, Heidelberg, 2005), pp. 4–23.
8. J. D. Cockcroft and E. T. S. Walton, *Proc. Roy. Soc. A* **136**, 619 (1932) and **137**, 229 (1932).
9. R. Hellborg, in *Electrostatic Accelerators: Fundamentals and Applications*, Ed. R. Hellborg (Springer Verlag, Heidelberg, 2005), pp. 104–109.
10. D. J. W. Mous, J. Visser, A. Gottdang and R. G. Haitsma, *Nucl. Instr. Meth.* **219/220**, 480–484 (2004).
11. R. J. Van de Graaff: *Phys. Rev.* **37**, 1919–1920 (1931).
12. R. Hellborg, Ed., *Electrostatic Accelerators — Fundamentals and Applications* (Springer Verlag, Heidelberg, 2005).
13. *Proc. 11th Int. Conf. on Heavy Ion Accelerator Technology (HIAT 09)*, online publication at http://epaper.kek.jp/HIAT2009/.
14. S. A. E. Johansson, J. L. Campbell and K. G. Malmqvist, Eds., *Particle Induced X-ray Emission Spectrometry (PIXE)* (Wiley & Sons, New York, 1995).
15. Th. B. Johansson, R. Akselsson and S. A. E. Johansson, *Nucl. Instr. Meth.* **84**, 141–143 (1970).
16. W. Maenhaut and K. G. Malmqvist, in *Handbook of X-ray Spectrometry: Methods and Techniques*, Eds. R. E. van Grieken and A. A. Marcowicz (M. Dekker, New York, 1992), pp. 517–582.
17. G. Calzolai *et al.*, *Nucl. Instr. Meth.* **B249**, 928 (2006).
18. J. L. Campbell, in *Handbook of Modern Ion Beam Analysis*, Eds. Y. Wang and M. Nastasi (Materials Research Society, Pittsburg, 2009), pp. 231–245.
19. A. Anttila, R. Hänninen, J. Räisänen, *J. Radioanal. Chem.* **62**, 441 (1981).
20. P. A. Mando, *Nucl. Phys. News Int.* **19**(1), 5–12 (2009).
21. I. Giles and M. Peisach, *J. Radioanal. Chem.* **50**, 307 (1979).
22. R. Mateus, A. P. Jesus and J. P. Ribeiro, *Nucl. Instr. Meth.* **B219/220**, 519–523 (2004).
23. J. Räisänen, in *Handbook of Modern Ion Beam Analysis*, Eds. Y. Wang and M. Nastasi (Materials Research Society, Pittsburg, 2009), pp. 147–173.
24. R. Hellborg and G. Skog, *Mass Spec. Rev.* **27**(5), 398–427 (2008).
25. *Proc. 12th Int. AMS Conference*, to be published in *Nucl. Instr. Meth. B*.
26. J. F. Ziegler, J. P. Biersack and U. Littmark, *The Stopping and Range of Ions in Solids* (Pergamon Press, New York, 1985).
27. J. F. Ziegler, http://www.srim.org/no.SRIM.
28. M. O. Thompson, http://www.genplot.com/doc/index.htm.
29. M. Mayer, in *Proc. 15th Int. Conf. on the Application of Accelerators in Research and Industry*, AIP Conf. Proc. Vol. 475 (American Institute of Physics, New York, 1999), p. 541.
30. C. Jeynes *et al.*, *J. Phys. D: Appl. Phys.* **36**, R97–R126, doi: 10.1088/0022-3727/36/7/201.

31. J. A. Leavitt, L. C. McIntyre and M. R. Weller, in *Handbook of Modern Ion Beam Analysis*, Eds. J. R. Tesmer *et al.* (Materials Research Society, Pittsburgh, 1995), pp. 37–81.
32. W.-K. Chu, J. W. Mayer and M.-A. Nicolet, *Backscattering Spectrometry* (Academic Press, New York, 1978).
33. G. Götz and K. Gärtner, *High Energy Ion Beam Analysis of Solids* (Akademie-Verlag, Berlin, 1988).
34. Y. Wang and M. Nastasi, Eds., *Handbook of Modern Ion Beam Analysis* (Materials Research Society, Pittsburg, 2009), Part 1.
35. M. L. Swanson, in *Handbook of Modern Ion Beam Analysis*, Eds. J. R. Tesmer *et al.* (Materials Research Society, Pittsburgh, 1995), pp. 231–300.
36. J. W. Mayer and E. Rimini, Eds., *Ion Beam Handbook for Materials Analysis* (Academic Press, New York, 1977).
37. Y. Wang and M. Nastasi, Eds., *Handbook of Modern Ion Beam Analysis* (Materials Research Society, Pittsburg, 2009), Part 2.
38. J. C. Barbour and B. L. Doyle, in *Handbook of Modern Ion Beam Analysis*, Eds. J. R. Tesmer *et al.* (Materials Research Society, Pittsburgh, 1995), pp. 83–138.
39. H. J. Whitlow, in *Proc. High Energy and Heavy Ion Beams in Materials Analysis Workshop* (Materials Research Society, Pittsburgh, 1990), pp. 243–256.
40. H. J. Whitlow, G. Possnert and C. S. Petersson, *Nucl. Instr. and Meth.* **B27**, 448–457 (1987).
41. H. J. Whitlow, in *Proc. High Energy and Heavy Ion Beams in Materials Analysis Workshop* (Materials Research Society, Pittsburgh, 1990), pp. 73–87.
42. A. Sagari *et al., Nucl. Instr. Meth.* **B216**, 719–722 (2007).
43. J. Kantele, *Handbook of Nuclear Spectroscopy* (Academic Press, London, 1995), p. 235.
44. H. J. Whitlow and R. Hellborg, in *Surface Characterisation*, Eds. D. Brune, R. Hellborg, H. J. Whitlow and O. Hunderi (Wiley-VCH, Weinheim, 1997), pp. 244–253.
45. F. A. Rickey, in *Handbook of Modern Ion Beam Analysis*, Eds. J. R. Tesmer *et al.* (Materials Research Society, Pittsburgh, 1995), pp. 21–44.
46. G. Blondiaux, J.-L. Debrun and C. J. Maggiore, in *Handbook of Modern Ion Beam Analysis*, Eds. J. R. Tesmer *et al.* (Materials Research Society, Pittsburgh, 1995), pp. 205–230.
47. K. Strijckmans, in *Surface Characterisation*, Eds. D. Brune, R. Hellborg, H. J. Whitlow and O. Hunderi (Wiley-VCH Weinheim, 1997), pp. 169–175.
48. G. Blondiaux, J.-L. Debrun and C. J. Maggiore, in *Handbook of Modern Ion Beam Analysis*, Eds. J. R. Tesmer *et al.* (Materials Research Society, Pittsburgh, 1995), pp. 619–641.
49. M. Laitinen *et al., Radiation Protection Dosimetry* **131**, 212 (2008), doi: 10.1093/rpd/ncn162.

50. H. J. Whitlow, in *Ion Beams in Nanoscience and Technology*, Eds. R. Hellborg, H. J. Whitlow and Y. Zhang (Springer, Berlin, Heidelberg, 2009), pp. 431–439, doi: 10.1007/978-3-642-00623-4.
51. P. M. DeLuca Jr, J. R. Tesmer and P. Rossi, in *Handbook of Modern Ion Beam Materials Analysis, Second Edition*, Eds. Y. Wang and M. Nastasi (Materials Research Society, Pittsburgh, 2009), pp. 425–434.

Chapter 6

PRODUCTION AND APPLICATIONS OF NEUTRONS USING PARTICLE ACCELERATORS

David L. Chichester
Idaho National Laboratory
2525 Fremont Avenue, Idaho Falls, ID 83415, USA
david.chichester@inl.gov

Advances in neutron science have gone hand in hand with the development of particle accelerators from the beginning of both fields. Early accelerator systems were developed simply to produce neutrons, allowing scientists to study their properties and interactions with matter, but people quickly realized that more tangible uses existed for them. Today, these systems are in use in such diverse applications as mineral assaying, oil well logging, radiography, and explosive detection to name a few. This chapter presents a technical introduction to the different ways particle accelerators are used to produce neutrons, an historical overview of the development of neutron-producing particle accelerators, a description of current industrial accelerator systems, narratives of the industrial fields where neutron-producing particle accelerators are used today, and comments on future trends.

1. Introduction

Within a few years of Cockcroft and Walton's pioneering work on particle acceleration and transmutation and Chadwick's discovery of the neutron in 1932, it was quickly recognized that neutrons could be produced using accelerators via the deuteron–deuteron (DD) disintegration ($^2H + {}^2H \rightarrow {}^3H + n$) reaction.[1–5] These discoveries, together with other related work at Cambridge University's Cavendish Laboratory, transformed the way experimental nuclear physics research was performed, and recognizing the usefulness of particle accelerators in nuclear physics experiments, more researchers soon followed this path.[6–8] With the basic building blocks in place, it didn't take long for these early researchers to recognize the potential commercial utility of having "electronic" sources of neutrons for use in academia, medicine, and industry.[9–13]

Since that time, particle-accelerator-based neutron sources (also referred to in this chapter as "electronic neutron generators" or ENGs) have been used in numerous applications. Most have been on a small scale, but one application stands out above the others — the use of small, vacuum-sealed (also known as "compact" or "sealed-tube") ENGs for geophysical exploration in the petroleum industry. Beyond this widespread application, other interesting but less frequently employed uses are for radiography, material characterization studies, on-line bulk material analyses, nuclear material assays, human and animal body-composition studies, and radiation-effects testing.

For many neutron applications, ENGs offer several advantages over radioisotope neutron sources, such as the spontaneous fission isotope californium-252 (^{252}Cf) or mixtures of alpha emitting radioisotopes with low-Z elements such as lithium or beryllium, that generate neutrons via the (α, n) reaction. Perhaps the most important advantage is that accelerators can be turned off, greatly simplifying issues in safety, transportation, and storage. Second, in most cases, accelerator-based sources can be operated to produce pulses of neutrons. Pulsing opens up additional measurement approaches that steady-state sources cannot readily accomplish, including neutron die-away time analyses. Pulsing also allows data to be collected during the time periods when neutrons aren't being produced, which can result in much stronger signal to noise ratios. Third, several reactions used in ENGs produce monoenergetic neutron beams, which again is a useful attribute in many applications. Finally, some ENGs make time-of-flight (TOF) measurements possible either through the use of extremely short pulses or the associated particle time-tagging technique.

Of course, there are several limitations associated with ENGs when compared to radioisotope sources. For example, the operational reliability of radioisotopes is essentially perfect — they don't stop producing neutrons. Accelerators by contrast are complicated devices that are vulnerable to multiple failure modes from both external causes (e.g., loss of power, shock and vibration, and extreme heat) and internal causes (e.g., component failures, vacuum leaks, and ion beam target deterioration). Also, for neutron intensities less than $\sim 10^9$ n/s, radioisotope sources are cheaper than accelerator-based systems — although for higher source strengths, the storage, packaging, and handling challenges associated with radioisotopes become significant.

Equally important is the fact that ENGs are much larger than any radioisotope source. This is immediately evident by examining the photograph in Fig. 1, which shows a typical multi-purpose portable ENG. In

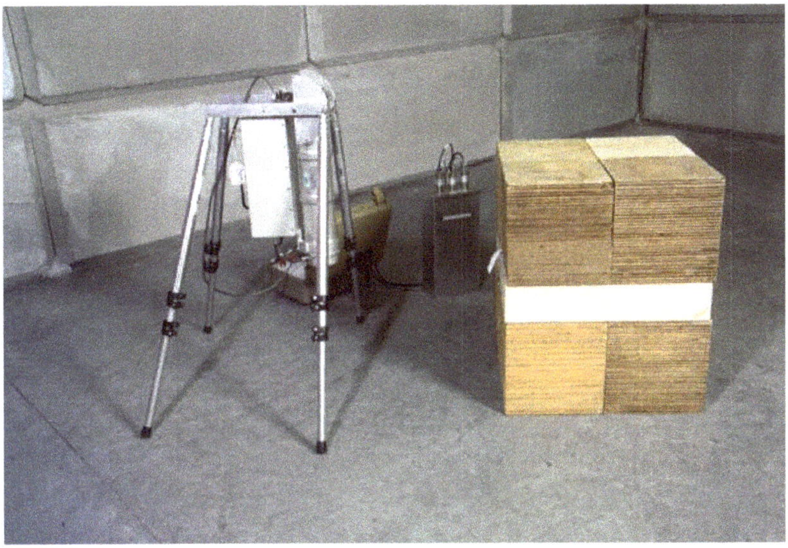

Fig. 1. Photograph of a compact ENG (left) in a prototypical setup for active neutron interrogation. A plywood cube to the right is being interrogated. This system produces $\sim 10^8$ n/s, is $\sim$15 cm × 25 cm × 60 cm in size, and weighs $\sim$12 kg. Directly behind it is a control box capable of operating the system for 16+ hours with a battery. Next to the control box is a cadmium-shielded, polyethylene-moderated neutron detector. Not shown is the laptop computer used to operate the system.

comparison, the volume of a ^{252}Cf source needed to produce the same neutron intensity is less than 2 cm^3. Neutron source size is of particular importance in the petroleum industry where instrumentation must be designed to fit within narrow-diameter boreholes. It is also important in other applications, especially when neutron moderating materials or more complicated assemblies designed to customize the emitted neutron energy spectrum are needed, because the size and weight of these assemblies can grow quickly, depending upon the external dimensions of the accelerator that must be accommodated.

Finally, the cost of ownership for radioisotope sources — based on either entry cost (cost per initial neutron emission rate) or total value of ownership (cost per total emitted neutrons) — has for years favored radioisotopes over ENGs. This has been the case for both short-lived sources such as ^{252}Cf and longer lived ones such as ^{241}AmBe. Operational lifetimes for compact ENGs are today only a few thousand hours of use (with a decaying output over time) before refurbishment and replacement parts are needed, and larger, higher-yield systems typically require more regular service. In contrast, radioisotope neutron sources operate with slowly decreasing outputs for years or decades before they need to be replaced.

Despite these disadvantages, ENGs have found applications in many niche areas. In the future, the use of these devices is likely to grow both in absolute and relative terms compared with radioisotope sources. Advances in radiation detection and measurement technology are steadily progressing, fueled primarily by applications in basic physics, medicine, and the life sciences. The new techniques developed in these areas are being adopted by the industrial sector and will inevitably expand the breadth and depth of neutron applications. Moreover, historical government subsidies for the production of radioisotope sources are being reduced, leading to higher prices and less predictable availability, thereby making their use economically unfavorable. At the same time, commercial accelerator-based solutions continue to become more efficient and cost effective. Lastly, government and public acceptance of the use of industrial radioisotopes is declining with perceived fears of poor security and the possibilities of theft, misuse, and potential use in radiological weapons (radiological dispersal devices). Recent reports by the US National Research Council and the North Atlantic Treaty Organization (NATO) underscore these concerns, and efforts are underway by many organizations to develop accelerator-based replacements for all current radioisotope sources.[14,15]

The main purpose of this chapter is to provide an overview of the technology used in accelerator-based neutron sources and to describe their industrial applications. This includes a fairly detailed discussion of the various nuclear reactions and types of accelerators that can be used to produce neutrons along with some examples of their use in industry. While several different types of accelerators have been used to produce neutrons in industrial settings, sealed-tube ENG systems are the most widespread and thus are of particular emphasis. Larger accelerator-based neutron systems, such as the Intense Pulsed Neutron Source (IPNS) at Argonne National Laboratory, the Los Alamos Neutron Science Center (LANSCE) at Los Alamos National Laboratory, and the Spallation Neutron Source (SNS) at Oak Ridge National Laboratory, are not included in this chapter. Their size and complexity make them unique research instruments far removed from the wider-scale commercial systems and industrial applications covered here. Also, while such neutron science facilities do support industrial research, their primary mission is basic science research. While some basic knowledge for understanding the design and operation of accelerator-based neutron sources will be provided, readers who are interested in more details can refer to several excellent review books on the underlying technology.[16–19]

2. Neutron Production

The neutron is different from all other forms of radiation used in industrial applications in that it is neutral and does not interact through electromagnetic interactions, making it insensitive to the interactions experienced by electrons, positrons, ions, and photons. Hence, neutrons interact primarily with atomic nuclei rather than the electron clouds that surround the nuclei. This makes them ideal probes of the nuclear structure of matter and especially sensitive to different isotopes of the same element. In these interactions, it is the kinetic energy of the neutron that determines the probability and type of interactions that will take place. Neutron inelastic scattering is the primary interaction at high (MeV) energies, while neutron capture is dominant at thermal (eV) energies.

An interesting fact is that in a free state (i.e., not bound in a nucleus) the neutron is unstable and will undergo radioactive decay into a proton, as shown in Eq. (1), with a lifetime of 886 s:

$$n^0 \rightarrow p^+ + e^- + \bar{\nu}_e. \tag{1}$$

An important and useful property that has a large influence on how neutrons interact with matter is that the neutron's mass (1.0087 u) is almost identical to that of the proton (1.0073 u). This means that neutron elastic scattering in hydrogenous materials can result in near complete energy transfer in just one collision, which quickly reduces a fast neutron population to thermal energies.

2.1. *Ion reactions*

Fusion of light nuclei is the most common method of producing neutrons with accelerators, and the most common of these reactions are listed in Table 1. While other reactions exist, these are the most efficient and easiest to employ, and each will be discussed in some detail in this section.

In principle, the direction and kinetic energy of the final products of these types of nuclear reactions are determined by the energy and momenta of the initial nuclei. In particular, the theoretical neutron energy can be determined as a function of the incident particle energy according to Eq. (2):[20]

$$E_3^{1/2} = \frac{(M_1 M_3 E_1)^{1/2} \cos\theta \pm \{M_1 M_3 E_1 \cos^2\theta + (M_3 + M_4)[M_4 Q + (M_4 - M_1)E_1]\}^{1/2}}{M_3 + M_4}. \tag{2}$$

Referring to the reaction in shorthand as "2(1, 3)4" where incident particle 1 reacts with stationary target nucleus 2 to produce daughter products 3

Table 1. Commonly used neutron production reactions.

Reaction	Shorthand	Q Value [MeV]	Threshold [MeV]	Minimum product energies [MeV]
$^{2}H + {}^{2}H \rightarrow {}^{3}He + n$	$^{2}H(d,n)^{3}He$	+3.269	NA	^{3}He: 0.82, n: 2.45*
$^{2}H + {}^{3}H \rightarrow {}^{4}He + n$	$^{3}H(d,n)^{4}He$	+17.589	NA	^{4}He: 3.54, n: 14.05
$^{1}H + {}^{7}Li \rightarrow {}^{7}Be + n$	$^{7}Li(p,n)^{7}Be^{\dagger}$	−1.644	1.880	^{7}Be: 0.21, n: 0.03
	$^{1}H(^{7}Li,n)^{7}Be^{\dagger}$	−1.644	13.094	^{7}Be: 10.0, n: 1.44
$^{2}H + {}^{7}Li \rightarrow {}^{8}Be + n$	$^{7}Li(d,n)^{8}Be$	+15.031	NA	^{8}Be: 1.68, n: 13.35
$^{1}H + {}^{9}Be \rightarrow {}^{9}B + n$	$^{9}Be(p,n)^{9}B$	−1.850	2.057	^{9}B: 0.18, n: 0.023
$^{2}H + {}^{9}Be \rightarrow {}^{10}Be + n$	$^{9}Be(d,n)^{10}B$	+4.361	NA	^{10}B: 0.40, n: 3.96

*The probability for $^{2}H + {}^{2}H$ (DD) producing a neutron is roughly the same as for producing $^{3}H + p$.

†There are competing end products for this reaction.

(a neutron) and 4, E_1 is then the incident particle energy, E_3 is the neutron energy, M_1 is the incident particle mass, M_3 is the neutron mass, M_4 is the associated daughter nucleus mass, and Q is the reaction energy (positive for exothermic reactions and negative for endothermic reactions). The direction of 1 towards 2 defines the neutron angle of emission θ after the reaction occurs. This relationship is plotted in Fig. 2 for the six reactions listed in Table 1 for neutrons emitted in the forward direction ($\theta = 0°$) and backward direction ($\theta = 180°$). Note that this plot does not show the probability that the reaction will take place, only the resultant neutron production if it does occur. Also, these neutron energies are theoretical estimates for infinitely thin targets. For thick targets, neutrons will be produced in each case with a continuum from the maximum energy down to the minimum energy.

For an exothermic reaction, only one neutron energy is possible at each observation angle for a given incident particle energy, and neutrons may be produced for all incident energies. However, the probability that a neutron production reaction will occur is not equal for all incident energies — the Coulomb repulsion generally makes the probability vanishingly small for low ion beam energy. In endothermic reactions, the beam energy must exceed the threshold value E_1^{th}, as defined in Eq. (3), before the reaction becomes possible. If the incident beam energy exceeds E_1^{th} but is less than a second critical value E_1', as given in Eq. (4), then two neutron energies are possible for each emission angle — thus the meaning of the "±" in Eq. (2). This double-valued neutron energy phenomenon exists only for endothermic reactions, reflecting the fact that until the velocity of the neutron in a center-of-mass coordinate

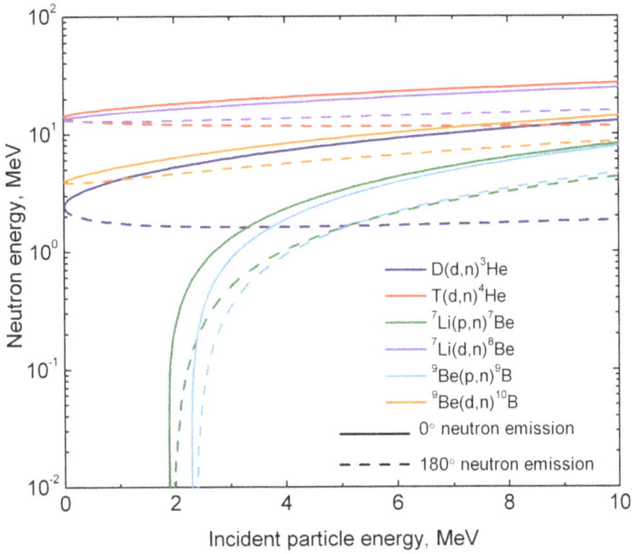

Fig. 2. Plot of the neutron energy produced from the reactions X(p/d,n)Y vs. beam (p or d) energy for emission at 0° and 180° with respect to the input beam direction.

system meets or exceeds the velocity of the coordinate system itself, forward going neutrons in the center of mass system may be observed in a backward direction in the laboratory system.[16] This effect is shown in Fig. 3 for the $^7Li(p,n)^7Be$ reaction.

$$E_1^{th} = \frac{-Q(M_3 + M_4)}{M_3 + M_4 - M_1}. \tag{3}$$

$$E_1' = \frac{-QM_4}{M_4 - M_1}. \tag{4}$$

The neutron yields from these reactions are based on their energy-dependent reaction probabilities as expressed in their reaction cross sections, σ. A plot of cross sections for several reactions is shown in Fig. 4. In practice, however, several other factors influence the neutron yields generated by accelerators. When an accelerated ion beam impacts the target, elastic scattering is much more likely to occur than nuclear capture. This causes the ions to lose energy as they traverse a target of appreciable thickness, which creates a continuum of energies at which fusion can take place from very low energies up to the incident beam energy. Known as the thick-target effect, this has a significant impact on the observed neutron yield versus beam energy.

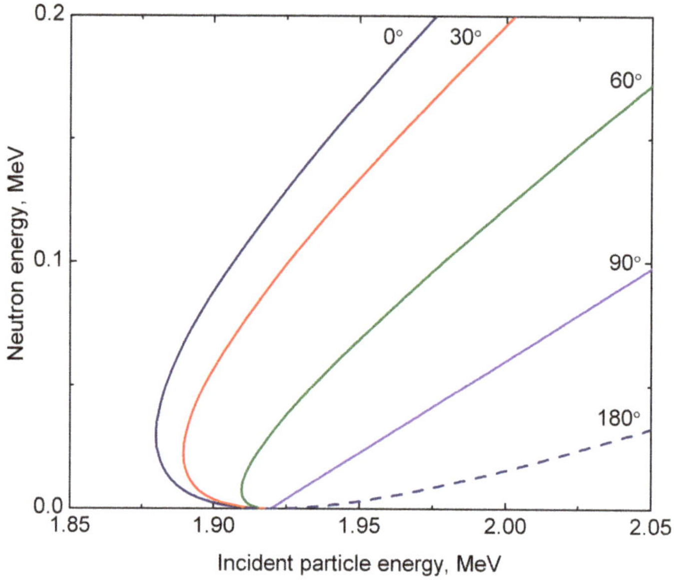

Fig. 3. The low-energy region of the $^{7}Li(p,n)^{7}Be$ reaction showing the double-valued nature of the emitted neutron energies for forward directed angles.

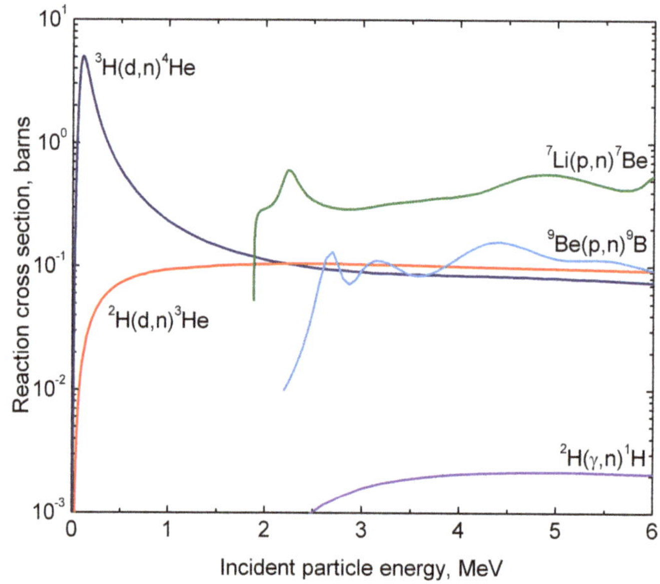

Fig. 4. Reaction cross sections for several accelerator-based neutron production reactions.[21]

Another factor influencing neutron yield versus beam energy is whether the accelerated beam consists of atoms, molecules, or some combination thereof. For example, accelerating both H and H_2 ions is possible using hydrogen gas in an ion source. In nearly all cases, some fraction of both species is present, with more atomic species present when hotter plasma sources are used (e.g., radiofrequency ion sources) and more molecular ion species present when cooler temperature ion sources are used (e.g., cold-cathode sources). For molecular ions, the interaction with the target may be approximated by assuming the ion splits into its constituents with each atom possessing its weight-fraction of the accelerating energy. Thus, if a H_2 molecule is accelerated to 100 keV and strikes a target, the neutron production can be approximated by two separate H atoms striking the target with a kinetic energy of 50 keV each. From Fig. 2, it is clear that two different neutron energies will be present at every emission angle, with the intensity depending upon the atomic/molecular ratio of the ion beam. A third factor that can influence the neutron yield is the target matrix and structure.[22] With deuterium and tritium reactions in compact ENGs, for example, it is common to use a hydriding metal occluder such as titanium, zirconium, or erbium that can store hydrogen within its lattice structure. These targets can have quite different yields as their densities can vary because they retain different concentrations of hydrogen isotopes depending on several factors, including preparation steps and target temperature.

Plots of thick-target neutron yield versus incident energy for the reactions listed in Table 1 plus the ^{9}Be photoneutron reaction are shown in Fig. 5. The DD and DT data are derived from Shope[22] and refer to a fully loaded titanium target, with deuterons incident on TiD_2 for the $^2H(d,n)^3He$ reaction and on TiT_2 for $^3H(d,n)^4He$. The other plots are taken from Hawkesworth.[23] Except for the $^9Be(\gamma,n)^8Be$ reaction, the solid plots refer to 100% atomic beams while the two dashed plots refer to 100% molecular ion beams.

2.1.1. $^2H + {}^2H \rightarrow {}^3He + n$

The DD fusion reaction is a straightforward and often used neutron production technique. The reaction is exothermic, and its cross section yields a useful neutron production rate with modest beam energies on the order of 100 keV and beam currents of $\sim$100 μA. Neutrons from this reaction start at $\sim$2.5 MeV. They are roughly monoenergetic at lower beam energies but less so at higher energies, as shown in Fig. 6. At higher energies, the $^2H(d,np)^2H$ breakup reaction, which occurs when the deuteron

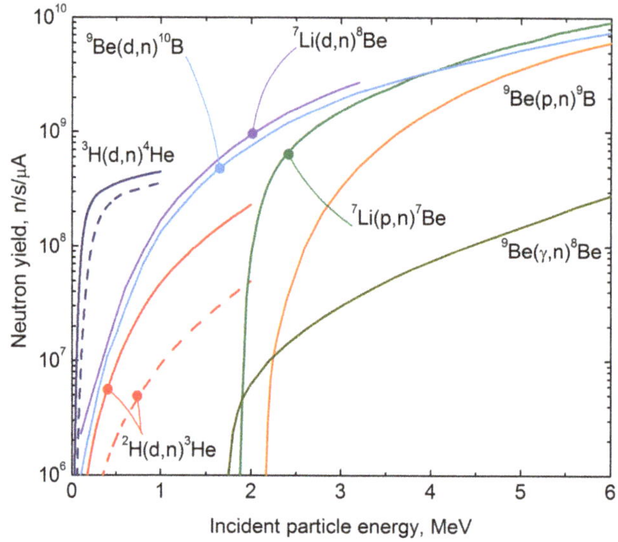

Fig. 5. Total neutron yield curves for several reactions as a function of beam energy.

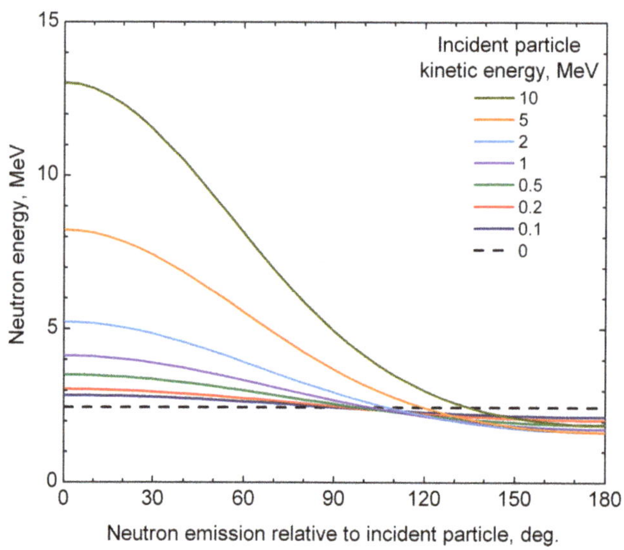

Fig. 6. Theoretical neutron energy spectra for the ^{2}H(d,n)^{3}He fusion reaction vs. emission angle for various incident beam energies.

energy exceeds 4.45 MeV, introduces a second energy group to the neutron spectrum. A further complication comes from the tritium made in the competing reaction ^{2}H(d, p)^{3}H, whose cross section is approximately the same as ^{2}H(d, n)^{3}He. This has been a problem in some high-yield, sealed-tube

ENGs because the build-up of tritium eventually leads to the introduction of higher energy neutron contamination from the DT reaction.[24]

The low Q value of the ^{2}H(d, n)^{3}He reaction results in a forward-directed anisotropic neutron yield at low beam energy, with the relative yield decreasing by a factor of four from 0° to 90° for a deuteron beam energy of 0.5 MeV.[19] The lower yield of this reaction versus DT fusion limits its use in many applications, but it is sometimes compensated by the lower energy of DD neutrons. This may be most advantageous for applications requiring thermal or epithermal neutrons.

2.1.2. $^2H + ^3H \rightarrow ^4He + n$

The DT fusion reaction is the one most often used for neutron production in compact ENGs. It has the highest neutron yield for low input beam energies (<1 MeV). As shown in Fig. 7, it produces roughly monoenergetic neutrons at beam energies of ~100 keV (a typical value for sealed-tube ENGs). The high Q value makes DT neutron sources a close approximation to isotopic sources at low beam energies.[19] However, in DT sealed-tube systems, the presence of both deuterium and tritium can introduce complexity to the neutron spectrum, and the use of molecular beam ion sources further complicates

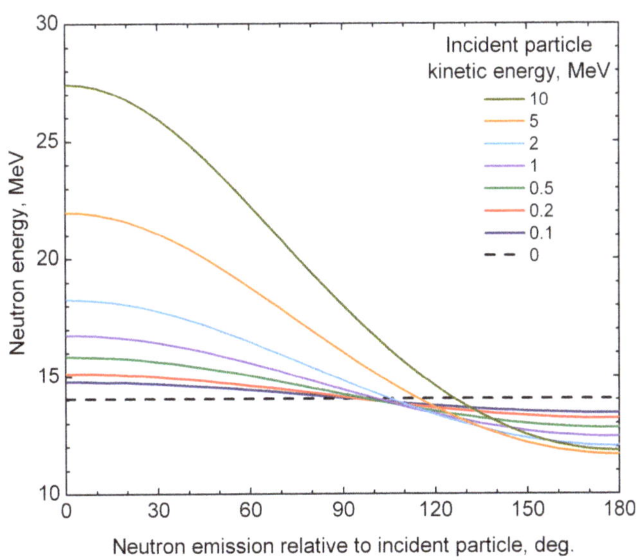

Fig. 7. Theoretical neutron energy spectra for the ^{3}H(d,n)^{4}He fusion reaction vs. emission angle for various incident beam energies.

this. As with the DD reaction, a breakup reaction adds a second neutron group to the output spectrum for deuteron energies exceeding 3.71 MeV.

Historically, the presence of tritium in accelerator systems led to challenges related to radiation safety and contamination control due to the potential for tritium leakage. While gas leakage can occur in sealed-tube systems, this has not proven to be a hurdle for commercialization of those systems and is no longer an issue in modern commercial sealed-tube ENGs. However, that is not the case for open-vacuum ENGs as will be discussed later.

2.1.3. $^1H + {}^7Li \rightarrow {}^7Be + n$

The forward-focused, relatively low-energy neutron spectrum produced by the ^{7}Li(p, n)^{7}Be reaction has motivated its use for some applications. As shown in Fig. 8, when the incident proton energy is only slightly higher than the reaction's threshold energy (1.88 MeV), neutrons are produced in a 60° forward emission cone with energies <0.1 MeV. This yields an improvement in efficiency over isotropic sources by allowing more neutrons to be directed towards the objective in interrogation applications. The forward-directed neutron beam also limits the back-angle radiation, which can reduce shielding requirements. The reaction produces neutrons of two energies at small

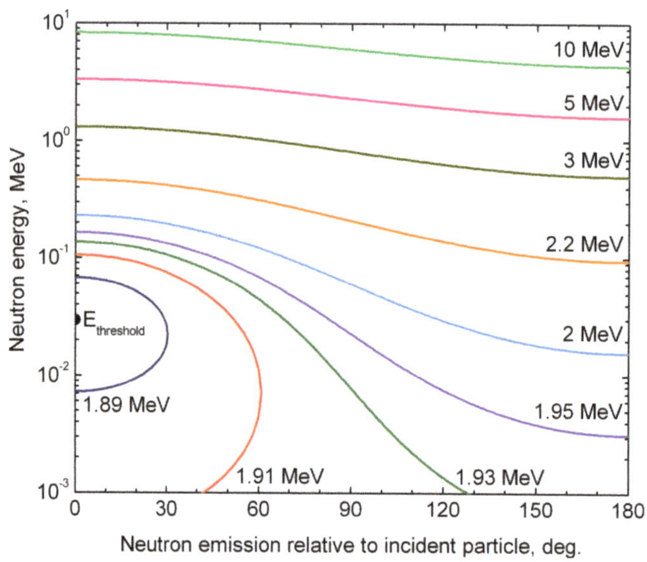

Fig. 8. Theoretical neutron energy spectra for the ^{7}Li(p,n)^{7}Be fusion reactions vs. emission angle for various incident beam energies.

angles, and as mentioned earlier, in thick targets a continuum of lower-energy neutrons is present due to ion beam energy losses in the target. A good presentation of the neutron spectra expected from this reaction can be found in the Ph.D. dissertation of C. L. Lee.[25]

The energy spectrum becomes more complicated when the incident proton energy exceeds 2.37 MeV. At this energy, the residual ^{7}Be nucleus can be left in its first excited state (0.429 MeV), and this results in a second, lower energy group of neutrons being generated as if the incident proton beam energy was lower by 0.429 MeV. Additional complexities occur as the energy is increased above 3.69 MeV due to the introduction of the ^{7}Li(p,n^{3}He)^{4}He reaction channel, and above 7.11 MeV with the opening of a channel for exciting the 4.57 MeV level of ^{7}Be. An engineering challenge associated with this reaction is to design a lithium target with sufficient cooling to prevent damage from the high power of the proton beam.[26]

2.1.4. $^2H + {}^7Li \rightarrow {}^8Be + n$

Another method for producing neutrons with a ^{7}Li target is to use the ^{7}Li(d,n)^{8}Be reaction. This is an exothermic reaction with a high Q value similar to the DT reaction. At lower deuteron beam energies, it produces a generally monoenergetic ($\pm$0.5 MeV) high energy neutron spectrum for all emission angles, as shown in Fig. 9. However, the cross section for the reaction increases less strongly with energy than the DT reaction, and deuteron energies up to 1 MeV are needed to approach comparable neutron yields (refer to Fig. 5). Nonetheless, when high energy neutrons are required, this reaction presents an attractive alternative to DT fusion as it avoids the use of tritium. However, the lithium target cooling challenges still remain a drawback.

2.1.5. $^1H + {}^9Be \rightarrow {}^9B + n$

Neutron production using ^{9}Be(p,n)^{9}B, an endothermic reaction with a threshold energy of 2.057 MeV, is similar to ^{7}Li(p,n)^{7}Be. When the proton beam energy is just above this threshold, the reaction is double-valued and emits neutrons in a forward direction as shown in Fig. 10. Even though slightly higher proton energies or larger beam currents are needed to achieve comparable neutron yields (see Fig. 5), beryllium's superior mechanical properties and higher melting temperature make it able to withstand higher beam power densities. It is easier to fabricate a target from beryllium and incorporate forced cooling if needed, making it more practical for routine

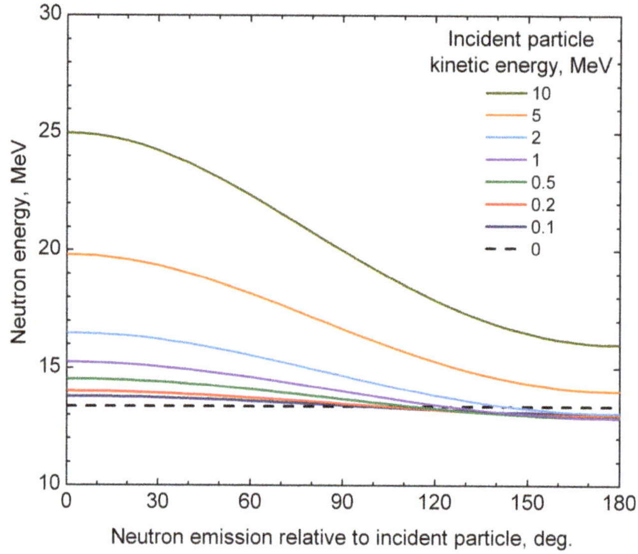

Fig. 9. Theoretical neutron energy spectra for the $^{7}Li(d,n)^{8}Be$ fusion reaction vs. emission angle for various incident beam energies.

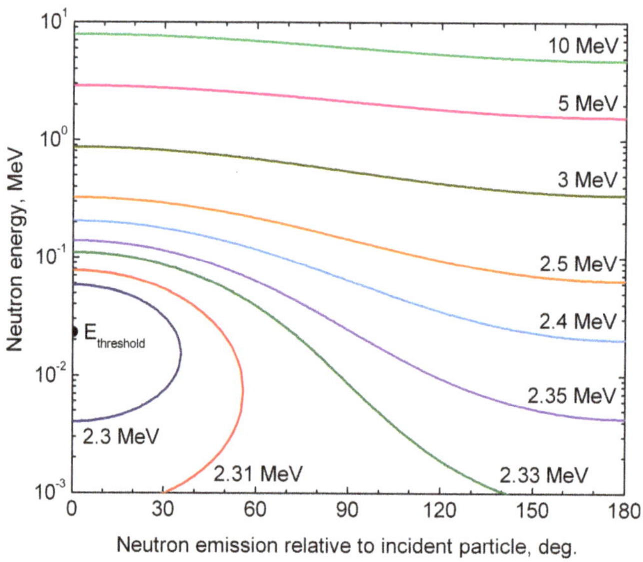

Fig. 10. Theoretical neutron energy spectra for the $^{9}Be(p,n)^{9}B$ fusion reaction vs. emission angle for various incident beam energies.

industrial applications. As with lithium, beam straggling (i.e., energy losses) in a beryllium target produces a continuum of neutron energies below the peak energy.

2.1.6. $^2H + {}^9Be \rightarrow {}^{10}B + n$

Rounding out the most commonly used neutron production reactions is ^{9}Be(d,n)^{10}B. Mirroring the similarities between lithium and beryllium reactions with protons, the energy dependence of the neutron yield from this exothermic reaction is similar to that of d + ^{7}Li. However, with a 4.361 MeV Q value (roughly one-third that of the lithium reaction), the neutron energy spectra start near 4 MeV as seen in Fig. 11. In some cases, the lower-energy neutron output, which is again a close approximation to a monoenergetic isotropic neutron source, is a desirable characteristic for applications requiring thermal or epithermal neutrons.

2.1.7. *Other two-body ion reactions*

In addition to the light ion reactions described above, there are many other two-body ion reactions that can be used to generate neutrons with accelerators. Table 2 lists a few of those that produce monoenergetic neutrons with varying energy characteristics and, depending upon beam–target choice, varying degrees of forward focusing. The first reaction uses a tritium target and is perhaps the most versatile in this category.[27] When used with thin gas targets, the ^{1}H + ^{3}H reaction generates near-monoenergetic neutrons over a wide energy range. However, the negative Q value removes it from

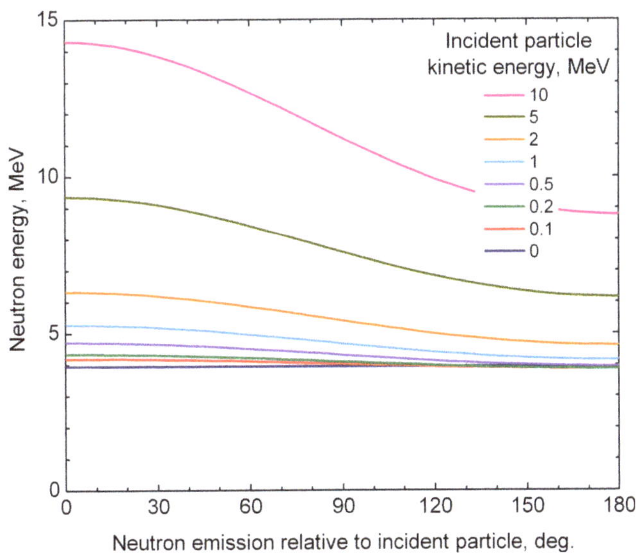

Fig. 11. Theoretical neutron energy spectra for the ^{9}Be(d,n)^{10}B fusion reaction vs. emission angle for various incident beam energies.

Table 2. Additional reactions potentially useful for neutron applications.[33,34]

Reaction	Q Value [MeV]	Threshold energy [MeV]
$^{1}H + {}^{3}H \rightarrow {}^{3}He + n$	−0.764	1.019
$^{3}H + {}^{3}H \rightarrow {}^{4}He + 2n$	11.332	NA
$^{1}H + {}^{6}Li \rightarrow {}^{6}Be + n$	−5.071	5.920
$^{2}H + {}^{6}Li \rightarrow {}^{7}Be + n$	3.381	NA
$^{1}H + {}^{10}B \rightarrow {}^{10}C + n$	−4.430	4.876
$^{2}H + {}^{10}B \rightarrow {}^{11}C + n$	6.465	NA
$^{1}H + {}^{11}B \rightarrow {}^{11}C + n$	−2.765	3.018
$^{2}H + {}^{11}B \rightarrow {}^{12}C + n$	13.732	NA
$^{1}H + {}^{12}C \rightarrow {}^{12}N + n$	−18.121	19.642
$^{2}H + {}^{12}C \rightarrow {}^{13}N + n$	−0.281	0.328
$^{1}H + {}^{13}C \rightarrow {}^{13}N + n$	−3.003	3.235
$^{2}H + {}^{13}C \rightarrow {}^{14}N + n$	5.326	NA
$^{1}H + {}^{36}Cl \rightarrow {}^{36}Ar + n$	−0.073	0.075
$^{1}H + {}^{51}V \rightarrow {}^{51}Cr + n$	−1.535	1.565

consideration in traditional low-voltage sealed-tube ENGs, and the need for a tritium handling infrastructure in an open-vacuum system adds an extra level of complexity not usually acceptable in industrial systems.

The reactions in this table, most notably the proton and deuteron reactions on boron, are all capable of generating useful neutron spectra.[28,29] In all cases, and in particular as the mass and complexity of the parent nuclei increase, care must be taken to understand the presence and impact of competing break-up reactions that complicate what would otherwise be monoenergetic spectra.[30] Also, as the target nuclei become heavier, high-energy γ-rays are produced along with the neutrons. A good example is the $^{2}H+{}^{11}B$ reaction in which 15.09 MeV photons are generated. These effects may be seen as benefits or hindrances depending upon the application.[31,32]

2.2. *Reactions using photons*

As an alternative to ion beams, high energy photons produced with electron accelerators can be used with some target nuclei for neutron production via photonuclear (γ, xn) or photofission (γ, f) reactions. These reactions occur when the incident photons have sufficient energy to overcome the neutron binding in the target nuclei. Table 3 lists the commonly used photoneutron targets and their neutron emission threshold energies, with beryllium and deuterium being the most widely employed.

Table 3. Commonly used target materials for producing neutrons with photons.

Target	Threshold energy [MeV]
^{2}H	2.22
^{6}Li	5.39
^{7}Li	7.25
^{9}Be*	1.66
^{210}Pb	5.18
^{235}U	5.31
^{238}U	5.08

*Multiple breakup reactions occur with the ^{9}Be photonuclear reaction.

The photon beams are most often produced by either linear or betatron accelerators. High energy electrons from the accelerator are directed towards a converter target (usually a high-Z material such as tungsten or tantalum), and as the electrons scatter and lose energy in the target, bremsstrahlung X-rays (or photons) are produced in a continuum spectrum extending up to the maximum energy of the electron beam. A suitable photoneutron target is then positioned in front of the convertor target. The neutrons are produced with a continuous energy spectrum, with the maximum energy being equal to the highest photon energy minus the threshold energy for the reaction.

The actual neutron yields from these reactions are dependent on the reaction cross sections and the specific parameters of the system, i.e., beam energy, beam current, X-ray converter target material and thickness, and neutron production target material and thickness. The cross sections for three of the reactions are shown in Fig. 12 for comparison. In one system, a thermal neutron flux of 1.23×10^{8} n/cm^{2}/mA/s was reported using a 5 MeV electron beam, a 1.698 g/cm^{2} thick tungsten converter, a 14 cm thick beryllium deuteride (BeD_2) neutron production target, and an elaborate neutron reflector.[35] In a second case, neutron production yields from LiD, Be, depleted U, and natural Pb targets were reported to be 5.2×10^{7}, 5.7×10^{7}, 4.7×10^{7}, and 8.0×10^{6} n/rad of beam dose, respectively, using a 10 MeV electron beam (the bremsstrahlung converter target was not defined).[36] Note that because of the dependence of the yield on the various parameters, reported photoneutron yields are often normalized to the photon beam dose rather than to the electron beam current.

One useful property of pulsed photoneutron sources is the inherent generation of broad-energy, "white spectrum" neutrons that can be used with

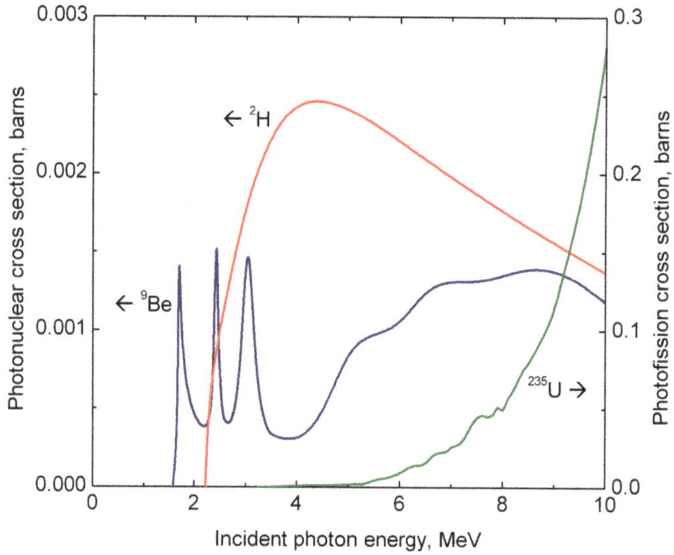

Fig. 12. Photonuclear neutron production cross sections for beryllium and deuterium and for the photofission cross section for ^{235}U.[37]

time-of-flight (TOF) techniques for energy-dependent measurements.[38–41] Accelerator systems designed for this purpose must be capable of generating pulses with widths of ~10 ns or less and typically incorporate beamlines with lengths of ~5 m longer. Neutrons are generated with a broad energy spectrum in a neutron production target, such as beryllium or deuterated water, during the accelerator pulse. The high energy neutrons leave the production target with different velocities v according to their kinetic energy E_n, as shown in the first half of Eq. (5) (neglecting relativistic effects). The velocity term in this equation can be rewritten as L/t, where L is the distance from the neutron production source to the detector (i.e., the length of the beamline) and t is the transit time from source to detector. Hence, the energy of neutrons measured in the neutron detector can be determined by recording the flight time using the accelerator beam pulse as the starting time, as shown in the second half of Eq. (5):

$$E_n = \frac{1}{2}mv^2 = \alpha^2\left(\frac{L}{t}\right)^2. \tag{5}$$

For an accelerator system with an infinitely short pulse width, the uncertainty associated with the timing measurements would be the primary factor impacting the energy resolution of the neutron measurements. Finite duration pulses further blur this energy resolution. Using 939.6 MeV/c^2 for the neutron rest mass m and 2.998×10^8 m/s for c, the constant α equals 72.3 $(\text{eV})^{1/2}(\mu\text{s/m})$. When the beam length is measured in meters and time is recorded in μs, the neutron energy is calculated in eV. Equation (6) gives an expression for the neutron energy resolution achievable in TOF measurements. In addition to minimizing timing uncertainties and reducing pulse widths, longer beamlines help to improve the precision of these measurements. The drawback is that the overall neutron intensity on target is diminished, which leads to longer measurement times.

$$\sigma_{\mathrm{E_n}} = 2\mathrm{E}_n \left(\frac{\sigma_{\mathrm{L}}^2}{\mathrm{L}^2} + \frac{\sigma_{\mathrm{t}}^2}{\mathrm{t}^2}\right)^{\frac{1}{2}} = 2\mathrm{E}_n \left(\frac{\sigma_{\mathrm{L}}^2}{\mathrm{L}^2} + \frac{\sigma_{\mathrm{t}}^2}{\alpha^2\mathrm{L}^2}\mathrm{E}_n\right)^{\frac{1}{2}}. \tag{6}$$

One possible disadvantage of photoneutron sources is the presence of high-energy photons. In situations requiring the measurement of a photon signature, this extraneous signal can degrade the measurement by reducing the signal to noise ratio or by rendering some of the neutron response signatures unusable. Further, the intense pulsed radiation fields present in these systems may temporarily disable (or blind) measurement equipment such as gas-filled counters, scintillators, and photomultiplier tubes.

3. Industrial Neutron Production Accelerators

Accelerators used to produce neutrons come in many different sizes and shapes. In laboratories and research environments, these systems often serve multiple purposes, generating both neutrons and ion beams (or even X-rays) for other research needs. In industrial settings, however, neutron production systems are often designed and built for a specific application. Considered in a broad context, these systems may be logically categorized in three classes: open-vacuum ion systems, sealed-vacuum (sealed-tube) ion systems, and X-ray systems. This section contains a brief review of the various types of accelerators in each of these classes and a discussion of other electronic systems that can produce neutrons and are often included in discussions of neutron sources.

3.1. *Open-vacuum systems*

The most familiar open-vacuum systems used to produce neutrons are large electrostatic (Cockcroft–Walton and Van de Graaff) systems, radio-frequency linear accelerators (RF linacs), and cyclotrons that employ the light-ion reactions of Table 1. Several accelerating arrangements based on these structures have been developed and deployed in commercial neutron production systems, and several references have been written that describe the scientific principles behind these devices as well as the engineering challenges involved in their design, construction, and operation.[42–44] Interested readers are encouraged to consult these references for more information on the underlying details of these systems.

While electrostatic and cyclotron systems have been used to generate neutrons, RF ion linacs based on the radio frequency quadrupole (RFQ) structure are the smallest and most versatile in the open-vacuum class. Several commercially produced RFQ neutron generators have been delivered in recent years for industrial neutron radiography and neutron activation applications ranging from nuclear material and mineral assay to cargo/luggage inspection studies for homeland security.[45] One system in particular has been operating for more than 12 years for the automated assay of spent nuclear fuel in a harsh industrial environment.[46]

At the present time, only a few open-vacuum ENGs are commercially sold each year for industrial applications; however, potential widespread deployment for security applications could result in an increase in that number. Prices for these systems can range from ∼US$500 k to several million, depending on the neutron output flux and beam characteristics required for the specific application.

These systems, similar to those commonly found in low-energy physics laboratories, include (1) an ion source to generate charged particles, (2) an extraction and acceleration section to remove ions from the source and accelerate them to the energies needed for the specific nuclear reaction chosen, and (3) a neutron production target. These three components are the key parts of both open-vacuum and sealed-vacuum neutron generators. However, as the name implies, open-vacuum systems require an active vacuum system that can vent to the environment. An active cooling system is also generally required due to the higher beam energy and larger size of the accelerator. The main purpose of the vacuum system is to keep the ion extraction and acceleration system at a low background pressure. The ion beam is normally generated by injecting hydrogen or deuterium gas into the ion source, which typically operates at pressures in the 1 Pa (∼10 mTorr) range or higher.[47] The accelerator section is isolated from the ion source by one or more small orifices in the ion extraction assembly that restrict the leakage of gas into the accelerator. Any gas that escapes from the ion source must be pumped by the vacuum system in order to minimize corona, direct electrical discharges (sparking), and beam losses from collisions with gas molecules in the accelerating section.

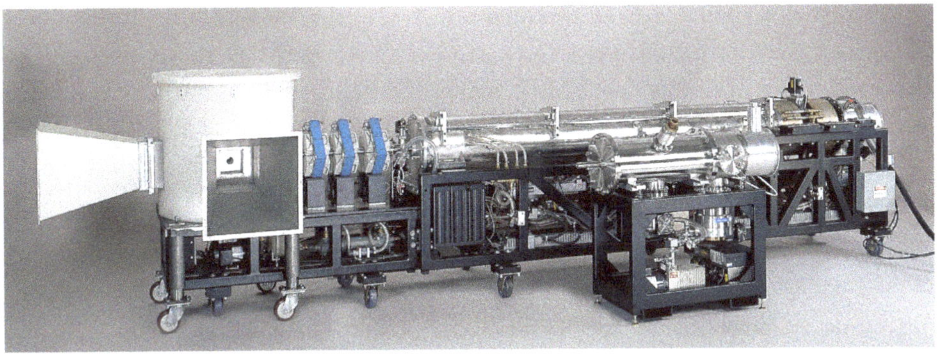

Fig. 13. Photograph of two RFQ linac neutron generator systems (photo courtesy of AccSys Technology, Inc.).

Open-vacuum systems are typically employed when ion energies in the MeV range are needed to produce a higher neutron yield or a neutron spectrum with specific characteristics. To achieve these energies, RFQ accelerators $\sim$1–4 m in length are common. Van de Graaff and Pelletron type accelerators are also used. To demonstrate their versatility in size and neutron production capability, Fig. 13 shows two RFQ neutron generators. The larger system in the background is a $\sim$7 m long, 4 MeV proton linac (shown with a moderator assembly) designed to produce a thermal neutron flux of 1×10^6 n/s/cm^2 at 2 m with a ^{7}Be target for neutron radiography. The smaller system in the foreground is a $\sim$1 m long, 1 MeV deuteron linac (shown with a short beamline and ^{7}Be neutron target assembly) that can produce a total neutron yield of 1×10^{10} n/s.

One advantage of open-vacuum linacs is their flexibility for service and maintenance. They can also be reconfigured for alternate missions through the use of a different gas in the ion source or a different target. RFQ linacs can generate up to 1 mA of ion beam currents and can produce neutron yields ranging from 10^8 to 10^{13} n/s, depending on the beam and target combination. Dynamitrons can accelerate slightly higher beam currents and can produce up to 10^{14} n/s in the largest systems. Most RFQ-based systems use water-cooled ^{7}Be targets that are rugged enough to be deployed in harsh environments and have much longer life-times than targets used in conventional sealed-tube ENGs. Tritium beams or targets are rarely used in open-vacuum systems due to the possibility that tritium gas could be accidentally vented to the environment. The health and safety hazards associated with tritium contamination have imposed too steep a barrier for its widespread use in open-vacuum ENGs for industrial applications. Only a few such systems have been deployed, mostly in research facilities.[48,49]

The major disadvantage of most open-vacuum systems in applications requiring portability is their size and supporting infrastructure needs. However, a small prototype 1 MeV proton RFQ neutron generator has been packaged and operated in two man-transportable trunk-sized containers.[50] In addition, 7 MeV RF ion linacs are now being installed and operated in trucks for medical radioisotope production, but these could also be used to generate neutrons.[51]

3.2. *Sealed-tube systems*

While open-vacuum accelerators are used in many research laboratories and in some select industrial settings, sealed-tube systems are found far more often in present day industrial applications. These systems are usually less complicated than their larger open-vacuum cousins and, in their simplest forms, share many technological similarities with light bulbs, vacuum tubes, and switch tubes.[52–54] Sealed-tube ENGs are also generally limited to accelerating voltages of less than 300 kV and therefore must generate neutrons via the DD or DT fusion reactions. However, the neutron yields are sufficient for most industrial applications, and for systems operating in the 100-kV range, their small size and low power requirements make them ideal for field applications.

Sealed-tube ENGs have four main subsystems: (1) a vacuum-sealed accelerator system ("neutron tube"), (2) a high-voltage power supply to generate the accelerating potential in the tube, (3) a high-voltage accelerator housing, and (4) a control console for operation. Sometimes the neutron tube and high-voltage power supply are collocated within the accelerator housing; sometimes only the neutron tube is in the accelerator housing and the high-voltage power supply is an external assembly connected to the housing and tube by special-purpose high-voltage cables.[55] The accelerator housing helps to protect the tube from external physical, electrical, and magnetic interferences. The housing also often contains a high-voltage insulating medium such as oil, liquid fluorocarbon, or sulfur hexafluoride gas to allow operation at voltages higher than could otherwise be sustained in open air.

A typical commercial neutron tube, the Zetatron, is shown in Fig. 14. All neutron tubes contain the main components shown in the schematic — a vacuum envelope (made of a combination of metal, ceramic, and/or glass components sealed together), an ion source, an ion acceleration section, and a target. An electron suppression device or Faraday cage (usually a metallic ring that is maintained at a negative bias of a few kV with respect to the target) is often included. Its purpose is to deflect electrons produced in the

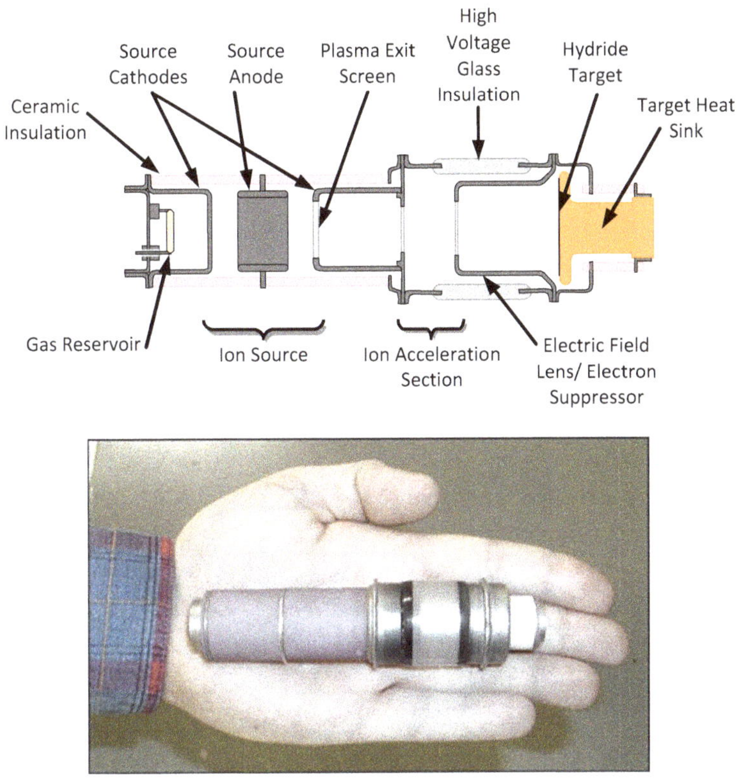

Fig. 14. Schematic cross section (top) and photo (bottom) of a Zetatron sealed neutron tube with a cold-cathode ion source, designed by Sandia National Laboratories and sold for commercial applications by Thermo Scientific, Colorado Springs, CO. (Schematic adapted from Ref. 56.)

target by ion bombardment and prevent them from reaching the positive electrodes or the vacuum envelope.

Another component frequently used is a gas reservoir, or "getter", consisting of a hydriding metal such as zirconium connected to a thin-wire heating element. Deuterium and/or tritium is stored in the reservoir in solid form when not in use and released into the tube by heating during operation. The electrical current that heats the reservoir is often used in a feedback loop that monitors the ion beam current on the target and adjusts it to maintain a constant beam current.

3.2.1. *History*

Reviewing early design concepts and acknowledging the adage that form follows function, it is interesting to note that the earliest generation of sealed-tube ENGs incorporated glass vacuum envelopes, targets, electron suppressors, and support components and had the same general size, shape, and layout still found in many of today's commercial cold-cathode ion source neutron tubes.[57–61] The first patent for the concept of a semi-sealed-tube ENG was filed in 1938 by Penning (see Ref. 9). This was followed by W. W. Salisbury of the Collins Radio Company in 1947.[62] It is unclear, however, if either of these designs was ever manufactured for commercial sale. The first patent for a sealed-tube ENG was filed by R. E. Fearon and J. M. Thayer of Well Surveys, Inc. in 1949; however, this design did not include a gas pressure regulator to control the system's operation.[63] Subsequent patents by Schlumberger, Dresser Industries (now Baker Atlas), Well Surveys (later affiliated with Gearhart Industries, now a part of Halliburton), and Mobil Oil Corporation further advanced the well-logging ENG concept.[64–68]

Other early sealed-vacuum ENG designs using deuterium beams with deuterated or tritiated targets, but not specifically intended for use in the oilfield industry, were reported in the early 1960s by Lawrence Radiation Laboratory, General Electric Research Laboratory, Philips Research Laboratories, Services Electronics Research Laboratory (SERL), and Kaman Corporation.[69–73] Philips led the efforts to commercialize this technology for laboratory use and soon thereafter developed laboratory-scale instruments for research.[74,75] Interest in the technology quickly grew from that point, and General Electric soon afterwards developed an ENG to support scientific measurements on lunar space missions.[76]

During the decades that followed, many entries were made into commercial ENG manufacturing for use in the oilfield industry and other areas, including the development of products by Efremov Institute (USSR, no longer producing ENGs), Kaman Nuclear (now Thermo Fisher Scientific, USA), Marconi Avionics (UK, no longer producing ENGs), Norelco Philips (now SODERN, France), SAMES (France, no longer producing ENGs), SERL (UK, no longer producing ENGs), Texas Nuclear (USA, no longer producing ENGs), and VNIIA (All-Russian Research Institute of Automatics, Russia). Descriptions of many of the ENGs are given in an International Atomic Energy Agency (IAEA) technical document.[77] In most cases, these manufacturers exclusively served either the oilfield services market or the "other-than-oilfield" markets. A timeline summarizing the evolution of sealed-tube neutron generators is given in Fig. 15.

As of today, over fifty years of market forces have transformed the commercial sealed-tube ENG market through consolidation, old application market deterioration, new market development, and the expansion of former

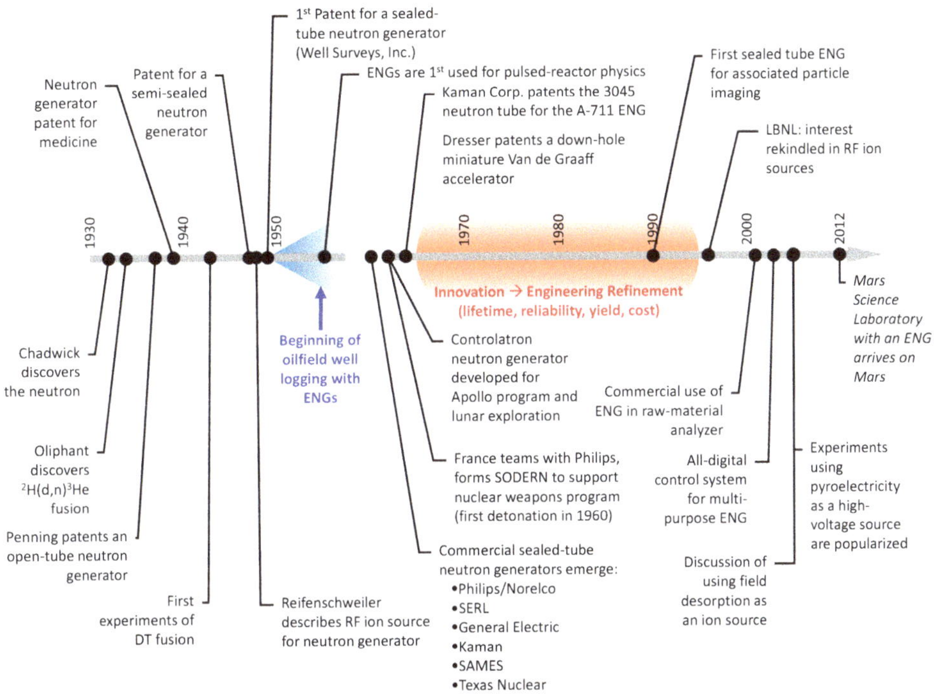

Fig. 15. Timeline for the evolution of sealed-tube neutron generators.

military ENG manufacturing facilities to civilian international sales. It is estimated that the current commercial manufacturers, including oil service companies, produce more than 200 new units annually. Prices can range from \$100 k to more than \$500 k, depending on the neutron output flux and type of system. In most cases, the DT fusion reaction is chosen over DD since higher energy neutrons are made with DT fusion (14.1 MeV versus 2.5 MeV from DD) and the maximum achievable neutron yield using both deuterium and tritium versus deuterium alone is 50–100 times greater.

3.2.2. *Industrial sealed-tube ENGs*

Commercial sealed-tube neutron generators can be classified in six categories as listed in Table 4. The LWD-ENGs are the latest category and perhaps the most technically sophisticated. They are designed to operate within the mechanism of a directional oilfield drill to provide real-time or near-real-time data on the geological formation in the vicinity of the drill bit. This information is used to help direct the drilling process and document the geological characteristics of the rock formations around the drilled hole.

Table 4. General categories of commercial sealed-tube electronic neutron generators.

Category	Primary applications	Performance characteristics
Logging-While-Drilling (LWD)	Oilfield drilling measurements	Extreme ruggedness and small diameter
Wireline (W)	Oilfield well surveys	Extreme ruggedness and small diameter
Portable (P)	Security, science	Light weight ($\sim$10 – 15 kg), small overall size, and operational flexibility
General Purpose (GP)	Bulk material analysis, security, medicine, general science	Operational flexibility, no active cooling
Active Cooling (AC)	Bulk and trace material analysis, security, radiation effects testing, general science	Higher neutron yield
Associated Particle (AP)	Security	AP detector performance, beam spot size (imaging)

Due to the very recent development of this technology and the competitive nature of the oilfield services industry, little public knowledge is available about these devices, but some general information can be found from the Schlumberger, Halliburton, and Baker Atlas websites. General descriptions of the technology related to W-ENGs can be found in nuclear science journals and journals dealing with wireline data analysis, but detailed descriptions remain elusive for the same reasons as with LWD-ENGs.[78–80] These instruments are highly-engineered devices capable of operating in pulsed modes under extreme conditions of shock and vibration and at temperatures often exceeding 250°C. Maximum neutron yields are $1 - 3 \times 10^8$ n/s (for DT), and typical operating lifetimes range from 500 to 1500 hours. Both of these parameters have improved 3- to 5-fold within the last decade.

Portable ENGs are another recent addition to the sealed-tube family. Technical advances in control systems and high-voltage systems made possible through the use of digital electronics have resulted in significant size reductions over previous generations that used analog electronics.[81] P-ENGs are optimized to balance neutronic performance with portability, with units available today weighing approximately 11–12 kg. They can be integrated into battery-powered systems using high-performance Li-ion batteries and small computers with only a marginal weight addition on the order of 5 kg.[55]

For applications requiring pulsed operation, they are capable of operating over a wide range of parameters with continuous pulse rates up to 20000 Hz, pulse duty cycles from 0–100%, and neutron yields from 0.1×10^8 to 2×10^8 n/s (DT). Development of these systems has been driven by customers in the security field, but their compact size, convenient use, operational versatility, and lower cost have served to expand their use into other applications for educational instruction and material analysis.

General purpose ENGs, as the name implies, are used in a multitude of application areas, including industrial bulk materials analyses, medicine, security, and general science research, by a wide range of users, including private companies, hospitals, universities, and national laboratories. GP-ENGs share operational flexibility with P-ENGs but are generally designed to optimize neutronic performance rather than achieve a lower weight.[81–84] In practice, this optimization results in systems with pulsing capabilities similar to portable systems but with maximum neutron yields three to five times higher, approaching 1×10^9 n/s using DT fusion. This value is important because it is at this level that a general transition is made in present generation ENG designs from passive cooling to the need for active cooling (for the target and often the ion source). Since active cooling requires pumps, heat exchangers, and additional control systems, the higher yield AC-ENGs are more complicated, more expensive, and typically two times larger than their GP-ENG counterparts.

AC-ENGs have been developed using a variety of high current ion sources including cold-cathode, hot-cathode, arc switch, and RF excitation types. Besides generally requiring active cooling, they also require larger, higher performance power supplies for higher beam currents and voltages. While other categories of ENGs utilize low current versions of the cold- and hot-cathode sources, nearly all ENGs that use RF ion sources fall into the AC category due to their inherently higher beam currents. Although the GP-ENG neutron range is generally bounded at the 1×10^9 level (DT), most AC-ENGs begin at the 1×10^{10} level (DT), probably due to market-driven demands that the increased cost and complexity associated with active cooling be accompanied by significantly higher performance. AC-ENGs are primarily aimed at applications in radiation effects testing and activation analysis and are not typically designed for pulsed operation. Some very-high-yield AC-ENGs have been developed, with neutron yields in some instruments exceeding 1×10^{13} n/s for short time periods. In the past, it was not uncommon for these very high neutron yield AC-ENGs to release tritium into the local environment during operation.[85,86] However, environmental standards regarding the release of radiation are much more stringent today, and current AC-ENGs operating in the 10^{10} range do not exhibit this behavior.

Interest in neutron generators that incorporate detectors to measure the recoil helium atoms associated with the fusion process is not new but has grown over the past decade, primarily in relation to security programs.[87–90] In these AP-ENGs, an associated particle (AP) detector — either a scintillator or solid-state detector — provides both timing (with ns time resolution) and directional information on when the neutrons are created and what their paths from the target are. In some cases, the directional information is used to provide a general indication or cone-beam understanding of the neutron flight path.[91–95] In other cases, AP detectors with high spatial resolution (in systems with small ion beam spot sizes) are used to calculate high-resolution neutron trajectories.[96,97] In low-resolution operation, the AP-ENG can be used to improve the signal-to-background ratio in a measured signature; for example, prompt γ-rays from a region of interest can be discriminated from other γ-rays originating behind the ENG. In high-resolution operation, AP-ENGs can be used to produce 2D and 3D images using γ-ray or neutron signatures from objects and topographic data reconstruction tools. AP-ENGs usually operate in a continuous mode during AP data acquisition although some devices have been built to operate in a pulsed mode. Typical neutron yields are in the 10^7–10^8 n/s range (using DT). In low-resolution systems, the beam spots may be on the order of 1 cm^2, but in high-resolution systems beam spots must be 1–3 mm^2.

3.2.3. *Long-lived DD neutron generator design considerations*

Occasionally, laboratory-scale testing with the DD fusion reaction in an open-vacuum accelerator system is used to extrapolate performance expectations for both the DD and DT reactions in sealed-tube systems — a practice that can be misleading. First, while the engineering differences between open-vacuum and sealed-tube systems may seem superficially obvious, there are numerous practical challenges associated with the production of sealed-tube systems intended for long term use and storage. Subtle design considerations must be taken into account when developing long-lived vacuum systems, including special methods for preparing and cleaning parts to avoid slow outgassing, techniques to avoid virtual leaks, and bonding methods to withstand high processing temperatures and harsh conditions of shock and vibration.[52] Second, as discussed earlier, in addition to generating neutrons, high-yield DD ENGs also produce tritium. Indeed, the reaction cross section at 100 kV for ^{2}H(d, p)^{3}H is 94% of that for ^{2}H(d, n)^{4}He as shown in Fig. 16.

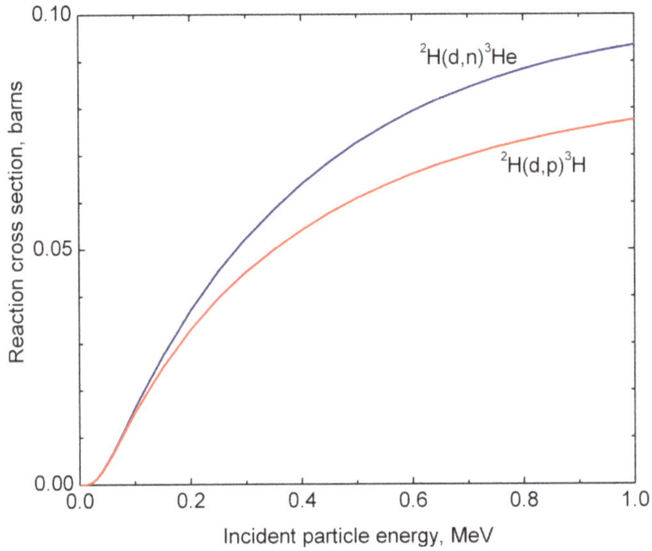

Fig. 16. A comparison of the ^{2}H(d,p)^{3}H and ^{2}H(d,n)^{4}He reaction cross sections.

The tritium contamination is dilute with respect to the initial deuterium loading but may reach non-trivial levels over time. For example, in a DD system producing 10^9 n/s, approximately 3.4×10^{14} tritium atoms (0.16 mCi) will be produced in the target after 1000 hours of operation. Although widely different target conditions exist in different neutron tubes, this tritium content might reasonably equate to a 0.1% tritium contamination that could contribute up to an extra 10% to a generator's output considering the higher yield from DT versus DD fusion.[24] Besides changing the expected yield, tritium contamination would also alter the emitted neutron spectrum making it "harder" (i.e., higher energy) than expected. Hence, tritium accumulation must be considered in sealed-tube DD ENGs designed for long lifetimes.

3.3. *Photoneutron systems*

Accelerator-based photoneutron sources have not yet achieved wide-scale industrial use.[35,98–100] Moderate energy (5–10 MeV) electron linacs and small-scale betatrons are most commonly proposed for these applications, but higher energy electron accelerators can also be used. The intense photon radiation accompanying neutron production in higher energy systems makes radiation shielding a challenge while also limiting their applicability when radiation measurements must be made. For radiation effects testing, these higher energy photoneutron sources serve a valuable role, but care must be taken to correctly assess dosimetric quantities.[101] Accelerator-based photoneutron sources have also been proposed for use in industrial-scale nuclear TOF analytical systems.[102,103] Beyond these few applications, however, few concrete descriptions exist of industrial-scale accelerator photoneutron systems.

3.4. *Other electronic neutron sources*

Two other electronic instruments that generate neutrons will be briefly discussed for completeness — the dense plasma focus (DPF) neutron source and the inertial-electrostatic confinement (IEC) neutron source. Strictly speaking, neither is a particle accelerator. However, they are often described in discussions of ENGs and hence warrant some mention here. Although neither has yet found practical application for industrial purposes, each possesses unique operating features that may someday lead to their routine use in areas where accelerator neutron sources are used today.

DPF neutron sources are pulsed-power instruments that discharge intense electrical currents to produce gaseous plasmas.[104–106] Deuterium is most commonly used, and the neutrons are generated from DD fusion occurring in the plasma. The formation of the plasma by electrical discharge is a highly energetic process that historically has resulted in substantial damage to the plasma generating electrodes. Cumulative damage from repetitive pulses leads to erosion of the electrodes, which eventually requires maintenance and refurbishment. In early generation systems, low operating lifetimes were not uncommon (from a few pulses, or "shots", to a few hundred pulses); however, more recent advances have led to systems capable of $\sim 10^8$ shots without maintenance. Neutron yields from DPF DD fusion sources have been reported over a wide range from 10^8 to 10^{11} n/shot, but the operating lifetime of higher yield systems is significantly reduced. Reported pulse rates for these inherently pulsed sources range from less than 1 Hz to several hundred Hz. DPF neutron sources require a substantial investment in equipment, including large capacitors, high-power switching systems, and occasionally water-cooling support. A unique attribute of these systems is that they generate very short neutron pulses, on the order of 10 ns. For short pulses of a few nanoseconds, it may be possible to use these sources for some TOF measurements.

IEC sources utilize hollow (usually spherical) electrode grids placed in the center of a vacuum chamber filled with deuterium gas to generate a low-density deuterium plasma by electrostatic discharge.[107–109] Fusion occurs when deuterium ions are accelerated toward the center region and collide during multiple transits through the plasma. This light-ion fusion is the result of the same reaction mechanisms that occur in accelerator-based sources. An often-cited advantage of IEC neutron sources over sealed-tube ENGs is that IEC sources do not require a metal hydride target. Because they do not use solid targets that erode with ion bombardment, long operating lifetimes on the order of 10^4 hours have been postulated for IEC sources. Reported neutron yields from IEC DD fusion sources cover a wide range from 10^5–10^{10} n/s. IEC sources are most often operated in a continuous mode, but pulsed versions have been proposed.[110,111] While most IEC sources use spherical electrodes and chambers, recent cylindrical designs may present new possibilities in some applications such as in-line conveyor belt analysis of coal or other minerals. IEC neutron sources have higher input power requirements than sealed-tube ENGs of comparable neutron yield (~1000 W versus ~100 W for a DT equivalent yield of 10^9 n/s). While DPF sources are only laboratory devices today, IEC sources are available commercially.

4. Industrial Applications

Within a decade after the neutron's discovery, patents were being filed for accelerator-based neutron sources, and hardware was being developed for medical applications. Following a nearly complete pause during World War II, diversified scientific and commercial interest in the technology quickly resumed in the 1950s and 1960s. Also during this period, the supporting technologies necessary for industrial-grade sealed-tube accelerators were perfected, resulting in the development of high-yield commercial systems for general-purpose applications and miniature systems for oilfield applications.

Evolutionary development continued in the 1970s through the 1990s with slow but steady improvements in manufacturing techniques and commensurately slow but steadily increasing improvements in performance and reliability. During this period, the total number of users of high-yield neutron generator systems actually decreased, but the industrial use of ENGs in oilfield applications grew. Going into the 21st century, decades of advances in digital electronics finally began to impact both open-vacuum and sealed-tube ENGs with the incorporation of digital systems and computer control. Concurrently, more sophisticated scientific modeling and simulation software led to a better understanding of the physics of particle accelerators and neutron generation. These advances resulted in the development of the higher performance, more efficient accelerator systems that are now commonly used in industrial applications. Applications in sub-surface geophysical exploration remain the single largest industrial use of ENGs, but other notable areas exist in the medical and security fields.

4.1. *Neutron interactions*

The neutron's high sensitivity to low-Z materials, good penetrability in mid-Z and high-Z materials, and exceptional sensitivity to isotopic species make it a versatile and powerful probe for analyzing materials.[16–19] The reason for its use in commercial applications is to exploit one or more of these properties. In some cases, the goal is simply to create an interaction in order to observe the effects, as in radiation hardness and effects testing of electronics. However, in most cases, the goal is to induce a reaction in a test material or sample and then observe reaction signatures to infer one or more physical properties of the subject material. These include, but are not limited to, isotopic composition, material mass, material thickness, material density, material geometry, and temperature.

The basic interactions of neutrons with matter — elastic scattering, inelastic scattering, and absorption — and their measurable signatures are well understood. Table 5 illustrates the use of these interactions and signatures to identify materials and their properties. In many applications, γ-rays produced from inelastic scattering or absorption (i.e., capture) are measured to infer information about the chemical composition of the material under interrogation. These γ-rays not only carry useful information about the isotopes from which they originated, but they can also provide information about the energy of the incident neutrons when reaction thresholds and resonances are considered. In measurements using pulsed neutron generators, time correlations between source pulses and reaction product γ-rays can provide information about the bulk nature of the surrounding media, including the density and average Z of nearby material. Secondary neutron detection also plays an important role in many applications, particularly those involving the detection, identification, or characterization of fissionable material. Time correlations between source and reaction product neutrons provide a highly sensitive signature for these materials. Time correlations among induced-fission neutrons (prompt and delayed) can also provide high-sensitivity signatures for identifying fissionable materials.

4.2. *Geophysical exploration*

Early uses of neutron generators for oilfield well logging were reported in the mid-1950s by Well Surveys, Inc. (Tulsa, OK), the McCullough Tool Company (Los Angeles, CA), and Schlumberger Well Surveying

Corp. (Houston, TX and Ridgefield, CT).[63,64,112] Since that time, scores of patents have been filed and research papers published on advances in the underlying technology and the oilfield applications that use neutron generators.[79,113–117] Common characteristics of these devices are (1) their small size, in order to fit within small diameter well holes from $< 5\,\text{cm}$ to $20\,\text{cm}$; (2) their simplicity of use and low power consumption, in order to operate up to several km below the Earth's surface and far from external control systems and power supplies; and (3) their extreme ruggedness and durability to withstand prolonged exposure to high temperatures and pressures, shock, vibration, and corrosive environments. These well-logging tools utilize sealed-tube DT fusion ENGs capable of generating $\sim 10^8$ n/s in pulsed

Table 5. Signatures of neutron interactions with matter and their use in materials identification.

Interaction	Signatures used to identify materials and their properties
Elastic Scattering	• Measuring the energy and direction of recoiling particles (e.g., inside a detector) gives the initial energy and direction of source neutrons, and with a well-characterized neutron source, the isotopic compositions and distributions within the sample. • Measuring the energy of scattered neutrons (in a TOF system) at one or more angles gives the isotopic composition of the sample, general physical parameters of the scattering environment, and possibly the physical distribution(s) in the sample of the scattering isotope(s). • Measuring the transmitted intensity (attenuation) or 2D transmitted distribution of neutrons (radiography) can infer the density and distribution of nuclei in the sample; computed tomography (CT) techniques, including rotation or irradiation from multiple angles, can generate 3D distribution data. (Note: this distribution is affected by all three interactions.)
Inelastic Scattering	• Making the same types of measurements as with elastic scattering gives similar information about the sample material. • Measuring the energy-dependent transmission of neutrons can identify and characterize unknown materials by comparing to known resonant characteristics of nuclei. • Measuring the γ-rays from the decay of excited nuclei, possibly together with neutron energy information, can infer the isotopic composition of the sample and general physical parameters of the scattering environment; using collimation and/or time correlation techniques can infer 2D and/or 3D isotopic distributions.

(*Continued*)

Table 5. (*Continued*)

Interaction	Signatures used to identify materials and their properties
Absorption	• Using the attenuation, radiography, and CT techniques described above gives similar information about the distribution of nuclei in the sample. • Measuring the energy-dependent transmission of neutrons can identify and characterize unknown materials by comparing to known resonant characteristics of nuclei. • Measuring the γ-rays or other particles emitted from nuclei following neutron absorption, or using products from the decay of extremely short-lived compound nuclei (prompt emissions) or from the decay of radioactive product nuclei, can yield information about the stoichiometry and absolute mass of isotopes in an analysis region. • Measuring the complex prompt products of nuclear fission (neutrons, photons, fission products, or other charged particles), analyzing the angular distribution and time correlations (including coincidences and higher-order multiplicity) of those emissions, and/or measuring the β-delayed neutron and photon emissions of fission products can identify fissionable materials.

operation ranging from 1000 to 20 000 Hz with duty cycles of ~10%. A photograph of a typical wireline well-logging tool with an internal sealed-tube generator is shown in Fig. 17. Such tools can be as long as 6.1 m (20 ft). It is estimated that 1000 (possibly more) of these devices are currently in use worldwide. In the simplest implementation, these ENGs may be used to replace radioisotope neutron sources such as ^{239}PuBe, ^{241}AmBe, or ^{252}Cf, operating in a continuous mode with one or more detectors. (In oilfield applications, radioisotope neutron sources are often referred to as chemical neutron sources.)

The primary challenge in well-logging measurements is to collect information about the properties of the rock formations near the well bore, taking into account the effects of the neutron generator, well structure, and borehole fluids. The most common approach is to use multiple detectors at different distances from the neutron generator. These detectors monitor thermal neutron flux at different distances above or below the neutron source in the borehole. Detectors close to the generator provide data related to the neutron moderating characteristics near the borehole while detectors further away provide more information about the bulk properties of the nearby rock, typically out to about 30 cm. Using detectors at different

Fig. 17. Oilfield workers preparing a small-diameter well-logging tool containing an internal ENG (suspended from the crane behind the truck). (Photograph courtesy of Schlumberger).

spacing, these tools are also able to eliminate the effects of a borehole casing (pipe). The thermal neutron intensity provides an often qualitative measure of moderating characteristics and, more specifically, hydrogen content (water or oil) in the rock. These measurements are considered a gauge of rock porosity. Ratios are taken of the data from the different detectors as a function of depth in the well to make the analysis more easily interpreted; these types of measurements are referred to as compensated neutron porosity logs. A schematic representation of a multiple-detector well-logging tool, the Schlumberger Accelerator Porosity Sonde, indicating the orientation of the ENG and five detectors, is shown in Fig. 18.

A second important measurement in neutron well logging is determining the rock formation's macroscopic neutron capture cross section, sigma (Σ). This is done using a pulsed neutron generator by measuring the decay rate of the thermal flux signature in between pulses, usually with a photon detector. This can also be used to determine, in combination with other data, water saturation in the rock for high-salinity environments. Another

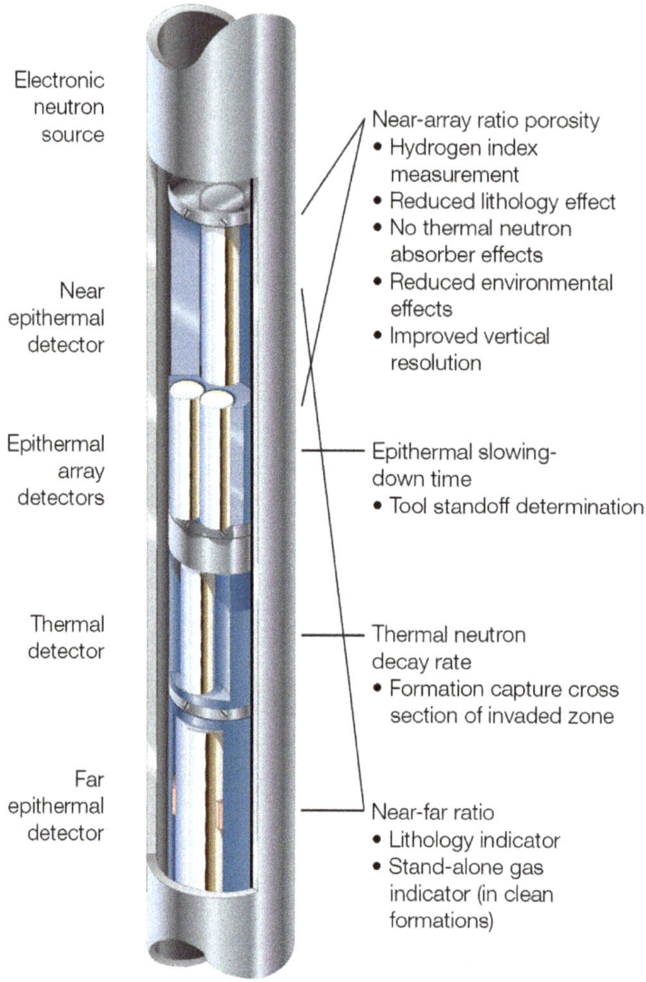

Fig. 18. Schematic representation of a five-detector well-logging tool using an ENG for assessing rock porosity. (Courtesy of Schlumberger.)

important measurement using pulsed neutrons is the detection of γ-rays from inelastic neutron scattering on carbon ($E_\gamma = 4.43\,\text{MeV}$) and oxygen ($E_\gamma = 6.13\,\text{MeV}$). Called the C/O log, this measurement takes advantage of the reasonably strong probabilities for these two reactions and the large difference in atomic density of carbon and oxygen in water (carbon $= 0$, oxygen $= 3.37 \times 10^{22}$ atoms/cm^3) and oil (carbon $= 3.67 \times 10^{22}$ atoms/cm^3, oxygen $= 0$). By using simple energy windows with a moderate-resolution γ-ray spectrometer, the relative strength of the carbon and oxygen peaks can be determined. This ratio together with other measures of the formation's properties can provide discrimination between oil and water in the rock. One novel twist proposed by Chen *et al.*[91] involved incorporating an AP-ENG in the C/O tool to provide more accurate ratios.

Pulsed ENGs can also be used in water-activation tests where water is flowing in the borehole. In these instruments, a γ-ray detector is placed farther away from the generator to measure the decay of ^{16}N (7.13 s half-life) produced when high-energy DT neutrons from the generator induce the $^{16}O(n,p)^{16}N$ reaction in water. This information is of interest in cases where it is suspected that water is flowing between the formation and the wall of a borehole casing. Yet another neutron logging measurement performed using pulsed ENGs (but also performed using radioisotope neutron sources) is full-fledged γ-ray spectroscopy using prompt γ-ray neutron activation analysis (PGNAA). In this measurement, a complete γ-ray energy spectrum is collected and then analyzed using a library of known response functions for each of the major rock elemental constituents expected in the formation (H, Si, S, Cl, Ca, Ti, Fe, etc.). Estimates can be made of the rock mineralogy in the formation by combining this information with other logging data.

Beyond oil and gas exploration, geological applications of ENGs also include the exploration for minerals and other subsurface materials. One simple example is the use of PGNAA for determining the salinity of water in drill holes.[118] The company EADS Sodern (France) has reported using down-hole PGNAA analysis to map copper in borehole logging. Other materials that have been considered for exploration by this technique include coal and precious metals. A notable application in geophysical exploration is uranium logging where the fission process is exploited.[119,120] A pulsed ENG is used to generate short, intense pulses of neutrons, and then one or more signatures from the fission process are detected, including the decay rate of thermal and epithermal neutrons and the intensity of beta-delayed neutrons and γ-rays from the decay of short-lived fission products.

4.3. *Gauging and radiography*

Gauging and radiography are early applications of fast and thermal neutrons for which a variety of ENGs were developed.[121–124] Gauging is a one-dimensional measure of attenuation often used in the oil industry to determine such properties as material density, fill level in tanks, or moisture content (through neutron moderation). Both direct attenuation measurements and back-scatter measurements detecting neutrons and capture γ-rays have been used for neutron gauging.[125–127] Neutron radiography shares many similarities with gauging measurements, except that radiography generates two-dimensional images compared with the one-dimensional neutron transmission measured in gauging. While neutron gauging is widely used for quantifying moisture content in a variety of materials, accelerator-based sources are rarely employed for this purpose due to the present availability of small, low cost radioisotope sources that are more appropriate for such a simple measurement.

In principle, neutron radiography is similar to X-ray radiography (see Chapter 7). However, the image contrast depends less on bulk density (average-Z) for neutron radiography than it does with X-ray imaging. Instead, it is dependent on the specific isotopic composition in the transmission region.[128] For example, low-Z inclusions or contamination from hydrogen are determined much more easily with fast neutrons than with X-rays. In thermal neutron radiography, the familiar thermal-neutron capture agents — boron, gadolinium, and cadmium — are most easily imaged and quantified. For fast-neutron radiography, accelerators may be used without much target infrastructure, but for thermal neutron radiography, a hydrogenous assembly must be used to moderate the fast neutrons to form a thermal energy beam.[129–131]

In radiographic imaging, the neutrons are most often measured using 2D scintillating screens coupled to charge-coupled-device (CCD) imaging cameras with a bent-optics setup to remove the CCD camera from direct exposure to the irradiating neutrons.[132] In high ambient γ-ray background environments that would also damage radiation-sensitive CCD cameras, shielding may be used to protect the cameras, but an alternate technique that is occasionally employed is to use neutron-sensitive foils and the activation transfer method.[133,134] In this approach, an activation sheet made of a neutron absorbing material that yields an activation product is placed behind a neutron-irradiated object. Depending on the composition of the object under test, neutrons are either scattered out of the incident neutron beam (e.g., fast neutrons interacting with hydrogen) or absorbed in the object (e.g., slow neutrons absorbed by boron), leaving a negative contrast image of the object's contents on the activation sheet. At the end of exposure, the activation sheet is then placed in close contact with a radiography film. As the neutron-activated nuclei in the sheet decay (e.g. copper for fast neutrons or dysprosium for thermal neutrons), they leave an image in the film that can be used for analyses. Storage-phosphor imaging plates coupled with transfer media (to improve their sensitivity to fast neutrons) for use with the activation transfer technique have also been employed for neutron radiographic imaging.[134]

Examples of industrial neutron radiography using accelerators are numerous and cover many different fields. In general, these applications seek to identify void spaces in man-made materials, detect and quantify damage in structural composite materials, or detect low-Z or high-Z objects among mid-Z materials.[135–139] A few specific examples include the use of fast neutron radiography for the inspection of composite helicopter rotor blades, the analysis of lithium-aluminum alloy rods, the analysis of voids in castings, and the inspection of air cargo to detect contraband.[140–143] A new technique in the radiography field is the use of AP-ENGs to provide spatial data for 2D image reconstruction.[144]

A noteworthy variant of the more familiar transmission radiography is resonant radiography where image data is collected at multiple neutron energies to exploit resonant absorption effects and provide contrast.[145] One method uses one of the light-ion reactions to produce monoenergetic neutrons at specific angles relative to the ion beam and then scans an object through different angles, reflecting on-resonance and off-resonance energies. Another approach is to use time-of-flight techniques with a time-gated detector to generate the on-resonance and off-resonance images.[146] On-resonance transmission images provide contrast to the isotope(s) in the energy ranges analyzed, while off-resonance images provide a method of correcting for non-isotope-specific neutron scattering and absorption.[147–149] Industrial measurements with high-energy neutrons have primarily focused on using resonances in carbon, oxygen, and nitrogen for homeland security to detect contraband such as explosives and narcotics and in the mining industry to locate precious minerals. For slow neutrons generated from a photonuclear neutron source in a TOF arrangement, prototype experiments have also demonstrated the feasibility of using resonant radiography to assay the isotopic composition of transuranic elements in spent fuel.[102,103]

4.4. *Laboratory activation analysis*

Activation analysis can be used to make practical analytical measurements of all of the elements in the periodic table except ^{4}He. A comprehensive review of these capabilities has been presented by Nargolwalla and Przybylowicz in Ref. 17. At one time, accelerator-based neutron sources were widely used in applied chemistry laboratories as an accurate tool for general-purpose chemical assay measurements.[150–153] With the progressive introduction of simpler (and often more accurate) laboratory analytical instruments, the use of neutron generators for these applications has diminished, but it continues in a few select areas.[154] Electronic neutron generators continue to be used as the primary tool for laboratory analyses in those areas where other competing technical approaches do not provide the same capabilities.[155,156]

Of these, the determination of trace quantities of oxygen in metal matrices remains the most important.[157,158]

4.5. *Biomedical applications*

Sealed-tube neutron generators have been used by several investigators as a tool for determining the elemental composition of the human body through PGNAA. In some cases, measurements are made to directly quantify the presence of multiple elements in the body.[159,160] In other cases, measurements are made to infer other aspects of body composition, such as body fat, protein, and water content.[161–164] Beyond human analyses, neutron generators have been used to scan livestock to assess lean body mass, using AP-ENGs in some cases.[165–167] For humans, body-composition studies provide general information on metabolism and health and, more specifically, can be used to examine diseases and other conditions leading to the loss or gain of weight. This information is particularly useful for nutritional studies involving the young and the elderly and for studying the effectiveness of diets designed to counter undesired weight loss.

4.6. *Bulk material analysis*

A rapidly growing industrial application for ENGs is bulk material analysis (BMA).[168–173] In this application, a neutron generator, often a pulsed ENG producing on the order of 10^8 n/s, is used with a specially-engineered moderating structure to generate a mixed-energy neutron spectrum. In most cases, the ENG and moderator assembly together with one or more γ-ray detectors is located near a conveyor belt carrying a raw material, such as mineral ore, coal, scrap metals, or cement. For high-hydrogenous materials, such as coal, the source is typically located above the conveyor belt, while for low-hydrogenous materials, such as metals, the source is typically located below the conveyor belt, although other options are possible.[171] The choice of where to place the neutron source is dictated by the need to adequately moderate the neutrons while ensuring the highest possible neutron flux uniformity throughout the measurement volume. The γ-ray detectors are usually placed on the opposite side of the conveyor belt from the ENG(s) in order to afford protection from high radiation fields and to keep interfering signals as low as possible. A photograph of an industrial system for assaying cement, installed at a plant, is shown in Fig. 19. The blue structures contain shielding material to reduce the external radiation fields at the sides of the BMA system to acceptable levels, while the yellow fencing is used to prevent personnel from getting too close to the conveyor belt entrance and exit.

Fig. 19. An industrial BMA system from EADS SODERN, erected to surround a conveyor belt transporting bulk raw materials. (Photo courtesy of Richard Kahn, EADS SODERN.)

BMA measurements using neutron generators are most often performed in pulsed mode, collecting fast neutron inelastic scattering data during pulses and thermal neutron PGNAA data in between pulses. The measurement precision of these types of devices is often considered proprietary information, but they are generally capable of determining the elemental composition of over a dozen elements in the material on the belt to better than 1% relative uncertainty in rolling averages of a few minutes or less.

A noteworthy industrial BMA application that has been investigated is the use of fast-neutron resonance radiography to detect and localize carbon (diamond) in kimberlite ore.[147,148,174,175] In order to process materials on a conveyor belt at a high rate using neutron resonance radiography, a very intense neutron source is needed together with exceptionally capable imaging and data processing resources. Two prototype systems using RF ion linacs were built for diamond detection in South Africa, and while specific performance data are not available, it is known that the concept was successfully demonstrated. However, questions regarding the economic viability of this approach on a commercial scale and the underlying technology readiness still need to be addressed. For the very high neutron yields ($>10^{12}$ n/s) required for this application, extreme challenges exist in cooling a conventional solid neutron-production target due to the high power of the ion beam.

To address this, alternative approaches have been developed, including windowless gas targets using high-speed rotating apertures and plasma-window apertures.[176,177]

4.7. *Radiation effects testing*

Accelerator radiation sources are frequently used to test the capability of modern electronic circuitry to withstand ionizing radiation and neutron radiation fields. These tests are of particular importance for electronics used outside of the Earth's atmosphere where highly energetic solar and cosmic radiation has the potential to deposit significant quantities of energy within the microscopic structures of modern integrated circuits. For neutron-susceptible circuitry, however, care must also be taken when using equipment within the Earth's atmosphere where cosmic-radiation-induced spallation reactions in the upper atmosphere can generate neutrons with GeV energies. As a whole, problems in integrated circuits caused by the interaction of single radiation quanta in the circuits are called single event effects (SEE).[178–180] Within the SEE category, there are several sub-categories of errors, including single event upsets (SEUs), single event latchups (SELs), and single event burnouts (SEBs).

Many resources are dedicated toward studying SEEs to understand their causes and to develop methods to avoid, mitigate, or deal with them. An important aspect of this work is the use of ENGs to test electronic circuitry and to investigate the causes of and solutions for dealing with SEEs. Several different neutron production reactions and many different accelerator types have been used for SEE testing.[181–183] The neutron energies of interest for these tests range from a few MeV to several hundred MeV. A particular challenge associated with these types of studies is the need to collect accurate dosimetric data in order to fully understand SEE initiation thresholds. Familiar dosimetry tools used in health physics, such as tissue-equivalent sensors, aren't well suited for radiation effects tests: silicon-equivalent sensors are often used instead.

4.8. *Detection of contraband, high explosives, and chemical weapon agents*

Starting in the early 1990s, a great deal of work has been done exploring neutron interrogation methods for detecting and identifying contraband and hazardous materials including narcotics, currency, alcohol and liquor, high explosives, unexploded ordinance, and chemical weapon agents.[184–194] The usual design for these applications includes a detector to measure inelastic and PGNAA γ-rays resulting from neutron scattering and neutron absorption, seeking first to detect and then often to quantify characteristic γ-rays from elements associated with the contraband or hazardous material. For materials such as narcotics, high explosives, and chemical weapon agents, simple generalizations for the relative density of a handful of elements can be used to distinguish them from benign material. In some cases, the presence of one or two key elements alone may be sufficient for detection. For narcotics detection, the approach generally seeks to quantify the relative strength of carbon, oxygen, and hydrogen in the interrogation volume, seeking to match particular ratios for these elements with known ratios for the targeted materials.[195–198] For explosives, the presence of nitrogen in high densities is used as a key indicator, while full analysis of N, C, O, and H is used to confirm detection and in some cases to identify the actual material.[188,189] Table 6 lists the key stoichiometric and elemental data that can be used to identify some important explosive materials.[199]

To demonstrate the small size that can be achieved for a portable system, Fig. 20 shows an ENG-based unit developed at Idaho National Laboratory for detecting and identifying chemical weapon agents and high explosives. It consists of a sealed-tube DT ENG (vertical aluminum cylinder in the center), a high-purity germanium γ-ray spectrometer (horizontal, in front of ENG), and a battery and control system (foreground). The only component not shown is a laptop computer for data analysis. The vertical cylinder in the

Table 6. Density and elemental stoichiometry of common explosive compounds.

Explosive Name	Molecular Weight	Stoichiometry				Density [g/cm^3]	Elemental Ratios			
		C	H	N	O		C/O	H/N	C/N	O/N
TNT	227.13	7	5	3	6	1.65	1.17	1.67	2.33	2
RDX	222.26	3	6	6	6	1.83	0.5	1	0.5	1
HMX	296.16	4	8	8	8	1.96	0.5	1	0.5	1
Tetryl	287.15	7	5	5	8	1.73	0.88	1	0.88	1.6
PETN	316.20	5	8	4	12	1.78	0.42	2	1.25	3
NG	227.09	3	5	3	9	1.59	0.33	1.67	1	3
EGDN	152.10	2	4	2	6	1.49	0.33	2	1	13
AN	80.05	—	4	2	3	1.5	0	2	0	1.5
TATP	222.23	9	8	—	6	1.2	1.5	NA	NA	NA
DNB	168.11	6	4	2	4	1.58	1.5	2	3	2
Picric Acid	229.12	6	3	3	7	1.76	2	1	2	2.33

Fig. 20. Photograph of the author with a prototype compact, portable DT neutron generator system for assaying chemical weapons agents and high explosives.

center to the left of the ENG is a safety cylinder holding a World-War-I-era phosgene ($COCl_2$) chemical weapon. Chemical weapon analysis is performed using a process similar to that used for high explosives but with screening of additional elements, including arsenic, chlorine, fluorine, phosphorous, and sulfur.[200,201]

When performing neutron inelastic scattering γ-ray spectroscopy and PGNAA in the field, an ever-present challenge is the presence of background interferences. The key elements used for detecting contraband — C, O, and H — are ubiquitous in the natural environment and in legitimate commercial cargo. A standard method for dealing with these legitimate signals is to measure the background in the vicinity of the area or object to be screened. This is not always successful since a representative background environment may not exist. Also in some cases such as cargo screening, nearby legitimate cargo clutter may not be easily separated from an interrogation area. To improve the sensitivity of these measurements, some detection approaches use the associated particle technique to "electronically" collimate the inspecting neutron flux to a known region of interest.[202–206] The goal is to use the spatial information afforded by the AP method to improve the signal to noise ratio and to provide a higher level of confidence for the identification of contraband in a smaller region of the interrogation volume.

In addition to inelastic neutron scattering and neutron absorption γ-ray spectroscopy, several other neutron-based approaches for contraband detection have been proposed, including elastic and inelastic neutron scattering, dual-mode photon and neutron radiography, and resonant neutron transmission radiography.[147,149,207–213] Dual-mode radiography provides a quick information-rich method for analyzing air cargo for smuggling and customs inspection. Nuclear resonance radiography has been shown to be a very promising technique for screening airline luggage and cargo to detect high explosives.

While high explosive screening for aviation security is the single largest application in this category, nuclear-based techniques are also being developed for explosive detection in other areas, including large-mass explosive detection for vehicle-borne explosives and stand-off explosive detection for improvised explosive devices.[193,199,214–215] Interest also exists in using ENGs to detect buried landmines.[216–218] Small anti-personnel land mines and larger anti-vehicle land mines are an insidious hazard in many parts of the world, with the potential to injure and kill decades after their emplacement. While real-world acceptance for these sensors has been inconclusive to date, continuing work may one day lead to neutron-based inspection systems that can help to solve this serious problem.

4.9. *Fissionable material analysis for safeguards*

In the realm of nuclear safeguards analysis, active neutron interrogation techniques have been used for several decades to measure and account for fissionable material in the nuclear fuel cycle.[219–221] In most of theses applications, the techniques are based on inducing fission within an interrogation volume and then measuring one or more of its signatures, i.e., prompt or delayed neutrons and γ-rays. One category of measurements looks for fissionable material left in the waste streams of the fuel cycle, mostly at fuel reprocessing facilities.[222–225] Assays are made of liquid streams in the reprocessing lines, liquid and solid waste byproducts from different stages of the recycling process, and solid wastes generated as residual wastes from nuclear material handling. Sealed-tube ENGs are most commonly used in these systems, but open-vacuum accelerators, such as RFQ linacs and betatrons, have also been used.[226,227] A particularly large number of projects have sought to develop large-scale scanning systems to assay waste drums and waste crates.[228–231] In the well-defined geometry and constraints of these systems, fissile material assay detection limits as low as 1 mg have been achieved.[232] In addition to gross material quantification, neutron-based safeguards techniques have also been developed to generate tomographic images of fissionable materials for security, control, and accountancy.[233]

Pulsed accelerators, usually sealed-tube ENGs, have also been used for performing neutron interrogation of both fresh and irradiated nuclear fuels for safeguards measurements.[234–236] A wide variety of techniques has been used for this work. Traditional fission signature analyses comprise a significant fraction of this work, but other approaches have been tried, most notably nuclear resonance techniques. In the resonant reaction category, one approach that has seen increased interest is the use of the "lead slowing down spectrometer" (LSDS) technique. LSDS exploits the resonant structure of the neutron fission cross sections of actinide elements for assaying plutonium in spent fuel.[237–240] Similarly, neutron resonance transmission analysis (NRTA) has been proposed as a useful approach for assaying plutonium within spent nuclear fuel.[102,103]

4.10. *Fissionable material detection for screening and security*

Neutron probing techniques have an obvious role in the detection of fissionable material for security purposes, as indicated by the ongoing research.[241–245] However, developers have approached the field as if they were ordering from an *à la Carte* menu of nuclear techniques for particular measurement challenges. Different accelerator types have been chosen including open-vacuum RFQ linacs, many different types of sealed-tube ENGs (some with associated particle imaging), and photoneutron systems. The most commonly used detectors are ^{3}He-based proportional counters, usually moderated with polyethylene and often wrapped in cadmium and/or boron to eliminate sensitivity to room-scattered thermal neutrons. Liquid scintillators and capture-gated liquid scintillators doped with boron, lithium, or gadolinium have been used for detecting both neutrons and γ-rays originating from induced fission. A variety of γ-ray spectrometers, including solid-state sensors and inorganic scintillators, have been used for this work as well.

One elegant approach for detection — an evolutionary development from radioisotope techniques using AmLi neutron sources — is the use of an accelerator to generate low kinetic energy (average of 60 keV) neutrons as a means to induce fission in an interrogation volume.[246] In this approach, a liquid scintillator is used to detect prompt fission neutrons during irradiation to take advantage of the inherent sensitivity of these scintillators to neutrons at the probe source energy. Another approach frequently applied is to measure prompt fission neutrons via the neutron die-away technique.[247–250] In this case, a pulsed neutron source, usually a sealed-tube DT ENG, is used to irradiate an object while detectors are used to measure neutrons between ENG pulses. The premise is that fast neutrons entering the system will lose energy and remain resident within the volume for long periods after each pulse, especially in hydrogenous matrices, and that if fissionable material is present, fission will occur during this period to generate fast-fission neutrons at time periods much longer than the time needed for neutrons from the original pulse to have scattered to low energies. A key to this approach is to use neutron sensors that are insensitive to thermal neutrons, such as shielded ^{3}He detectors or liquid scintillators. Figure 21 shows an example of the type of data generated from one of these measurements. The cumulative neutron count data are shown in a time spectrum for the first 3000 μs after each pulse

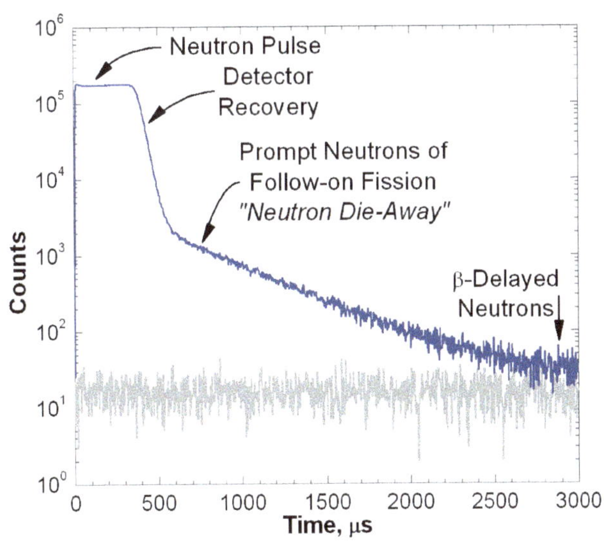

Fig. 21. Neutron die-away data taken of 9.4 kg of highly enriched uranium shielded in a 60 cm plywood cube using a DT ENG pulsed at 300 Hz with a 10% duty cycle. Data were collected using a 10 atm moderated and shielded ^{3}He proportional counter. The lower trace is the background.[251]

with the data from each pulse added together for a 600 s acquisition. During the ENG pulses, a plateau is observed when the ENG output is the dominant neutron source. After each pulse, the count intensity quickly drops until the prompt fission die-away signal becomes a dominant feature. The die-away signal decays with a pseudo-exponential dependence until background neutrons and neutrons from the decay of β-delayed neutron-emitting fission products are the only remaining source terms.

Rather than measure the die-away neutron fission signature as described above, another approach often employed in active neutron interrogation systems is to measure neutrons and γ-rays emitted from the decay of short-lived fission products generated within the volume being interrogated. Since nearly all of these fission products decay via β-particle emission, they are sometimes called β-delayed emissions. Measuring β-delayed neutrons is advantageous because of the strong signal that can be generated and the low natural background.[252–256] A time dependence analysis of the 200+ fission products that decay with neutron emission yields a six-component exponential decay profile, with half of the total emission probability for uranium having a half-life of 2.23 s or less.[257] In a parallel approach, methods to measure β-delayed γ-rays from short-lived fission products have been investigated in order to use this as a signature to detect shielded fissionable material.[258–260] While γ-rays continue to be emitted from fission products for long time periods following the fission, a time dependence similar to the β-delayed neutrons is observed at short time periods. An advantage of this approach is that the intensity of the β-delayed γ-ray signature is stronger than that of the β-delayed neutrons. However, data collection can be complicated at very short periods following the neutron pulses due to the presence of thermal neutrons that are captured and produce multi-MeV capture γ-rays. Both methods can be impacted by the presence of benign interferences when DT neutrons are used, either as a result of the $^{17}O(n,p)^{17}N$ reaction in which ^{17}N decays with a 4.2 s half-life and emits neutrons or the $^{16}O(n,p)^{16}N$ reaction in which ^{16}N decays with a 7.1 s half-life and emits high energy γ-rays.[261]

Other signatures from active neutron interrogation that are more challenging to measure and use for detecting shielded fissionable material include prompt emissions during high-energy neutron pulses, neutrons and associated γ-rays from inelastic scattering, and neutron-capture γ-rays. One important signature familiar from passive screening is the detection of correlated neutron emissions by the measurement of neutron coincidences and analysis of neutron multiplicities. This technique has been shown to be effective for both detecting and characterizing fissionable material.[262–264] However, analysis times may be long in some geometries.[265]

4.11. *Other applications*

In addition to the applications discussed above, ENGs have been utilized for analysis in several other fields, demonstrating that neutron-based analytical techniques can play an important role in solving many challenges in the areas of industrial measurement, gauging, and quality assurance. Sealed-tube ENGs are the most commonly chosen tool for many reasons, including their smaller size, ease of use, portability, and lower cost when compared with larger open-vacuum systems. These applications include:

- *Environmental Analysis*. ENGs have been used in non-destructive measurements with neutron inelastic scattering activation analysis and PGNAA to detect and quantify hazardous elements in the soil at many contaminated locations.[266–268] In the US, an important list of hazardous elements is defined by the Resource Conservation and Recovery Act (RCRA) guidelines for dealing with hazardous waste. Included in this list are the heavy metal elements As, Ba, Cd, Cr, Pb, Hg, Se, and Ag. Neutron-based techniques have also been developed to assay RCRA elements in waste barrels using PGNAA methods.[269–271]
- *Soil Analysis*. Portable ENGs have been used for field measurements to assay carbon content in soil.[272,273] The aim of this work has been the development of a method to quantity carbon present in soil to evaluate long term trends in global climate change, to understand natural carbon concentration variations in soil, and to evaluate man-made methods for sequestering carbon.
- *Flow Analysis in Pipes*. Techniques have been developed to use pulsed ENGs to analyze the flow of fluids in pipes, including water activation using the $^{16}O(n, p)^{16}N$ reaction. In addition to mud and water flow analysis in the petroleum industry as mentioned earlier, analysis techniques have been considered for other industrial processes in order to understand flow rates, residence times, and mixing. A comprehensive overview of these techniques has been presented by Perez-Griffo *et al.*[274]
- *Extraterrestrial Exploration*. Beginning with the Apollo program and the development of the Controlatron neutron generator for lunar soil analysis, there has been a continued interest in placing small ENGs on space probes destined for the Earth's moon and Mars.[76,275–278] Although no project has yet progressed to full mission deployment of an ENG on the Moon,

a current US National Aeronautics and Space Administration (NASA) program, the Mars Science Laboratory, is now underway to land a small ENG on Mars in 2012. The use of the ENG on this mission is part of the Dynamic Albedo of Neutrons (DAN) experiment, which aims to detect and quantify hydrogen in the Martian soil.

5. Summary and Future Trends

There is a long history of using accelerator-based neutron sources in industrial applications. Sealed-tube electronic neutron generators are the most commonly used type, but others — especially RFQ accelerators — have found commercial interest as well. By all accounts, the most prevalent use of these devices is in geophysical exploration for the petroleum industry, but neutron generating accelerators are also found in a diverse array of other fields and applications. In measurement applications, the unique nature of the neutron and how it interacts with matter make it an invaluable tool for understanding the physical world, while in radiation effects testing, accelerator neutron sources provide a realistic method for testing electronics for vulnerabilities to cosmic radiation and atmospheric spallation neutrons.

Given the current state-of-the-art in accelerator-based neutron generator technology and knowledge of ongoing and proposed research activities from information in the public domain, it is possible to make a reasonable forecast of technological developments expected in the next 5–10 years. Such a forecast, however, is necessarily limited for developments in the oilfield services industry due to its proprietary nature. Some reasonable expectations in commercial ENG development over this time period, exclusive of the oilfield sector, are given below in no particular order.

- *Expansion of ENG Use to Replace Radioisotope Neutron Sources.* Following the recent trends observed in the US and abroad related to security, it is reasonable to expect additional pressure will be placed on radioisotope neutron source users to switch to ENGs. The use of tritium may not be a concern in these cases: the comparative risks of tritium being used in radiological dispersal devices versus americium or plutonium is small.[279]
- *Very-High-Yield ENGs Using RF Ion Sources.* Ongoing research in the US and South Korea is progressing towards the development of practical high-yield ENGs using modern RF ion source technology. High-yield sealed-tube DT ENGs of this type were manufactured by SERL (England) in the 1960s, but the technology did not advance from those early efforts. It is reasonable to expect that with this new research there will be a reintroduction of commercial ENGs reaching yields of $\sim 10^{12}$ n/s

(DT) with the ability to operate in pulsed modes. These next-generation systems will most likely use active cooling, and their lifetimes could approach 10^4 h. Motivations for this development will include the need for neutron sources for basic research and education (following the closing of more university research reactors), for electronic radiation hardness testing (with further miniaturization of microelectronic circuitry), and for security screening.[280,281]

- *High-Efficiency ENGs Using Advanced Ion Sources.* Current research into the use of microelectromechanical-system (MEMS) nanoemitter arrays has shown that quantum field ionization and desorption can be used to generate ions much more efficiently than the plasma-based ion sources used in today's ENGs. If further laboratory successes and research occur in this field, it is realistic to expect it to lead to the development of smaller P-ENGs in the 5 kg range with neutron yields approaching 10^9 n/s (DT) without active cooling. Further, the electrical efficiency of these systems should be significantly better than current generation systems, thus leading to much longer operating periods when powered by batteries. Research in this area will be motivated mostly by security applications.[282,283]
- *Better Performance AP-ENG Detectors.* Research is underway to improve the performance of high-resolution scintillator-based particle detectors and solid-state detectors for use in AP-ENGs. In addition, work is underway to develop better data readout electronics for these systems. Continuing research should contribute to the development of next-generation AP-ENG instruments with better performance (lifetime, ruggedness, and beam spot size) than present-generation devices. Developments in this area will be driven by nuclear safeguards research, security applications, and medical investigations.[284,285]
- *Improvements in the User Interface.* Mirroring developments in the underlying ENG technologies, advances in control system technology will likely be made to improve ease of use and incorporate more sophisticated self-diagnostics capabilities by building upon the "smart sensor" concept for process instrumentation. For example, rather than setting operating parameters such as high voltage or beam current, the user should only have to set a "neutron yield" parameter and let the instrument determine the optimal combination of voltage and current. Also, technology should allow the ENG to automatically indicate when a component is in need of service, such as replacement of the neutron tube, in order to streamline use in continuous-duty process monitoring applications.
- *Novel ENG Power Source Development.* The fact that piezoelectric crystals (exploiting pyroelectric and/or ferroelectric properties) can be used

for ENG power sources has been known for decades. However, present small-scale research efforts in this and other areas may lead to interesting ENG design concepts in the future, perhaps in response to the need for low-cost long-lived electronic alternatives to radioisotope neutron sources. Regular and ongoing advances in solid-state microelectronics may also be expected to yield noteworthy performance improvements over current state-of-the-art compact power systems.[286,287]

- *Longer-Lived ENG Neutron Tubes.* The operating lifetime of today's commercial ENGs is primarily limited by the degradation of the hydride target within the sealed neutron tube, which results in a reduction of the overall neutron yield and the need to replace the tube. Research can be performed to improve the resistance of hydride targets to this damage mechanism. It is reasonable to expect operational lifetimes exceeding 10^4 h in the 10^9 n/s (DT) range with relatively stable neutron yields (i.e., the yield decreases by <50% of the initial yield) through a combination of improved target design and automated control systems.
- *Advanced Open-vacuum Accelerator Concepts.* In a broader view of the future of accelerator-based neutron sources, there are several research and design efforts underway for innovative open-vacuum systems that may have a great impact on future industrial neutron applications. Some of these technology advances include the development of high-efficiency superconducting cyclotrons, advanced linear accelerator and accelerator storage ring technology, high-gradient accelerating structures, and other exotic ion source and acceleration concepts.

Acknowledgements

This text has been adapted from Idaho National Laboratory report INL/EXT-09-17312. I would like thank Marianne and Bob Hamm for their patience and perseverance for editing this chapter and working to assemble and coordinate this book. Work supported by the U.S. Department of Energy, National Nuclear Security Administration, Office of Nonproliferation and Verification Research and Development, under DOE Idaho Operations Office Contract DE-AC07-05ID14517. This manuscript has been authored by Battelle Energy Alliance, LLC under Contract No. DE-AC07-05ID14517 with the U.S. Department of Energy. The United States Government retains and the publisher, by accepting the article for publication, acknowledges that the United States Government retains a nonexclusive, paid-up, irrevocable, worldwide license to publish or reproduce the published form of this manuscript, or allow others to do so, for United States Government purposes.

Disclaimer

References

1. J. Chadwick, *Proc. R. Soc. Lond.* A **136**, 692 (1932).
2. J. D. Cockcroft and E. T. S. Walton, *Proc. R. Soc. Lond.* A **137**, 229 (1932).
3. J. Chadwick, *Proc. R. Soc. Lond.* A **142**, 1 (1933).
4. M. L. E. Oliphant, P. Harteck and E. Rutherford, *Proc. R. Soc. Lond.* A **144**, 692 (1934).
5. B. Cathcart, *The Fly in the Cathedral* (Penguin Books Ltd., London, 2004).
6. M. L. E. Oliphant and E. Rutherford, *Proc. R. Soc. Lond.* A **141**, 259 (1933).
7. M. L. E. Oliphant, B. B. Kinsey and E. Rutherford, *Proc. R. Soc. Lond.* A **141**, 722 (1933).
8. W. H. Zinn and S. Seeley, *Phys. Rev.* **50**, 1101 (1936).
9. F. M. Penning, US Patent 2, 211, 668 (1940) (The Netherlands, 1937).
10. W. Schutze, US Patent 2, 240, 914 (1941) (Germany, 1938).
11. H. I. Kallmann, US Patent 2, 251, 190 (1941) (Germany, 1938).
12. H. I. Kallmann, US Patent 2, 287, 619 (1942) (Germany, 1939).
13. H. I. Kallmann, US Patent 2, 287, 620 (1942) (Germany, 1939).
14. *Radiation Source Use and Replacement, Abbreviated Version*, National Research Council of the National Academies (The National Academies Press, Washington, DC, 2008).
15. *International Approaches to Securing Radioactive Sources Against Terrorism*, NATO Science for Peace and Security Series, Eds. W. D. Wood and D. M. Robinson (Springer, Dordrecht, 2009).
16. J. B. Marion and J. L. Fowler, *Fast Neutron Physics, Part I: Techniques & Part II: Experiments and Theory* (Wiley and Sons, New York, 1963).
17. S. S. Nargolwalla and E. P. Przyblowicz, *Activation Analysis with Neutron Generators* (Wiley and Sons, New York, 1973).

18. S. Cierjacks, Ed., *Neutron Sources for Basic Physics and Applications* (Pergamon Press Ltd., Oxford, 1983).
19. J. Csikai, *CRC Handbook of Fast Neutron Generators*, Vols. I and II (CRC Press, Boca Raton, 1987).
20. K. S. Krane, *Introductory Nuclear Physics* (Wiley and Sons, New York, 1988).
21. *Evaluated Nuclear Data File (ENDF) ENDF/B-VI.8*, US National Nuclear Data Center (Brookhaven National Laboratory, 2009), www.nndc.bnl.gov.
22. L. A. Shope, *Report SC-TM-66-247* (Sandia National Laboratories, NM, 1966).
23. M. R. Hawkesworth, *Atomic Energy Rev.* **15**, 169–220 (1977).
24. F. E. Cecil and E. B. Nieschmidt, *Nucl. Instr. Meth. Phys. Res.* **B16**, 88 (1968).
25. C. L. Lee, Ph.D. Thesis, Massachusetts Institute of Technology (1998).
26. B. Blackburn, Ph.D. Thesis, Massachusetts Institute of Technology (2002).
27. M. Drosg, in *SPIE Conf. Proc.* **2339**, 145 (1995).
28. S. Takayanagi, N. H. Gale, J. B. Garg and J. M. Calvert, *Nucl. Phys.* **28**, 494 (1961).
29. G. U. Din, M. A. Nagarajan and R. Pollard, *Nucl. Phys.* **A93**, 190 (1967).
30. M. Drosg, *Contribution to Int. Workshop on Fast Neutron Phys.* (2002), http://homepage.univie.ac.at/manfred.drosg/dresdenb.htm.
31. A. Antolak, B. Doyle, D. Morse and P. Provencio, *Report SAND2006-0995* (Sandia National Laboratories, Albuquerque NM, 2006).
32. M. Goldberg, United States Patent 7,381,962 (see Table I) (2008).
33. *U.S. National Nuclear Data Center* (2009), http://www.nndc.bnl.gov/qcalc.
34. M. Drosg, *DROSG-2000: Neutron Source Reactions, Data Files with Computer Codes for 59 Monoenergetic Neutron Source Reactions*, Report IAEA-NDS-87 (International Atomic Energy Agency, Austria, 2005).
35. L. Auditore *et al.*, in *Proc. EPAC 2004*, 2347 (2004).
36. Z. W. Bell, V. L. Chaklov and V. M. Golovkov, *Report Y/DW/1730 R1*, (Oak Ridge Y-12 Plant, Oak Ridge, TN 1998).
37. *Handbook on Photonuclear Data for Applications, Cross Sections and Spectra*, Report IAEA-TECDOC-Draft No 3 (International Atomic Energy Agency, Vienna, Austria, 2000), www.nds.iaea.org/photonuclear.
38. R. L. Macklin, R. B. Perez, G. de Saussure and R. W. Ingle, Report ORNL/TM-10666 (Oak Ridge National Laboratory, Oak Ridge, TN, 1988).
39. L. C. Leal, Report ORNL/TM-11547 (Oak Ridge National Laboratory, Oak Ridge, TN, 1990).
40. M. E. Overberg, B. E. Moretti, R. E. Slovacek and R. C. Block, *Nucl. Instr. Meth. Phys. Res.* **A438**, 253 (1999).
41. M. Flaska, Ph.D. Thesis, Technical University of Delft, Delft, The Netherlands (2006).
42. W. Scharf, *Particle Accelerators and Their Use, Parts 1 and 2* (Harwood Academic Publishers GmbH, Switzerland, 1991).

43. S. Humphries, *Principles of Charged Particle Acceleration* (Wiley and Sons, Hoboken, 1999).
44. *Electrostatic Accelerators: Fundamentals and Applications*, Ed. R. Hellborg (Springer-Verlag, Berlin, 2005).
45. R. W. Hamm, in *SPIE Conf. Proc.* **4142**, 6 (2000).
46. M. J. Wakem and M. Brownridge, in *Proc. Adv. for Future Nucl. Fuel Cycles (ATALANTE 2004)*, 01-01 (2004).
47. B. Wolf, Ed., *Handbook of Ion Sources* (CRC Press, Boca Raton, 1995).
48. J. Kollár, M. Tatara and D. Chorvát, *Int. J. Appl. Rad. Iso.* **26**, 635 (1975).
49. *Radiation Protection and Measurement for Low-Voltage Neutron Generators*, NCRP Report No. 72 (National Council on Radiation Protection and Measurements, Bethesda MD, 1983).
50. D. Dietrich *et al.*, *Nucl. Instr. Meth. Phys. Res.* **B241**, 826 (2005).
51. E. Gimshi, A. Tsechanski and E. Shalom, in *Proc. 12th Int. Workshop on Targetry and Target Chemistry*, 63 (2008).
52. C. Faure, *Acta Electronica* **13**, 317–364 (1970).
53. G. Lindin and J. Massieux, *Sealed Neutron Tubes*, Report SAND83-6003 (Sandia National Laboratories, Albuquerque NM, 1985).
54. W. H. Kohl, *Handbook of Materials and Techniques for Vacuum Devices* (American Institute of Physics, New York, 1995).
55. D. L. Chichester, J. D. Simpson and M. Lemchak, *J. Radioanal. Nucl. Chem.* **271**, 629 (2007).
56. L. A. Shope *et al.*, *Int. J. Appl. Rad. Iso.* **34**, 269 (1983).
57. D. A. Von Kessar, *Helv. Phys. Acta* **8**, 601 (1935).
58. F. M. Penning and J. H. A. Moubis, *Physica* **4**, 1190 (1937).
59. A. Bouwers, F. A. Heyn and A. Kuntke, *Physica* **4**, 153 (1937).
60. O. Reifenschweiler, *Electrotechink und Maschinenbaus* **74**, 96 (1957).
61. J. D. Gow and L. Ruby, *Rev. Sci. Instr.* **30**, 315 (1959).
62. W. W. Salisbury, US Patent 2,489,436 (1949) (filed 1947).
63. R. E. Fearon and J. M. Thayer, US Patent 2,712,081 (1955) (filed 1949).
64. H. B. Frey, US Patent 2,769,096 (1956) (filed 1952).
65. C. Goodman, US Patent 2,943,239 (1960) (filed 1954).
66. C. Goodman, US Patent 2,991,364 (1961) (filed 1954).
67. C. Goodman, US Patent 2,996,618 (1961) (filed 1954).
68. A. H. Frentrop, US Patent 3,246,191 (1966) (filed 1961).
69. O. Reifenschweiler, US Patent 3,124,711 (1964) (filed 1959).
70. J. D. Gow and H. C. Pollock, *Rev. Sci. Instr.* **31**, 235 (1960).
71. O. Reifenschweiler, *Nucleonic* **18**, 69 (1960).
72. P. O. Hawkins and R. Sutton, *Rev. Sci. Instr.* **31**, 241 (1960).
73. B. J. Carr, US Patent 3,393,316 (1966) (filed 1964).
74. J. S. Buhler, *The Norelco Reporter* **8**, 87 (1961).
75. O. Reifenschweiler, *Philips Res. Repts.* **16**, 401 (1961).

76. *Development of the Controlatron Neutron Source Tube for Lunar Surface Analysis*, Report GEX-332 (General Electric Company, Milwaukee, 1962).
77. *Manual for Troubleshooting and Upgrading of Neutron Generator*, Report IAEA-TECDOC-913 (International Atomic Energy Agency, Vienna, 1996).
78. C. Goodman, US Patent 2,943,239 (1960) (filed (1954).
79. R. C. Smith, C. H. Bush and J. W. Reichardt, *IEEE Trans. Nucl. Sci.* **35**, 859 (1988).
80. H. G. Pfutzner and M. Mahdavi, *SPIE Conf. Proc* **2339**, 201 (1995).
81. H. G. Pfutzner, J. L. Groves and M. Mahdavi, in *IEEE Nucl. Sci. Symp. & Med. Imaging Conf., 1994* (IEEE, New York, 1994), p. 812, doi: 10.1109/NSSMIC.1994.474489.
82. O. Reifenschweiler, *Philips Res. Repts.* **16**, 401 (1961).
83. P. Bach, *et al.*, *J. Radioanal. Nucl. Chem.* **168**, 393 (1993).
84. V. D. Aleksandrov, *et al.*, *Appl. Rad. Iso.* **63**, 537 (2005).
85. B. E. Leonard, in *Proc. Workshop on High Intensity Neutron Generators*, 44 (1972).
86. J. E. Jobst, Z. G. Burson and F. F. Haywood, in *Proc. Workshop on High Intensity Neutron Generators*, US Atomic Energy Commission, Eds. H. A. Lamonds, B. E. Leonard and E. J. Story (EG&G, 1972), p.159.
87. J. D. L. H. Wood, *Nucl. Instr. Meth. Phys. Res.* **21**, 49 (1963).
88. R. McFadden, P. W. Martic and B. L. White, *Nucl. Instr. Meth.* **92**, 563 (1971).
89. C. M. Gordon and C. W. Peters, *Appl. Rad. Iso.* **41**, 1111 (1990).
90. *Associated Particle Imaging (API)*, Report DOE/NV/11718-223 (Special Technologies Laboratory, Santa Barbara, 1998).
91. Z. Chen *et al.*, US Patent 6,297,507 (2001).
92. A. V. Kuznetsov *et al.*, *Appl. Rad. Iso.* **61**, 51 (2004).
93. B. Perot *et al.*, in *Proc. Int. Topical Mtg. Nucl. Res. Appl. and Utilization of Accelerators*, Paper SM/EN-07 (International Atomic Energy Agency, Vienna, 2009).
94. D. N. Vakhtin *et al.*, in *Proc. Int. Topical Mtg. Nucl. Res. Appl. and Utilization of Accelerators*, SM/EN-11 (International Atomic Energy Agency, Vienna, 2009).
95. P. Le Tourneur, in *Proc. Int. Topical Mtg. Nucl. Res. Appl. and Utilization of Accelerators*, Paper SM/EN-17 (International Atomic Energy Agency, Vienna, 2009).
96. D. L. Chichester, M. Lemchak and J. D. Simpson, *Nucl. Instr. Meth. Phys. Res.* **B241**, 753 (2005).
97. P. A. Hausladen, J. S. Neal and J. T. Mihalczo, *Nucl. Instr. Meth. Phys. Res.* **B241**, 835 (2005).
98. J. A. Deye and F. C. Young, *Phys. Med. Biol.* **22**, 90 (1977).
99. R. E. Morgado, *Report LA-11393-C*, 10–33 (Los Alamos National Laboratory, Los Alamos, 1987).

100. Z. W. Bell, V. L. Chaklov and V. M. Golovkov, *Report Y/DW-1730 R1* (Oak Ridge Y-12 Plant, Oak Ridge, 1998).
101. B. Mukherjee, D. Makowski and S. Simrock, *Nucl. Instr. Meth. Phys. Res.* **A545**, 830 (2005).
102. R. A. Schrack, J. W. Behrens, R. G. Johnson and C. D. Bowman, in *Neutron Radiography, Proceedings of the First World Conference*, Eds. J. P. Barton and P. Von Der Hardt (D. Reidel Publishing, Dordrecht, 1983), p. 495.
103. R. A. Schrack, *Radiation Effects* **95**, 1309 (1986).
104. M. J. Bernstein, *Rev. Sci. Instr.* **40**, 1415 (1969).
105. V. Nardi and J. Brzosko, *Report LA-11393-C Supplemental* (Los Alamos National Laboratory, Los Alamos, 1987).
106. R. A. Krakowski, J. D. Sethian and R. L. Hagenson, *J. Fusion Energy* **8**, 269 (1989).
107. R. L. Hirsch, *J. Appl. Phys.* **38**, 4522 (1967).
108. T. H. Rider, S. M. Thesis, Massachusetts Institute of Technology (1991).
109. G. H. Miley *et al.*, *IEEE Trans. Plasma Sci.* **25**, 733 (1997).
110. Y. Gu, G. Miley and S. DelMedico, in *Digest of Tech. Papers, Tenth IEEE Int. Pulsed Power Conf.*, Vol. 2 (IEEE, New York, 1995), p. 1500.
111. J. H. Sorebo, G. L. Kulcinski, R. F. Radel and J. F. Santarius, *Fusion Sci. Tech.* **56**, 540 (2009).
112. J. A. Clark, *The Chronological History of the Petroleum and Natural Gas Industries* (Clark Book Co, Houston, 1963).
113. A. H. Youmans, C. W. Millis, E. C. Hopkinson and W. D. Bishop, *Proc. Symp. Soc. Prof. Well Log Analysts*, J. Petroleum Technology **16**(3), 319 (1964).
114. R. Ethridge, J. Valentine and E. Stokes, in *Conf. Record IEEE Nucl. Sci. Symp. 1990*, (IEEE, New York, 1990), p. 820, doi: 10.1109/NSSMIC. 1990693467.
115. R. C. Odom, *et al.*, in *Proc. Symp. Soc. Prof. Well Log Analysts, 49th Annual Logging Symp.* (Society of Petrophysicists & Well Log Analysts, 2008), 2008-O.
116. D. E. Johnson and K. E. Pile, *Well Logging in Nontechnical Language, 2nd Edition* (Penn Well Publishing Co., Tulsa, 2002).
117. D. V. Ellis and J. M. Singer, *Well Logging for Earth Scientists, 2nd Edition* (Springer, Dordrecht, 2007).
118. M. Borsaru *et al.*, *Appl. Rad. Iso.* **64**, 630 (2006).
119. W. W. Givens, W. R. Mills, C. L. Dennis and R. L. Caldwell, *Geophysics* **41**, 468 (1976), doi: 10.1190/1.440627.
120. F. D. Thibideau, *The Log Analyst* **18**(6) (1977), document id:1977-vXVIIIn6a2.
121. H. I. Kallmann and E. Kuhn, Reichspatentamt Patentschrift Nt. 726,278, Deutsches Reich (1942).
122. L. Holland and M. R. Hawkesworth, *Nondestructive Testing* **4**, 330 (1971).
123. H. Berger, *Appl. Rad. Iso.* **61**, 437 (2004).

124. I. A. Anderson, R. L. McGreevy and H. Z. Bilheux, Eds., *Neutron Imaging and Applications: A Reference for the Imaging Community*, (Springer Science + Business Media, New York, 2009).
125. B. D. Sowerby, US Patent 4, 884, 288 (filed 1989) (Australia, 1985).
126. G. Xiang *et al.*, *J. Isotopes* **18**, 26 (2005).
127. J. S. Schweitzer and R. C. Lanza, in *AIP Conf. Proc.* Vol. 475 (American Institute of Physics, Melville, 1999), p. 675.
128. E. Lehmann, G. Frei, A. Nordlund and B. Dahl, *IEEE Trans. Nucl. Sci.* **52**, 389 (2005).
129. A. J. Cox and P. E. Francois, *Nucl. Instr. Meth.* **92**, 585 (1971).
130. A. R. Spowart, *Nucl. Instr. Meth.* **92**, 613 (1971).
131. D. G. Vasilik, R. L. Murri and G. P. Fisher, *Nucl. Tech.* **14**, 279 (1972).
132. E. Bogolubov *et al.*, *Nucl. Instr. Meth. Phys. Res.* **A542**, 187 (2005).
133. J. P. Barton, *Trans. Amer. Nucl. Soc.* **10**, 443 (1967).
134. V. Mikerov, V. Samosyuk and S. Verushkin, *Nucl. Instr. Meth. Phys. Res.* **A542**, 192 (2005).
135. D. E. Wood, *Trans. Amer. Nucl. Soc.* **10**, 443 (1967).
136. W. E. Dance, S. Cluzeau and H.-U. Mast, *Nucl. Instr. Meth. Phys. Res.* **B59/57**, 907 (1991).
137. S. Cluzeau and P. Le Tourneur, in *Proc. Regional Meeting: Nuclear Energy in Central Europe Present and Perspectives* (Nuclear Society of Slovenia, 1993), p. 576, ISBN 96-90004-1-2.
138. K. H. Kim, R. T. Klann and B. B. Raju, *Nucl. Instr. Meth. Phys. Res.* **A422**, 929 (1999).
139. C. L. Hollas and L. E. Ussery, *Report LA-UR-04-0989* (Los Alamos National Laboratory, Los Alamos, 2004).
140. J. J. Antal *et al.*, *Report MTL TR 89-52* (US Army Materials Technology Laboratory, Watertown, 1989).
141. J. J. Antal *et al.*, *Report MTL TR 90-18* (US Army Materials Technology Laboratory, Watertown, 1990).
142. D. A. Allen *et al.*, *Nucl. Instr. Meth. Phys. Res.* **A353**, 128 (1994).
143. J. E. Eberhardt *et al.*, *Appl. Rad. Iso.* **63**, 179 (2005).
144. C. J. Evans and Q. B. Mutamba, *Appl. Rad. Iso.* **56**, 711 (2002).
145. J. I. W. Watterson and R. M. Ambrosi, *Nucl. Instr. Meth. Phys. Res.* **A513**, 367 (2003).
146. V. Dangendorf *et al.*, in *Proc. Seventh World Conf. Neutron Radiography* (2002), available at http://arxiv.org/ftp/nucl-ex/papers/0301/0301001.pdf.
147. G. Chen, Ph.D. Thesis, Massachusetts Institute of Technology (2001).
148. D. Vartsky, in *Int. Workshop on Fast Neutron Detectors and Applications*, Proceedings of Science, FNDA2006-084 (2006).
149. B. W. Blackburn *et al.*, *IEEE Nucl. Sci. Symp. Conference Record*, NSS '07 (IEEE, New York, 2007), p. 1016, doi :10.1109/NSSMIC.2007.4437185.

150. R. Gijbels, A. Speecke and J. Hoste, *Anal. Chim. Acta* **43**, 183 (1968).
151. G. Breynat, J. P. Bocquet and J. P. Garrec, *Nucl. Instr. Meth.* **92**, 499 (1971).
152. R. E. Wainerdi, R. Zeisler and W. A. Schweikert, *J. Radioanal. Nucl. Chem.* **37**, 307 (1977).
153. J. W. Mitchell, S. Yegnasubramanian and L. Shepherd, *J. Radioanal. Nucl. Chem.* **112**, 425 (1987).
154. R. Lupu, A. Nat and A. Ene, *Nucl. Instr. Meth. Phys. Res.* **B217**, 123 (2004).
155. D. L. Perry *et al.*, *Nucl. Instr. Meth. Phys. Res.* **B213**, 527 (2004).
156. P. R. Renne *et al.*, *Appl. Rad. Iso.* **62**, 25 (2005).
157. W. D. James, *J. Radioanal. Nucl. Chem.* **219**, 187 (1997).
158. W. D. James and C. D. Fuers, *J. Radioanal. Nucl. Chem.* **244**, 429 (2000).
159. K. Boddy, I. Holloway and A. Elliott, *Phys. Med. Biol.* **19**, 379 (1974).
160. E. D. Williams, K. Boddy and J. K. Haywood, *Phys. Med. Biol.* **22**, 1003 (1977).
161. J. J. Kehayias *et al.*, *Amer. J. Clin. Nutr.* **53**, 1339 (1991).
162. I. E. Stamatelatos *et al.*, in *Proc. Int. Conf. Appl. Nucl. Tech.: Neutrons in Research and Industry*, Proc. SPIE 2867, Eds. G. Vourvopoulos and T. Paradellis (Society of Photo-optical Instrumentation Engineers, 1996), p. 379, doi: 10.1117/12.267937.
163. J. J. Kehayias *et al.*, in *SPIE Conf. Proc.* **3769**, 224 (1999).
164. K. J. Ellis, in *Human Body Composition*, Eds. S. B. Heymsfield *et al.* (Human Kinetics, Champaign, 2005).
165. C. Chung and Y.-Y. Chen, *J. Radioanal. Nucl. Chem.* **169**(2), 333 (1993).
166. S. Mitra *et al.*, *Asia Pacific J. Clin. Nutr.* **4**, 187 (1995).
167. J. E. Wolff *et al.*, in *Proc. New Zealand Soc. Animal Prod.* **56**, 1 (1996).
168. P. Bach *et al.*, *J. Radioanal. Nucl. Chem.* **168**, 393 (1993).
169. G. Beurton, B. Ledru and P. LeTourneur, *SPIE Conf. Proc.* **2339**, 424 (1995).
170. P. Lebrun *et al.*, in *AIP Conf. Proc.* Vol. 475 (American Institute of Physics, Melville, 1999), p. 695.
171. C. S. Lim, *J. Radioanal. Nucl. Chem.* **262**, 525 (2004).
172. C. S. Lim and D. A. Abernethy, *Appl. Rad. Iso.* **63**, 697 (2005).
173. B. D. Sowerby, *Appl. Rad. Iso.* **67**, 1638 (2009).
174. R. M. Ambrosi, Ph.D. Thesis, University of Witwatersrand, Johannesburg (1999).
175. R. M. Ambrosi *et al.*, in *SPIE Conf. Proc.* **4142**, 331 (2000).
176. E. B. Iverson, Ph.D. Thesis, Massachusetts Institute of Technology (1998).
177. A. de Beer *et al.*, *Nucl. Instr. Meth. Phys. Res.* **B170**, 259 (2000).
178. E. Normand, *IEEE Trans. Nucl. Sci.* **43**, 461 (1996).
179. E. Normand, *IEEE Trans. Nucl. Sci.* **43**, 2742 (1996).
180. R. D. Schrimpf and D. M. Fleetwood, Eds., *Radiation Effects and Soft Errors in Integrated Circuits and Electronic Devices* (World Scientific Publishing, Singapore, 2004).

181. A. H. Taber and E. Normand, *Report DNA-TR-94-123* (Defense Nuclear Agency, Alexandria, 1995).
182. D. L. Oberg *et al.*, *IEEE Trans. Nucl. Sci.* **43**, 2913 (1996).
183. P. J. McMarrr *et al.*, *IEEE Trans. Nucl. Sci.* **50**, 2030 (2003).
184. E. M. A. Hussein, in *Proc. Int. Soc. Opt. Eng.* **1736**, 130 (1992).
185. J. C. Overley *et al.*, *Nucl. Instr. Meth. Phys. Res.* **B99**, 728 (1995).
186. T. Gozani, *Nucl. Instr. Meth. Phys. Res.* **A353**, 635 (1994).
187. A. Buffler, *Rad. Phys. Chem.* **71**, 853 (2004).
188. S. Steward and D. Forsht, *Appl. Rad. Iso.* **63**, 795 (2005).
189. D. Strellis and T. Gozani, *Appl. Rad. Iso.* **63**, 799 (2005).
190. K. Wilhelmsen *et al.*, *J. Radioanal. Nucl. Chem.* **271**, 725 (2007).
191. D. Koltick *et al.*, *Nucl. Instr. Meth. Phys. Res.* **B261**, 277 (2007).
192. T. Gozani and D. Strellis, *Nucl. Instr. Meth. Phys. Res.* **B261**, 311 (2007).
193. S. M. McConchie, Ph.D. Thesis, Purdue University (2007).
194. E. H. Seabury *et al.*, in *AIP Conf. Proc.* **1099** (American Institute of Physics, Melville, 2009), p. 928.
195. T. G. Miller *et al.*, in *SPIE Conf. Proc.* **2867**, 215 (1996).
196. L. Zhang and R. C. Lanza, in *IEEE Nucl. Sci. Symp. Conf. Record, 1998*, Vol. 3 (IEEE, New York, 1998), p. 1532.
197. P. C. Womble *et al.*, in *AIP Conf. Proc.* **475** (American Institute of Physics, Melville, 1999), p. 691.
198. P. A. Dokhale *et al.*, in *AIP Conf. Proc.* **576** (American Institute of Physics, Melville, 2001), p. 1061.
199. C. Bruschini, *Subsurface Sensing Tech. Appl.* **2**, 299 (2001).
200. A. J. Caffrey *et al.*, *IEEE Trans. Nucl. Sci.* **39**, 1422 (1992).
201. E. H. Seabury *et al.*, in *AIP Conf. Proc.* **1099** (American Institute of Physics, Melville, 2009), p. 928.
202. E. Rhodes *et al.*, *IEEE Trans. Nucl. Sci.* **39**, 1041 (1992).
203. G. Nebbia *et al.*, *Nucl. Instr. Meth. Phys. Res.* **A533**, 475 (2004).
204. S. Blagus, D. Sudac and V. Valković, *Nucl. Instr. Meth. Phys. Res.* **B213**, 434 (2004).
205. M. Lunardon *et al.*, *Nucl. Instr. Meth. Phys. Res.* **B213**, 544 (2004).
206. B. Perot *et. al.*, in *Proc. IAEA-TM-29225*, (International Atomic Energy Agency, Vienna, 2006), A-05.
207. E. M. A. Hussein, P. M. Lord and D. L. Bot, *Nucl. Instr. Meth. Phys. Res.* **A299**, 453 (1990).
208. B. D. Sowerby *et al.*, in *Proc. Int. Topical Meeting on Nuclear Research Applications and Utilization of Accelerators* (International Atomic Energy Agency, Vienna, 2009), SM/EN-01.
209. C. B. Franklyn, in Ref. 208, SM/EN-12.
210. *The Practicality of Pulsed Fast Neutron Analysis for Aviation Security* (National Research Council, Washington DC, 2002).

211. D. Strellis, T. Gozani and J. Stevenson, in Ref. 208, SM/EN-05.
212. R. C. Lanza, in *Proc. IAEA-TM-29225* (International Atomic Energy Agency, Vienna, 2006), A-02.
213. M. B. Goldberg *et al.*, in Ref. 208, SM/EN-13.
214. *Existing and Potential Standoff Explosives Detection Techniques* (The National Academies Press, Washington DC, 2004).
215. E. L. Reber *et al.*, *Nucl. Instr. Meth. Phys. Res.* **B241**, 738 (2005).
216. A. V. Kuznetsov *et al.*, *Appl. Rad. Iso.* **61**, 51 (2004).
217. M. Lunardon *et al.*, *Appl. Rad. Iso.* **61**, 43 (2004).
218. P. Le Tourneur, in *Proc. IAEA-TM-29225* (International Atomic Energy Agency, Vienna, 2006), B-03.
219. T. Gozani, *Nucl. Tech.* **13**, 8 (1972).
220. T. Gozani, *Report NUREG/CR-0602* (US Nuclear Regulatory Commission, Washington DC, 1981).
221. M. Clapham *et al.*, in *Proc. 19th ESARDA Symp. Safeguards and Nuclear Material Management*, ESARDA 28, Eds. C. Foggi and F. Genoi (1997), p. 359.
222. P. Jover, *IEEE Trans. Nucl. Sci.* **17**, 517 (1970).
223. B. J. McDonald, G. H. Fox and W. B. Bremner, *Report IAEA-SM-201/61* (International Atomic Energy Agency, Vienna, 1976), p. 589.
224. J. Romeyer-Dherbey, C. Passard and J. Cloue, in *Proc. 15th ESARDA Annual Symposium*, Rome, Italy (1993), p. 543.
225. A.-C. Raoux *et al.*, in *Proc. Int. Conf. Nuclear Waste, Safewaste-2000*, (Société Francais d'Energie Nucléaire, 2000), p. 420.
226. M. J. Waken and M. Brownridge, in *Proc. ATALANTE 2004*, Nîmes, France (2004), 01-01, available at http://www-atalante2004.cea.fr/home/liblocal/docs/atalante2000/01-01.pdf.
227. V. M. Golovkov, V. L. Chakhlov and M. M. Shtein, *Atomic Energy* **96**, 127 (2004).
228. B. D. Rooney *et al.*, in *IEEE Nucl. Sci. Symp. Conf. Record, 1998*, Vol. 2 (IEEE, New York, 1998), p. 1027.
229. S. G. Melton, R. J. Estep and E. H. Paterson, *Report LA-UR-00-2468* (Los Alamos National Laboratory, Los Alamos, 2000).
230. A.-C. Raoux, *Nucl. Instr. Meth. Phys. Res.* **B207**, 186 (2003).
231. W. Hage, *Nucl. Instr. Meth. Phys. Res.* **A551**, 396 (2005).
232. W. E. Kunz *et al.*, in *Proc. 3rd Annual ESARDA Symp. Safeguards and Nuclear Materials Management*, Karlsruhe, Germany (1981), p. 119.
233. P. A. Hausladen, *Nucl. Instr. Meth. Phys. Res.* **B261**, 387 (2007).
234. S. J. Tobin *et al.*, *Report LA-UR-08-03763* (Los Alamos National Laboratory, Los Alamos, 2008).
235. S. McConchie *et al.*, in *AIP Conf. Proc.* Vol. 1099 (American Institute of Physics, Melville, 2009), p. 643.

236. D. L. Chichester and E. H. Seabury, in Ref. 235, p. 851.
237. H. Krinninger, S. Wiesner and C. Faber, *Nucl. Instr. Meth.* **73**, 13 (1969).
238. H. Krinninger, E. Ruppert and H. Siefkes, *Nucl. Instr. Meth.* **117**, 61 (1974).
239. Y. D. Lee *et al.*, *Nucl. Instr. Meth. Phys. Res.* **A459**, 365 (2001).
240. A. Gavron, L. E. Smith and J. J. Ressler, *Nucl. Instr. Meth. Phys. Res.* **A602**, 581 (2009).
241. S. Fetter *et al.*, *Sci. Global Security* **1**, 225 (1990).
242. S. Koonin *et al.*, *Report JSR-03-130*, JASON Study Group (The MITRE Corporation, McLean, 2003).
243. C. E. Moss *et al.*, *IEEE Trans. Nucl. Sci.* **53**, 2242 (2006).
244. *Detecting Nuclear and Radiological Materials*, Royal Society Policy Document 07/08 (The Royal Society, London, 2008).
245. *American National Standard Minimum Performance Criteria for Actvie Interrogation Systems Used for Homeland Security*, Standard ANSI N42.41-2007 (IEEE, New York, 2008).
246. P. Kerr *et al.*, *Nucl. Instr. Meth. Phys. Res.* **B261**, 347 (2007).
247. Y. P. Bogolubov *et al.*, *Nucl. Instr. Meth. Phys. Res.* **B213**, 439 (2004).
248. K. A. Jordan, Ph.D. Thesis, University of California, Berkley, (2006).
249. D. L. Chichester and E. H. Seabury, *IEEE Trans. Nucl. Sci.* **56**, 441 (2009).
250. K. A. Jordan, T. Gozani and J. Vujic, *Nucl. Instr. Meth. Phys. Res.* **A589**, 436 (2008).
251. D. L. Chichester and E. H. Seabury, in *Proc. Int. Topical Meeting on Nuclear Research Applications and Utilization of Accelerators* (International Atomic Energy Agency, Vienna, 2009), SM/EN-0.
252. W. Rosenstock and T. Köble, in *Proc. Joint ESARDA/INMM Workshop on Science and Modern Technology for Safeguards*, Arona, Italy (1996), p. 29.
253. W. Rosenstock, T. Köble and P. Hilger, in *Proc. 21st ESARDA Symp. on Safeguards and Nuclear Material Management*, Seville, Spain (1999), p. 321.
254. C. E. Moss *et al.*, *IEEE Trans. Nucl. Sci.* **51**, 1677 (2004).
255. C. E. Moss *et al.*, *Nucl. Instr. Meth. Phys. Res.* **B241**, 793 (2005).
256. W. Myers, C. A. Goulding and C. L. Hollas, *Report LA-UR-06-3984* (Los Alamos National Laboratory, Los Alamos, 2006).
257. G. R. Keepin, T. F. Wimett and R. K. Zeigler, *Phys. Rev.* **107**, 1044 (1957).
258. E. B. Norman *et al.*, *Nucl. Instr. Meth. Phys. Res.* **A51**, 608 (2004).
259. D. R. Slaughter *et al.*, *Nucl. Instr. Meth. Phys. Res.* **A579**, 349 (2007).
260. J. M. Hall *et al.*, *Nucl. Instr. Meth. Phys. Res.* **B261**, 337 (2007).
261. A. Dougan *et al.*, *Report UCRL-TR-202775* (Lawrence Livermore National Laboratory, Livermore, 2004).
262. C. L. Hollas, C. Goulding and B. Myers, *Report IAEA-SM-367/17/06* (International Atomic Energy Agency, Vienna, 2001).
263. J. T. Mihalczo *et al.*, *Nucl. Instr. Meth. Phys. Res.* **B213**, 378 (2004).
264. J. A. Mullens *et al.*, *Nucl. Instr. Meth. Phys. Res.* **B241**, 804 (2005).

265. J. M. Verbeke *et al.*, *Report UCRL-PROC-231582* (Lawrence Livermore National Laboratory, Livermore, 2007).
266. D. V. Ellis *et al.*, in *Trans. SPWLA 36th Annual Logging Symposium* (Society of Petrophysicists and Well Log Analysts, 1995), available at http://www.onepetro.org/mslib/servlet/onepetropreview?id=SPWLA-1995-C&soc=SPWLA.
267. J. B. Shapiro, W. D. James and E. A. Schweikert, *J. Radioanal. Nucl. Chem.* **192**, 275 (1995).
268. J. K. Shultis *et al.*, *Appl. Rad. Iso.* **54**, 565 (2001).
269. R. J. Gehrke and G. G. Streier, *Report INEL/EXT-97-00141* (Idaho National Engineering and Environmental Laboratory, Idaho Falls, 1997).
270. A. R. Dulloo *et al.*, in *Proc. 1999 Industry Partnerships to Deploy Environmental Technology Conference* (National Energy Technology Laboratory, Morgantown, 1999), paper 1.2.
271. A. R. Dulloo *et al.*, *Nucl. Instr. Meth. Phys. Res.* **B213**, 400 (2004).
272. L. Wielopolski *et al.*, *IEEE Trans. Nucl. Sci.* **47**, 914 (2000).
273. S. Mitra *et al.*, *IEEE Trans. Nucl. Sci.* **54**, 192 (2007).
274. M. L. Perez-Griffo, R. C. Block and R. T. Lahey, *Nucl. Sci. Eng.* **82**, 19 (1982).
275. R. L. Caldwell *et al.*, *Science* **152**, 457 (1966).
276. R. G. Johnson, L. G. Evans and J. I. Trombka, *IEEE Trans. Nucl. Sci.* **26**, 1574 (1979).
277. M. W. Busch and O. Aharonson, *Nucl. Instr. Meth. Phys. Res.* **A592**, 393 (2008).
278. M. L. Litvak *et al.*, *Astrobiology* **8**, 605 (2008).
279. *Radiation Source Use and Replacement* (National Academies Press, Washington DC, 2008).
280. J. Reijonen, K. N. Leung and G. Jones, *Rev. Sci. Instr.* **73**, 934 (2002).
281. I. J. Kim and H. D. Choi, *Nucl. Instr. Meth. Phys. Res.* **B241**, 917 (2005).
282. D. L. Chichester *et al.*, *Nucl. Instr. Meth. Phys. Res.* **B261**, 835 (2007).
283. I. Solano *et al.*, *Nucl. Instr. Meth. Phys. Res.* **A587**, 76 (2008).
284. P. A. Hausladen, J. S. Neal and J. T. Mihalczo, *Nucl. Instr. Meth. Phys. Res.* **B241**, 835 (2005).
285. J. S. Neal *et al.*, *IEEE Trans. Nucl. Sci.* **55**, 1397 (2008).
286. B. Naranjo, J. K. Gimzewski and S. Putterman, *Nature* **434**, 1115 (2005).
287. J. A. Geuther and Y. Danon, *J. Appl. Phys.* **97**, 074109 (2005).

Chapter 7

NONDESTRUCTIVE TESTING AND INSPECTION USING ELECTRON LINACS

William A. Reed
Varian Medical Systems
Security & Inspection Products
6811 Spencer Street,
Las Vegas, Nevada 89119, USA
williamreedphd@gmail.com

This chapter focuses on the role of electron accelerators that produce X-rays in NDT applications and emphasizes topics of interest to those using accelerator-based inspection products. In addition, it highlights the developments in digital detector technology that have become the precursor to more sophisticated digital radiography and computed tomography applications as well as the emerging security inspection market. Because the demand for accelerators for cargo inspection has vastly exceeded all other industrial NDT applications, this market segment, its special requirements, and future trends are comprehensively discussed. Accelerator systems that are specifically designed to produce neutrons can also be used for some of the same applications, but these are covered in Chapter 6.

1. Introduction

Nondestructive testing and inspection (NDT) refers to the examination of materials and equipment with techniques that do not alter, destroy, or otherwise interfere with the object under test. While many technologies have been developed for this purpose, radiographic techniques were among the earliest and remain ubiquitous in the industry. Because radiography offers unique capabilities for revealing the internal structures of dense materials, this method of inspection has been increasingly adopted and customized for a variety of new applications. The first radiographs taken for NDT utilized X-ray tubes to expose film behind the object of interest. However, the relatively low energy X-rays from these devices resulted in little penetration through many materials of interest. With the advent of commercially available electron linear accelerators (commonly referred to as "linacs") in the late 1950s, radiographers began to use these more powerful sources to create clear radiographs of significantly thicker and denser materials, allowing manufacturers better quality control for large manufactured products and complex assemblies. Concomitantly, the radiography industry was becoming increasingly organized and formally established, resulting in more professionally trained practitioners. This capability of testing larger objects along with the availability of more sophisticated inspection methods resulted in a significant demand for industrial accelerators. During the intervening decades, new applications and improved methods have continued the trend.

From its beginnings in 1941, the American Society of Nondestructive Testing (ASNT), which was originally the American Industrial Radium and X-ray Society, led industry efforts in educating its members, coordinating general information on radiography, and improving radiographic techniques. Today the ASNT publishes *The Handbook for Radiological Testing* (Bossi, *et al.*),[1] which provides authoritative information on virtually all radiographic inspection topics. Along with the association's conferences and training, the Handbook serves to standardize radiographic practices for the industry. Therefore, readers are directed to this handbook as an excellent resource that details accepted practices and techniques for a wide range of radiographic testing applications.

1.1. *History of X-rays for NDT*

The first radiographs were reported as early as 1896, shortly after the discovery of X-rays by Wilhelm Conrad Roentgen in 1895. Prior to 1912, however, radiographs were used mainly in medical practice for imaging broken bones.[2] During those early years, X-ray tubes were refined to produce stronger and more reliable outputs. These tube innovations included metal targets, curved cathodes, and more robust gas pressure systems. Researchers also had begun to investigate the practical use of radioscopy in industrial applications and construct machines that could inspect armaments destined for deployment in World War I. This technology also evolved into systems for detecting contraband using the fluoroscopy of packages, similar in purpose to today's airport baggage inspection machines.[3]

By the 1920s, X-ray developments primarily centered on medical applications. However, scientists in Germany and the United Kingdom continued to explore industrial radiography applications, which resulted in the development of both laboratory and field-based radiographic testing equipment. In the United States, Horace Lester was best known for establishing the groundwork for industrial radiography by documenting defects in castings and welds and relating those discontinuities to premature failure.[4] As a noted metallurgist, his work at the Watertown Arsenal demonstrated the value of industrial radiography and eventually led to the formation of the ASNT.[2]

While X-rays were being produced from increasingly sophisticated tubes, their physical characteristics limited their use to energies below ~500 keV. Researchers were experimenting with other devices such as resotrons, betatrons, and Van de Graaff generators that could produce much higher energies. These employed different techniques to accelerate electrons to very high energies, with some exceeding 25 MeV. However, their disadvantages included size and weight, along with relatively low dose outputs and undesirably large (effective) spot sizes. A small beam spot size is desirable for imaging because it results in a more point-like X-ray source, which reduces the shadowing effects (or penumbra) in the image.

For applications requiring much less flux, radioactive isotopes became commonplace sources of high energy X-rays. Isotopes such as cobalt-60 and iridium-192 produce gamma rays exceeding 1 MeV and have useful dose rates in relatively small packages. For example, cobalt-60 has a robust photon rate with output spectra at both 1.17 and 1.33 MeV. Frequently, manufacturers create pellet sizes in the 1 mm range to meet their flux requirements while achieving small spot sizes. When greater flux is needed, multiple pellets can often be combined in the same source holder.

General Electric is credited with developing the first commercial non-isotope X-ray source that exceeded 1 MeV. Introduced in 1939, their resonant transformer design became available in 1 and 2 MeV versions. Higher energy X-ray machines based on Van de Graaff generators and betatrons were manufactured with outputs of 15 to 25 MeV. However, the limited flux of those machines encouraged researchers to investigate other technologies, leading eventually to the development of the radio frequency (RF) electron linac.

Electron linacs were manufactured starting in the 1950s to address very specific military needs for X-ray sources with energy and flux combinations significantly higher than available with isotope sources or other types of X-ray generators. For example, in 1959 Varian Associates (now Varian Medical Systems) delivered a 10 MeV accelerator to the US Navy for the inspection of the Polaris missile and other munitions. Since then, a modest but growing commercial and military market for high energy X-ray systems has driven the development and productization of a broad range of commercial accelerators. This effort was led by Varian Associates, which produced the Linatron® family of accelerators in 1968 with energy ranges from 1 to 15 MeV. Linatron accelerators are still produced by Varian Medical Systems through its industrial group known as Security & Inspection Products. While some of the original systems are still operating in the field, new systems with contemporary electronics and more advanced functionality are available from Varian and other manufacturers worldwide. The newer systems utilize more automated features, and many have interactive touch screen displays that allow operators to monitor and control parameters relating to operation. For example, the Varian Linatron K-15 has a human–machine interface (HMI) that displays a schematic of the subsystems, allowing operators to get more information just by touching the appropriate area. Colors are used to indicate status — parameters in their normal ranges are shown in green and faults are shown in red. The integral computer also maintains a digital log of each event for future review.

1.2. *X-ray tubes vs. electron linacs*

Traditional X-ray tubes are commonly used to image small objects, and they provide an excellent solution for many NDT applications. At energies between 80 keV and 450 keV and operating with modest flux rates, X-ray tubes are relatively small and cost-effective with adequate durability for industrial use. However, the physical properties of most X-ray tubes limit their energy to ~450 keV and a continuous dose rate of 15–30 rad/min (0.15–0.30 Gy/min) at 1 m. At these maximum kilovolt energies and flux combinations, effective penetration through steel remains well under 4 in thicknesses. For example, a typical 250 keV energy equates to a half-value layer (HVL, defined as the thickness of material at which half the incident flux has been attenuated) through steel of 0.25 in (6.35 mm), allowing for maximum effective imaging of a steel section that is 2.5 in (63.5 mm) thick.[5] This estimate is based on a rule of thumb for maximum penetration of approximately 10 HVLs of steel at these energies.

For industrial applications, especially in the production environment, penetration and flux rates are important beyond the requirement for a good quality image. Within limits, greater penetration and more flux can reduce exposure times and allow faster imaging. This becomes especially important when estimating the production cost of using accelerator technology for the inspection of thicker or denser materials such as turbine blades or heavy castings. In addition, accelerators are often considered a safer alternative to isotope sources[6] (see p. 52 of the reference) because they produce X-rays only when power is applied.

Table 1. Maximum penetration vs. energy in steel.

Technology	X-ray tubes			Linear accelerators		
Energy	120 keV	250 keV	400 keV	3 MeV	6 MeV	9 MeV
Steel Thickness @ 10 HVL	1.0 in	2.5 in	3.5 in	9.1 in	11 in	12 in

Electron linacs used for generating X-rays (sometimes referred to as "X-ray linacs") are unique in their ability to produce X-rays with high dose rates (flux) and energy outputs in the MeV range. This allows them to operate significantly beyond the capacity of X-ray tubes or isotope sources to effectively penetrate steel and other dense materials and produce images in short time periods. Table 1 illustrates the maximum penetration in steel that can be expected from X-ray tube versus linear accelerator technology. The ASTM Standard E94-04[5] gives the equivalent penetration for other materials. The penetrations shown in Table 1 are based on the standard of imaging through a maximum of 10 HVLs of steel. At megavolt levels, the penetration values shown can be multiplied by a factor of 1.5 to 2.0 depending on the imaging conditions. This additional penetration capability is due primarily to beam hardening effects of imaging through thick sections but also includes optimizing the imaging chain with the use of low scatter techniques and sensitive detection. This standard is explained in more detail later.

While 1 to 9 MeV accelerators will meet most inspection requirements, higher energies may be needed to image very thick and dense objects such as heavy equipment castings, some composite materials, and large welded assemblies. Other examples include NASA's requirement to examine solid rocket boosters as well as military applications for inspecting ship and aircraft components. For these applications, accelerators with outputs of 15 MeV or more and flux rates exceeding 100 Gy/min at one meter are available as standard products. These systems can penetrate an 18 inch (45.72 cm) thickness of steel or the equivalent in other materials. Even with such massive energy outputs, these large X-ray machines are designed to operate continuously for hours without interruption and may have effective service lives exceeding 10 years.

2. Market Overview

Beginning in the 1950s, RF linacs emerged as a significant inspection tool for military and civilian manufacturers of armaments, castings, pressure vessels, turbine blades, and other critical components. Early electron linac manufacturers included well known companies such as General Electric, Mitsubishi Electric (under the name "Melco"), and Varian Associates (now Varian Medical Systems). The current major manufacturers of electron linacs for security inspection and industrial NDT are Varian and Nuctech (China). There are also several other industrial accelerator vendors worldwide, including EuroMeV in France and Siemens in Germany, along with XScell (formerly Schonberg Research Corp) and L&W Research in the United States. Other linac manufacturers such as Accuray and Tomotherapy promote their accelerators primarily for the medical market in the US. A number of companies, such as Mevex, Iotron, and Advanced Cyclotron Systems, also provide equipment in support of accelerator technology.[7]

A significant accelerator market also exists in Asia. For example, in Japan, Sumitomo, Mitsubishi, and Hitachi provide synchrotron and cyclotron technology in support of radioisotope medicine. In addition, countries such as Russia, India, Korea, and China also have companies and institutions that supply the industrial, medical, and research markets. For a complete review of the Asian accelerator industry, see the recent presentation by C. Tang.[8] There are also a changing number of smaller manufacturers who specialize in modifying old medical accelerators for industrial use or who purchase and refurbish used industrial systems for new applications.

Industry-wide estimates are difficult to obtain, but the total number of industrial accelerators produced worldwide for nondestructive testing and inspection between the 1950s and 2010 has clearly exceeded 1500 systems. Nearly one half of these have been utilized by the rapidly emerging security and cargo inspection market. At the close of 2010, the global industrial inspection and testing market for 3–15 MeV electron accelerators can be estimated at 200–300 accelerators per year, representing a direct value of perhaps US$250M. New, smaller, and more portable accelerators in the 1–3 MeV range are just entering the market and have the potential to add significantly to the population over the next few years.

The total market value related to these products is much larger when considering that many of these accelerators are designed into complex digital imaging systems that include large detector arrays, high powered computers, and sophisticated software. In addition, they are often housed in specially designed facilities or customized vehicles. Furthermore, these systems must be operated and serviced by trained personnel throughout their product life cycles.

For the last three decades, the Varian Linatron L-6000 and the Melco MR-15R have been the industry standards for industrial NDT accelerators with outputs over 9 MeV. With some units operating for more than 30 years, these early accelerators proved to be robust and reliable systems for inspecting large objects. Both of these earlier products are now obsolete, primarily because replacement parts can no longer be obtained. Today, a number of technological innovations have led to more reliable and cost-effective systems to replace them. In addition, today's high energy accelerators are easier to use, are available with more energy options, have higher flux outputs, and can be purchased with more factory supplied accessories.

While the NDT market has seen a relatively consistent expansion in recent decades, the simultaneous trend for manufacturers to consolidate their operations has resulted in a more modest overall net growth in this market segment. Consequently, the traditional NDT and inspection market continues with steady organic growth from a combination of new and replacement systems. However, a number of new applications continue to drive significant growth in market demand for industrial electron accelerators and have become the dominant force for the industry. These new applications are discussed in the following sections.

2.1. *Computed tomography*

Computed tomography, commonly referred to as "CT" scanning, of industrial products using accelerators has evolved from the extensive medical use of this technique to address a diverse set of industrial material and component design problems. The utilization of industrial CT machines (typically with X-ray tube sources) has increased over the last decade as more powerful computers and control electronics have become available at much lower cost. This has led to new software tools that can utilize the resulting three-dimensional data generated by CT scans and apply it to numerous applications.

While accelerator-based industrial CT systems fill an important market niche, they are significantly less common than tube-based systems, primarily due to more limited applications and the cost of the large scale machinery necessary to image the correspondingly larger objects. However, some objects are highly dense and relatively small, creating the ideal scenario for CT inspection with accelerator sources. For example, modern turbine blades are critical structures in jet engines and require 100% imaging to assure quality. Their construction from dense alloys and composites calls for inspection with the high energy and high flux rates available from accelerators. This is especially important because these parts are typically manufactured in a production line environment where speed of inspection is an important factor in overall cost. Accelerator-based industrial CT inspection may become a much larger market in the near future as smaller and more portable accelerators become available at the same time that a robust emerging market for CT inspection of air cargo is unfolding. More information on this is given later in this chapter.

2.2. *Metrology and reverse engineering*

The three-dimensional examination capability of CT creates the opportunity for advanced software to not only detect but also quantify the internal characteristics and dimensions of structures within a part or assembly. By receiving complete three-dimensional data from CT systems, metrology software can reconstruct the images and calculate the actual dimensions between internal structures. Companies such as Carl Zeiss IMT (Germany) with their Metrotom series of products have incorporated X-ray-based CT to achieve this goal. Larger versions can be configured with linear accelerator sources when purchased as a custom product. Their functionality is the same as smaller tube-based systems, but additional care is taken to maintain tolerances over larger dimensions and control leakage radiation and its associated scatter. In addition, data files become much larger as objects grow in three dimensions. Consequently, these customized accelerator-based metrology systems tend to be larger and more complex, and feature more robust computer processing power.

2.3. *Hazardous waste identification*

Hazardous waste identification can be particularly problematic because of the danger associated with opening sealed containers (often 55 gallon barrels or larger) with unknown or unverified content. These steel barrels and larger vessels containing liquids or solids can be too dense for X-ray tubes to penetrate, creating a market opportunity for X-ray examination with accelerator sources. Consequently, a number of X-ray waste inspection systems have been deployed, typically consisting of 3 to 6 MeV electron accelerators with high energy digital detector arrays for rapid imaging and analysis.

The emerging market for hazardous waste inspection is beginning to take advantage of interlaced dual-energy accelerators, which can provide additional capability for material discrimination as an aid for identifying contents. First generally available to the industrial market in 2007, high energy material discrimination is a combination of hardware and software features that seek to identify specific materials based on the ratio of their X-ray absorption rates at two distinct energies. This technology was developed for the cargo screening market and is discussed in more detail later in the chapter.

2.4. *Portable inspection applications*

Industrial accelerators have traditionally been too large to readily accommodate field-portable applications such as bridge or pipeline inspection. However, a trend toward producing smaller, lighter, more modular X-ray accelerators was set in motion when the US homeland security customers established a demand for truck-mounted inspection systems and partially funded the necessary development. This trend is enabling the development of smaller and more portable inspection systems, paving the way for new field uses for high energy radiography. These new system solutions include self-contained portable detector panels, laptop computers with imaging software, and novel accelerator designs that are smaller and lighter than their larger counterparts.

The military has a particular interest in lightweight X-ray-based imaging systems for a number of tactical field applications, such as improvised explosive device (IED) identification and bomb disposal. While several manufacturers have provided deployable systems based on battery operated X-ray tube sources, their performance is limited by the available flux and energy output of small tubes. The miniaturization of industrial accelerators operating in the 1–2 MeV range will meet a strong need in this market. To accomplish this, manufacturers are incorporating innovations such as X-band RF accelerating structures, energy recovery technology, and solid-state modulators.

2.5. *Isotope replacement*

A large number of isotope sources have been licensed by industry to provide high energy X-rays for all types of applications. As of 2008, there were approximately 5000 devices containing nearly 55 000 high-activity radiation sources licensed for use in the United States.[9] As indicated earlier, two isotopes in particular (cobalt-60 and iridium-192) emit megavolt X-rays and are commonly used in the radiographic inspection industry. With concerns over the possibility that these sources might be used in radiologic dispersal devices ("dirty bombs"), the US Congress passed an energy policy law[10] in 2005, which included a provision to review civilian uses of radiation sources and identify alternatives. The committee empowered by this act, working under the auspices of the National Research Council, noted that the users of isotope sources do not incur the total cost of these devices (e.g., disposal costs) and recommended incentives to switch to alternative sources. For example, the committee estimated that the cost of returning used cobalt-60 sources to the manufacturer is typically in the tens of thousands of US dollars. Thus, the projected costs of isotope sources are expected to rise, and licensing will be more stringent as regulators attempt to incentivize alternatives.

During the period from 2005 through 2010 and in spite of the identified costs of radioactive sources, little progress was made in removing them from the public domain. According to a recent report from the US Department of Energy (Ref. 6, p. 52): "The deployment of radioactive sources represents a national security risk. Growing concerns exist about the supply, security, and transport of radioactive isotopes for use in areas such as industrial R&D, basic research, medical tracers, and security applications." Consequently, replacing isotope sources with accelerators has become a priority for cost, safety, and homeland security reasons.

2.6. *Homeland security*

In 1992, a workshop on the role of nuclear physics in combating terrorism reported that the application of particle accelerators for combating terrorism fell into three categories: (1) forensic investigation, (2) the accumulation of basic nuclear and materials properties data, and (3) inspection techniques.[11] These areas remained primarily a scientific interest until the terrorist attacks in the United States on September 11, 2001 ("9/11"), which catalyzed a market demand for these technologies. Since then, technology from each of the three categories has experienced a different outcome.

Forensic applications are intended to identify the constituents of suspect materials that include radioactive components. The workshop participants speculated that accelerators could play a significant role in determining the source of materials through identifying trace isotopic ratios that are unique to a specific material-processing path. Accelerator requirements for this application involve high energies and special equipment that best suit the laboratory environment, which would preclude routine field use. While semi-portable applications are possible, the designs would need to address issues such as significant power consumption, cooling requirements, and shielding safety concerns. Consequently, the commercial market for these systems has been very limited and is mostly driven by government regulatory and security authorities.

In the second category, accelerators were expected to prove useful in facilitating the classification of materials and establishing reference data for use in security initiatives. The authors of the report noted that more accurate gamma-ray attenuation coefficients were needed to develop better gamma-ray radiography techniques. Other examples included the need for improved photon attenuation coefficients, cross sections, neutron absorption scattering cross sections, and the like. These applications are also laboratory-based, calling for a small number of accelerators operating in specialized research environments.

The final category represents more traditional X-ray use for inspection. Accelerators that produce bremsstrahlung radiation were already available for industrial inspection of castings, welded assemblies, rocket motors, and other large objects. As the security market developed, this equipment was quickly adapted for high resolution imaging of cargo containers and large vehicles. Conceptually, these were scaled-up versions of traditional airport-style baggage screening systems. However, their design requirements addressed a number of challenges related to the need for increased size, faster speed, and higher energy.

Neutron-based inspection was described as another potential application of accelerators, incorporating techniques such as thermal neutron analysis (TNA), fast neutron analysis (FNA), and pulsed fast neutron analysis (PFNA). In practice, neutron-based inspection systems direct neutrons to a specific area and then look for the unique gamma ray signatures emitted from the object of interest. When sufficient quantities of the threat material are present and the signal strength is adequate, these systems hold the promise of producing excellent material identification results.

Manufacturers are currently offering neutron inspection products, primarily based on isotope-based neutron sources rather than accelerators. The initial challenges for commercializing accelerator-based neutron generators were reported to include high cost, low flux, and limited operating lifetimes. For more information on the use of neutrons for inspection, see Chapter 6 of this book.

Since the 1992 report was published (and particularly following 9/11), security inspection applications have become the dominant industrial use for industrial electron accelerators, resulting in more than 700 of these X-ray systems being deployed for cargo screening throughout the world. This began in earnest with the creation of the Cargo Security Initiative (CSI) by the US Customs Service, with the objective of preventing adulterated cargo from entering the US. This market's momentum picked up speed with the advent of the US SAFE PORT Act in 2006, which required 100% scanning of US-bound cargo containers by 2012.[12] Under the requirements of this Act, foreign port operators have a vested interest in choosing X-ray inspection systems that can effectively prescreen cargo bound for US ports.

As a consequence of this US law, other countries are considering similar screening requirements for their inbound cargo. In addition, the use of large containers for importing cargo is increasing and was projected to account for 83% of all US imports in 2010.[13] As a point of reference, The Journal of Commerce On-line reported that a total volume of 12.7 million 20-foot equivalent containers were imported to the 10 largest US ports in 2009,[14] and this volume is estimated to continue to grow as the global economy improves.

2.7. *Air cargo*

Air cargo security is also a high priority for the US Transportation Security Administration (TSA) and sister organizations throughout the world. It was first widely recognized as an area of weakness in 1988 when a Pan Am 747 was bombed over Lockerby, Scotland. The bomb was hidden in a suitcase located in a cargo container in the forward cargo hold of the plane. Since then, standard airport screening has been expanded to include checked baggage, but other non-passenger air cargo remains a significant threat.

Air cargo represents a challenge for screening systems because it is packaged in different ways and may include a large percentage of "break-bulk" packaging (individual items and palletized material, often banded and shrink-wrapped for shipment). In other cases, cargo may be pre-loaded in large air cargo containers (referred to as ULDs), which vary in size by plane but can have cross sections exceeding 100 in × 200 in. Another challenge is that air cargo frequently arrives at collection points just prior to flight departures and must be processed quickly. Thus, effective screening of air cargo requires that equipment operate at high speeds, be able to fully penetrate the dimensions of the packaging, and be able to help the operator rapidly identify the contents. The effective implementation of such a system has generally been envisioned as a high speed conveyer system that would accept both individual boxes and palletized cargo.

The real challenge for obtaining market acceptance for such systems has been to achieve government certification. In August 2010, the Rapiscan division of OSI Systems was the first company to obtain US TSA approval for an accelerator-based air cargo screening system. Known as the Rapiscan Eagle A1000, this product is used to inspect air cargo pallets and containers using a 1 MeV X-ray accelerator. This energy is deemed sufficient to penetrate air cargo pallets without the associated shielding requirements of higher energy systems. While this is the first certified system of its kind, the US Department of Homeland Security (DHS) is seeking to advance this technology through additional research projects.

To that end, the Science and Technology Directorate of DHS has provided some guidance for their needs relating to future air cargo screening systems in the goals set forth under their research solicitation for an air cargo system (research opportunity BAA 10–19, published June 17, 2010). The key requirements listed in this solicitation refer to screening medium ($0.3\,\text{gm/cm}^3$) and high density ($>1\,\text{gm/cm}^3$) air cargo in pallets up to $40\,\text{in} \times 48\,\text{in} \times 60\,\text{in}$ and weighing up to 3500 lbs. The goal is to find IEDs "made with a variety of commercial, military, and improvised initiation circuitry, power sources, initiation devices (detonator), and main (explosives) charge material" (quote from p. 5 of the BAA). Successful systems are defined as those that can effectively image at least 20 pallets (or skids) per hour at a resolution defined by the detection of a 26 AWG wire under varying conditions.

Consequently, as this market matures, accelerator-based systems are expected to play the dominant role in meeting these stringent industry and government requirements while minimizing the disruption to legitimate commerce.

3. NDT Electron Linac Technology

Today, industrial electron linacs for NDT are produced with output energies in the range of $\sim$1–15 MeV. The lower bound is primarily economic. Below 1 MeV, traditional X-ray tubes often provide more cost-effective solutions with superb focal spot options and adequate flux for many applications. At the high end, only a few industrial applications require more than 15 MeV, making these systems a built-to-order specialty product. Most industrial applications fall between 3 and 9 MeV with flux outputs between 0.3 and 30 Gy when measured at 1 m from the target. In general, energy output is positively correlated to flux, i.e., higher flux rates usually correspond to higher energy accelerators.

Industrial NDT systems have traditionally been designed to operate in factory environments, protected from environmental extremes, and isolated from other equipment. They are installed in concrete shielded rooms (also known as bays or bunkers) with adequate radiation safety interlocks and manipulator equipment to locate the objects under examination and assist in moving them into and out of the room. The safety interlocks are typically connected directly to the X-ray machine controls so that X-rays cannot be initiated without a safe status at all measured points. Additionally, inspection bays are equipped with pre-warn alarms and emergency stop (E-stop) buttons that disable the X-ray output upon actuation should personnel be trapped inside during X-ray operation.

Contemporary NDT accelerators generate bremsstrahlung radiation from the rapid deceleration of high energy electrons when they penetrate targets composed of high atomic number (Z) materials such as tungsten and copper. The heat generated at the target from high and prolonged flux rates typically requires targets to be designed with liquid cooling. Tungsten is frequently the material of choice because of its combination of high density ($Z = 74$) and characteristically high melting point.

Neutrons are also produced in the target material by (γ,n) photonuclear reactions when the incident energy of a bremsstrahlung photon exceeds the binding energy of a neutron in a target nucleus. Neutrons are produced at a higher rate in tungsten than in lower-Z materials such as copper and steel. Since neutron production is generally an undesirable side-effect of X-ray production, there is a design tradeoff in material selection. In other words, higher-Z materials are desirable for their bremsstrahlung production but undesirable for their neutron production. For tungsten, the photonuclear effect occurs with photon energies at 6.18 MeV and higher. Therefore, this is not a significant concern for accelerators with output energies of 6 MeV or less.

While the production of bremsstrahlung X-rays and neutrons in the target is nearly isotropic, neutrons are also produced in the heavily collimated shielding materials that are often added to reduce X-ray scatter. Thus, neutron production can increase when adjustable collimators are at their closed positions. Because neutrons are subject to elastic collisions in high-Z shielding materials, those materials are not especially effective as neutron shields. Consequently, the manufacturers of accelerators above 6 MeV may include other shielding materials that are specifically selected for neutron absorption and may specify additional safety requirements for operating these systems.

3.1. *NDT accelerator basics*

Most modern industrial electron accelerators utilize a standing wave coupled-cavity structure for acceleration of an electron beam using microwave fields. First developed in 1964, this structure is significantly smaller, lighter, and more cost-effective than many of the traveling wave systems still used in scientific laboratories. For a more detailed technical description and theory of operation of these and other linear accelerator structures, see *Principles of*

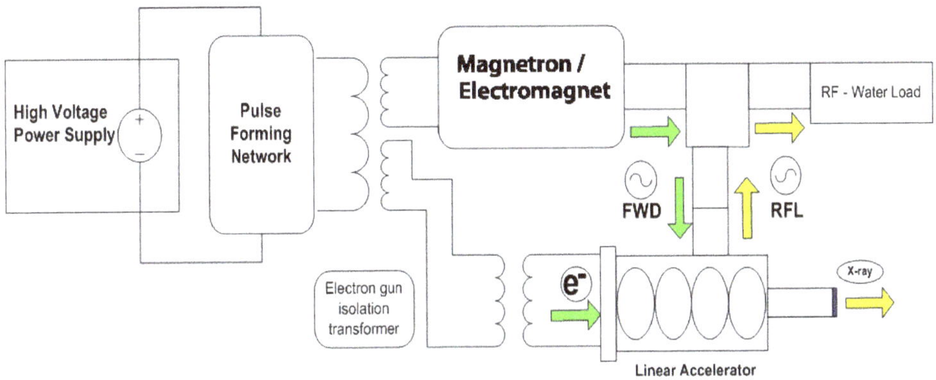

Fig. 1. Schematic layout of a typical standing wave industrial electron accelerator.

RF Linear Accelerators by Wangler.[15] A schematic layout of a typical compact standing wave electron accelerator for industrial NDT is shown in Fig. 1.

These electron linacs operate in a pulsed mode, such that the microwave field (typically about 3 GHz for S-band and approximately 9.3 GHz for X-band) in the accelerator cavities and the high voltage at the electron gun only exist for a few μs at a time. The very high voltages and currents required to create X-rays are most economically produced as pulses of energy rather than continuous outputs. This reduces the cost of many system components and in some cases, allows smaller components to be used. In most NDT systems, the pulse widths are typically a few μs with pulse repetition rates that vary from 50 to 500 Hz. This produces a beam duty cycle ranging from $\sim$0.02 to 0.2%.

Most of these accelerators rely on a magnetron to produce microwaves for the accelerator cavity. For magnetron-based systems, X-ray output stability is often a function of the magnetron's ability to start immediately and produce outputs that correspond to its pulsed high-voltage input. Because magnetrons are essentially mechanically resonant oscillators, there is a small variation in pulse startup from unit to unit and over their operating range. Also, there can be some normal variation in the input pulses that drive the magnetron into oscillation. Consequently, there are two notable pulse output artifacts inherent in contemporary magnetron-based systems. First, some output pulse jitter occurs that is primarily traceable to slight variations in magnetron oscillation during the first part of each output pulse. Linac manufacturers usually provide a pulse stability specification of about 5% to quantify this variation.

The second artifact is the probability of missing pulses at the output of the magnetron. Even with modern magnetron designs, occasional missing pulses occur at the output even when the input is within tolerance. Missing pulse rates vary depending on the magnetron used but typically occur at a rate of less than 0.1% when averaged over long time intervals. However, in practice, there is a statistical likelihood that when missing pulses do occur, they will sequentially occur in groups of two or three rather than being equally distributed over time.

With traditional film radiography, these artifacts have little or no consequence since many hundreds or thousands of pulses are integrated to produce an image. However, more recent computed tomography and digital radiography applications require that many fewer pulses be used to image a single "slice" of an object. In some applications, manufactures have specified one pulse per slice, so that a single missing pulse will produce a black (unexposed) line across the image. System designers can address this characteristic through their use of multiple pulses per scan line or software interpolation techniques. While these occasional missing pulses are undesirable, they should not be considered a defect in the system.

Some accelerator systems utilize klystrons rather than magnetrons for their source of RF power. Klystrons are generally larger, heavier, and more expensive than magnetrons but have a number of advantages. Their output frequency, phase, and power can be precisely controlled, and they don't suffer the occasional missing pulse phenomenon of magnetrons. Another advantage is that klystrons are available with much larger average power outputs than magnetrons, making them more common in higher power accelerators.

Interlaced (sometimes called interleaved) dual-energy accelerators are designed mostly for security inspection applications. These systems feature X-ray output pulses that alternate between two distinctive energies and generally operate near maximum repetition rates. The most significant functional difference for these dual-energy accelerators is that a minimum of two pulses is needed for each scan line, effectively reducing the scan rate by 50%. Because high scanning speeds are a valuable feature in cargo screening applications, manufacturers are addressing this limitation with new designs that allow higher overall pulse rates.

3.2. *Major subsystems*

Most industrial NDT X-ray systems are composed of four main subsystems. The *X-ray head* is located in the inspection bay and produces the X-rays. It contains the sealed accelerator "tube", which houses the electron gun, coupled-cavity structure, X-ray target, and supporting components. It also includes the high power source of microwaves that are fed into the accelerator tube to create the fields that accelerate the electrons through the cavities. The X-ray head contains power supplies, control electronics, and some degree of shielding to reduce stray radiation. It may also have a laser sighting option that is useful during radiographic setup procedures to identify the projected center of the X-ray beam. Filters, collimators, and other options may also be available from the manufacturer. These can be built-in or attached to the output of the head.

The *modulator cabinet* typically includes circuitry for a number of functions, most of which provide high voltage electrical and control signals to the X-ray head and *control console*. Traditionally, the modulator also produces the high power pulses that drive the electron gun and microwave power source. These pulses are sent to the X-ray head through high voltage cables, which limits the maximum distance between the control console and X-ray head to about 300 ft. Some newer designs have incorporated the generation of high voltage pulses into the X-ray head itself, thus eliminating the long run of high voltage cables and simplifying the connections between the modulator and the X-ray head.

Because modulators have traditionally included controls and indicators of accelerator operation, facility planners may locate them in close proximity to the operator and the control console. With older systems, operators placed an oscilloscope near the modulator to observe and measure reflected microwave power from the accelerator. This was useful because operators had to routinely adjust the magnetron frequency to keep the accelerator in tune and producing full output. Modern accelerators have excellent AFC (automatic frequency control) and no longer need this legacy feature for manual tuning.

The *control console* usually features a touch-screen panel and display system to control all aspects of the X-ray system operation. Most systems provide an intuitive interface that allows appropriate selections and identifies any problem conditions through a color-coded display of virtual buttons. Users may select operating parameters such as pulse rate and triggering but are prevented from adjusting parameters that would be detrimental to effective operation. For service personnel, there are password-restricted display and selection screens that assist in troubleshooting and identification of previous performance metrics. The control console also includes a key switch to prevent unauthorized personnel from operating the equipment. Consoles may have a second key switch that determines whether the console is in local or remote operation. In remote operation, the system control is switched to a second console with similar controls.

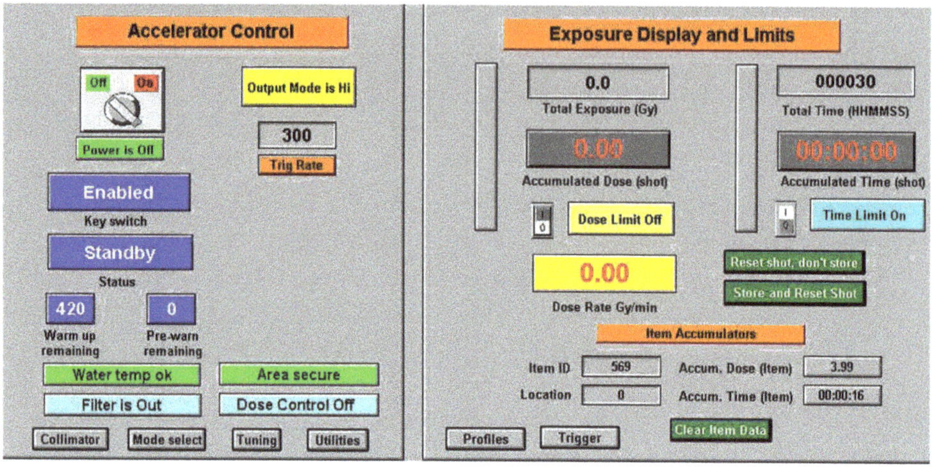

Fig. 2. GUI display from an industrial NDT accelerator computer-based operating system.

Graphical user interfaces (GUIs), such as the one shown in Fig. 2, that provide operation from a dedicated computer are also available. These GUIs provide the operator with more user-friendly information about the status of the machine and its controllable parameters. Computer interfaces generally replace the standard control console and may provide additional features such as long term storage of operating parameters.

The final major subsystem is the *cooling system* (chiller), which serves as a liquid-cooled temperature controller for the X-ray head and related parts. The chiller provides coolant at a precise temperature (usually 30°C) and pressure to circulate through heat exchangers in the critical heat generating components such as the accelerator tube, power supplies, magnetron, and target. Sensors in the X-ray head will alarm if coolant flow falls outside of specifications and will instruct the control console to shut off the accelerator. Chillers are the one major component that is usually located outdoors, so they are often designed for environmental exposure and extended temperature operation.

Newer NDT accelerator designs tend to be more modular with standardized controllers and more interchangeable components. With the advent of electron accelerators designed especially for homeland security applications, X-ray heads are available with numerous shielding choices and additional collimator options that are optimized to work with digital detector arrays. The mobile environment has also led to more robust mechanical frames and mounting techniques along with component selections that meet the extended temperature requirements consistent with outdoor operation and storage.

3.3. *Beam properties*

Specifications for NDT systems identify their energy and flux range along with beam characteristics and X-ray leakage patterns. Flux and energy combinations are related in a complex manner; however, higher energy outputs usually correspond to higher dose (flux) capability. Most NDT accelerators, like the example shown in Fig. 3, have output ranges similar to those listed in Table 2.

Pulse repetition rate (pps) directly affects the flux rate in these accelerators and is often the primary adjustment mechanism to vary the output. The typical adjustment range is 50–350 pps. Operators can also set X-ray exposure to terminate on elapsed time or by accumulated dose as measured by the accelerator's internal ion chamber. Because most accelerators utilize

Fig. 3. An example of a contemporary industrial accelerator for NDT applications, the Varian Linatron-M.

Table 2. Typical NDT accelerator output energy and flux rates.

Output e$^-$ energy	3 MeV	6 MeV	9 MeV
X-ray flux @ 1 m	0.3–3.0 Gy	0.3–8.0 Gy	0.3–30.0 Gy

a pancake style ion chamber to monitor the flux directly following the target, any additional collimation of the primary beam must be considered when calculating the flux that reaches the object of interest.

3.3.1. *Radiation leakage*

Leakage is defined as the radiation produced outside of the primary beam and is usually specified as radiation emanating at greater than 60° off-axis from the beam center. Typical linac leakage specifications are on the order of 0.1% of the primary beam, although the manufacturer may offer optional shielding that reduces off-axis radiation to $\leq 2 \times 10^{-5}$ of the primary beam. It is customary to measure and specify this leakage along the horizontal axis at 10° increments. For inspection systems operating in shielded rooms, leakage may only be a concern because of its ability to produce scatter artifacts on the X-ray images. However, for accelerators used in security applications, personnel safety concerns may create the need for additional internal shielding. In other cases, additional shielding may be added in order for digital detectors within a specified imaging chain to work at maximum sensitivity and maintain an adequate signal to noise ratio. This includes shielding and collimation at the X-ray source as well as at the detectors. Overall, manufacturers tend to offer a number of shielding options consistent with their intended applications, and customers routinely add additional temporary shielding when needed for particular test setups.

3.3.2. *Energy spectrum*

NDT systems produce unfiltered X-rays in a broad spectrum of energies up to the maximum electron energy at the target (i.e., "end point"). This X-ray spectrum is known as "white radiation" until it becomes filtered by obstacles in its path. When absorbing materials are placed in the path of the X-ray beam, they tend to absorb a greater proportion of lower energy photons. This "hardens" the beam and results in a narrower spectrum of emitted X-rays that are usable for imaging. Eliminating lower energy X-rays is desirable because they often do not fully penetrate the object under examination but can be a source of additional scatter.

3.3.3. *Beam flatness*

NDT systems in the 2–15 MeV range produce a significantly greater intensity of X-rays on the central beam axis than at angles just a few degrees off-axis. As a rule of thumb, higher energy accelerators have a more pronounced gradient, such that the peak for a 9 MeV system is a substantially greater proportion of the total output than that of a 3 MeV source from the same manufacturer. Beam flatness is traditionally defined as the percentage decline of the X-ray output over the center $15^\circ(\pm 7.5^\circ)$ of the beam. Figure 4 illustrates how the beam flatness can decrease significantly once the electron beam energy exceeds 1 MeV, as shown with a 22° cone collimator in place.

Since most radiography applications utilize film with reasonably uniform exposure characteristics, it is often desirable to add a compensator or beam flattener to produce a more uniform intensity across the field of interest. The disadvantage of a beam flattener is that it works by reducing the intensity of the peak, perhaps by as much as 50%. Consequently, some manufacturers offer automated beam flatteners that can be remotely inserted or removed from the beam path as needed. An example of an automated beam (or "field") flattener is shown in Fig. 5. It is pictured without its cover to show the internal mechanism. The device is remotely controllable with two

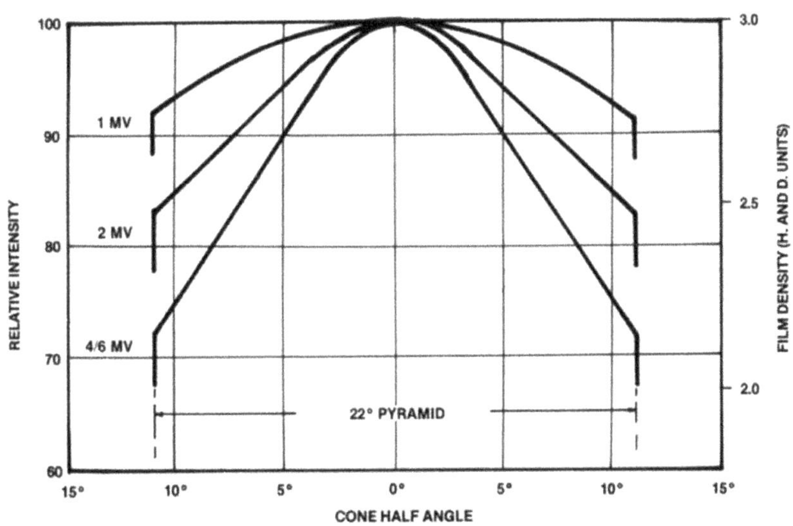

Fig. 4. X-ray output intensity profiles at different beam energies.

Fig. 5. Remotely controllable field flattener assembly.

settings. As shown, the field flattener is positioned directly in front of the target. The alternate setting is to have no field flattener in the path.

3.3.4. *Beam symmetry*

Beam symmetry is of similar importance to beam flatness since radiographers expect to see images that are consistent with the object under examination and not subject to deviations in illumination. Beam symmetry can be especially important when the objects examined are symmetric because operators compare each half-image referenced to a central axis. Beam symmetry can be influenced by a number of factors such as the target design, beam steering mechanisms, filters, and collimators. The degree of beam symmetry is usually verified during factory tests.

3.3.5. *Focal spot size*

The focal spot size of most industrial NDT accelerators is approximately 2 mm in diameter and is specified at the 50% intensity level of the output. That is, when a piece of X-ray film is exposed directly at the output of the target, the resulting image has a bright center with decreasing density moving away from the center. Through visual observation or software analysis, an outline is drawn around the center at the 50% intensity point. This is

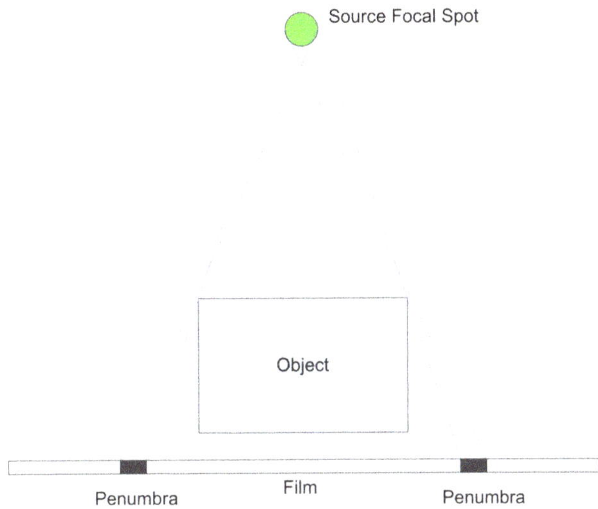

Fig. 6. Illustration of penumbra.

customarily specified as the spot size. Some manufacturers offer options that will narrow the focal spot to 1 mm or less, but there is a practical minimum primarily due to target heating. Note that the actual focal spot is seldom perfectly round, and therefore, it is specified in relation to its maximum dimension.

Focal spot size is important in many radiography applications because a large focal spot produces a shadowing effect on the image known as penumbra, as illustrated in Fig. 6. Thus, small spot sizes are desirable for creating crisp images, although the impact of spot size will be influenced by the geometry of the exposure and the depth of the object. In other words, moving the source farther from the object under examination also reduces the penumbra effects. Because the effective flux at the object is reduced by the square of the distance from the source, this can be a limiting factor.

3.3.6. *Collimation*

It is usually undesirable to allow the X-ray beam to cover an area larger than the desired exposure area because additional X-ray scatter can be produced, which appears as "fog" on film or "noise" on electronic detectors. Three types of collimation are available to limit these effects. First, *pre-collimation* is a term used to describe shielding that is placed between the X-ray source and the object. Radiographers may place lead bricks, shot bags, or other materials in the beam path to further collimate the X-ray beam prior to its contact with the object.

Radiographers also utilize *post-collimation* — the technique of placing shielding materials between the object and the recording media. Post-collimation is generally possible when there is substantial distance between the object and the recording media and is advantageous to reduce scatter from the inspected object itself or other items in close proximity.

The third type of collimation is referred to as *source* or *primary* collimation because it is internal to the accelerator itself as designed by the manufacturer. It may consist of both fixed and variable elements. To provide this, manufacturers select materials such as tungsten or lead for use around the target with a collinear opening after the target to produce the desired beam shape. These fixed elements may be cones, squares, or lines to match the intended use. For accelerator outputs above 6 MeV, collimation materials are selected based on both their X-ray and interaction characteristics.

The photograph in Fig. 7 shows an accelerator with an internal square-source collimator surrounded by an octagonal-shaped steel collimator. Note that this model includes a cylindrical receptacle for an additional fixed collimator that can be configured to meet customer requirements.

Another type of source collimation is an external motorized collimator, as seen in Fig. 8, that allows individual control over vertically and horizontally mounted "jaws" that further shape the beam. These are particularly valuable

Fig. 7. Standard fixed collimation.

Fig. 8. External collimator with cover removed.

when objects of varying size and shape are frequently radiographed. As shown in Fig. 9, these can be added to the outside of an accelerator to further limit the beam shape.

Externally motorized collimators are typically designed with independently operating jaws, and some include a rotation feature as well. They are remotely controllable at the system console and have position readouts to inform the operator of the rotation and jaw position. Typically, devices with the rotation feature allow about 10–15° of field size adjustment that is repeatable to <0.5°. Leakage through such devices is usually <5% of the incident X-ray beam.

4. Digital Detectors

Digital detectors are rapidly replacing film for a number of traditional NDT applications and are standard in the computed tomography and security inspection industries. Manufacturers offer proprietary detector systems that include the unique integration of X-ray sensitive crystals, scintillator materials, special shielding, and customized electronics. There are several design approaches representing tradeoffs for variables such as detection technique, pixel size, operating speed, cost, and durability.

Fig. 9. A Varian Linatron 15 MeV system with an external motorized collimator.

For example, pixelated digital detectors may rely on the intrinsic direct conversion of X-ray photons to electrical charge in the detector ("direct conversion"), or they may use a scintillator optically coupled to the detector. In this second case, the X-ray quantum first interacts with the scintillator, and the resulting visible-light photons generate photoelectrons in the digital detector ("indirect conversion"). Direct conversion is typically used for low energy X-ray detection. For indirect conversion, designers can select specific scintillator compounds to efficiently absorb X-rays and convert the energy to visible light. Scintillator crystals of various compounds and grain sizes can be used to produce a variety of resolutions and brightness. The visible light is then sensed by the photodiodes that comprise a typical pixilated digital detector. Detector engineers deal with a number of design trade-offs when selecting the scintillator material that best suits the application. Factors such as radiation hardness, crystal transparency, conversion efficiency, afterglow, durability, and cost must all be considered for a specific application. For example, afterglow refers to the ongoing illumination of the scintillated crystal after the photons have excited it. The afterglow duration limits the overall system speed since additional signals striking the detector will be incorrectly measured to the extent that afterglow is still present.

Digital detectors are available in both arrays and panels. The panel in Fig. 10 is typical of many digital detector panels used for medical and industrial applications. This panel utilizes an amorphous silicon detection

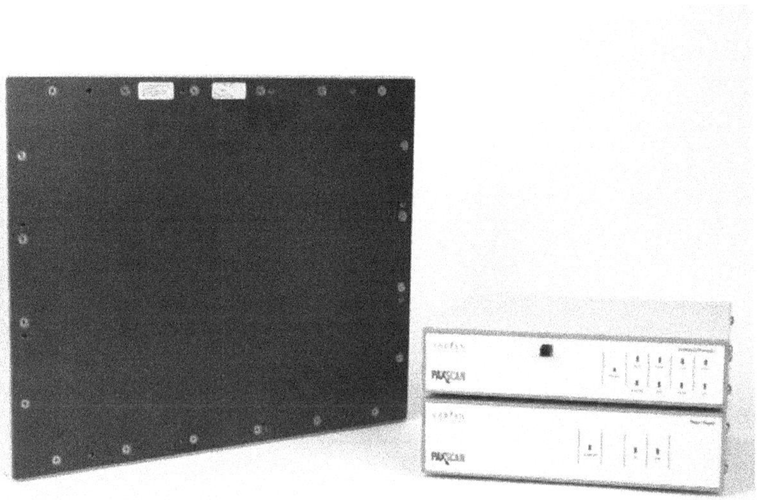

Fig. 10. A typical X-ray detector panel shown with electronic interface modules.

technology combined with cesium iodide scintillator material. It has a 194 μm pixel pitch and provides a resolution of 2048 $\times$ 1536 pixels with an overall active surface of 40 cm $\times$ 30 cm.

When film is an option, cost tradeoffs are important considerations for choosing between film or detector panels. The comparison involves considering the cost of the film, processing, storage, and labor for each exposure made with film. While the initial cost of detector panels can be quite high, the total cost of ownership often favors them, especially when the volume of exposures is high.

Digital radiographs also have other advantages. For example, they facilitate computer analysis and detailed comparison with historical images. The trend of organizations to switch from film to digital detectors is expected to accelerate as detector panels continue to become more economical and digital radiographs become more accepted by radiographers and quality-control auditors.

4.1. *High energy digital detectors*

High energy digital detectors differ from their lower energy counterparts because they are built to absorb much more energetic photons while simultaneously retaining sufficient sensitivity to record a very wide range of flux rates. Electronic components in the detector arrays must also be designed to withstand high energy X-ray absorption that would otherwise damage the electronics. Because gamma ray damage occurs primarily as a function of the cumulative flux that is absorbed by the detector electronics, engineers have developed a number of novel designs to address these issues. These include detector designs with alternative spatial orientation, special shielding, and fiber optics to reduce the quantity and impact of accumulated dose. Care also must be given to the effects of neutron bombardment on electronics for X-ray energies ≥ 9 MeV.

4.2. *Detector performance*

The optimal performance of digital detectors requires that a number of design parameters be carefully addressed. For example, the array shown in Fig. 11 is a cadmium tungstate ($CdWO_4$) scintillator-type detector assembly designed for use with megavolt energy X-rays. It is mechanically configured for placement at a specific distance from the source. Its curved profile establishes the optimal distance from the target such that direct path X-rays are equidistant to all detectors, allowing each discrete detector

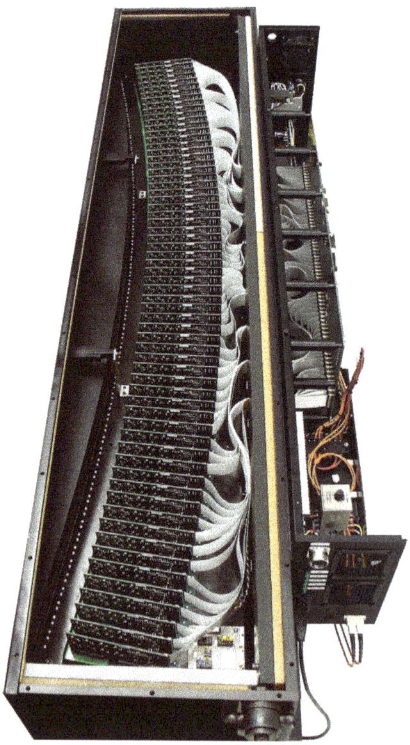

Fig. 11. Modular detector array.

to maintain a consistent focal distance. Additional design considerations include temperature control for noise reduction, the use of special detector shielding, and optimizing the detector pitch to match the collimated opening.

Similar to detector panels, the arrays must be specifically designed for the application and include a number of important design characteristics to enhance the performance of the imaging chain. These include:

- *Dynamic range.* Dynamic range can be defined as the difference between the maximum signal output from the detector and the minimum signal threshold above the noise. Because signals originate as analog voltages, they are converted to digital through an analog-to-digital converter. Typically, 18–20 bit converters are utilized, creating single bit resolutions equivalent to the $10\,\mu\mathrm{V}$ range. If the digital resolution of the converter exceeds the dynamic range, then a few bits may reflect noise rather than signal.
- *Sensitivity.* Sensitivity can be stated in either analog or digital terms depending on the context. For X-ray detectors, it is typically the minimum incremental value of the digital output, usually defined by the characteristics of the analog-to-digital circuitry in the detector circuit. It may also refer to the lowest level of signal that can be reliably detected. In both cases, sensitivity is established by design and limited by noise sources in the detector system.
- *Element size.* Detector elements range in size from ~2.5–10 mm and are selected to best match the desired application. Larger elements gather more photons, which facilitates faster scanning, and are generally more economical when they can adequately image the objects of interest. Finer pitched detectors have a cost premium and may have the flexibility to be used alone or in combination with adjacent cells to produce a larger effective element. When used as individual elements, the resulting images contain more data and can produce more detailed images if they are sufficiently illuminated.
- *Signal to noise performance.* Electrical noise limits the minimum detectable signal for a digital detector and occurs as the composite effect from a variety of sources. For fixed threshold systems, detector outputs are calibrated to provide the minimum signal at some level above the noise floor such that outputs are null when no signal is present. In this way, the detection threshold provides the best statistical probability of detecting a valid signal while rejecting most noise. With variable threshold systems, electronic circuitry or software algorithms can modify the threshold level to achieve the optimal performance during varying conditions. The

threshold level may be set during machine calibration or may be varied throughout the scanning process.

- *Scatter rejection.* Scatter is the generic term for photons arriving at the detector from secondary paths. As such, it is considered a source of interference or noise at the detector. Because photons received at the detector elements from scatter sources are indistinguishable from direct path photons, manufacturers utilize a number of techniques to eliminate as much scatter as possible. Scatter is discussed in more detail later in this chapter.
- *Artifact reduction.* Artifacts are visible elements in an output image, such as streaks, dark lines, or spots that can originate from a variety of sources or can be the result of signal variations in the imaging system. Software filtering is the primary remedy for occasional artifacts in scanned images. When these do occur, the nature of the artifacts can supply some diagnostic information to the operator. For example, in traditional cargo screening systems, solid horizontal lines that are consistent for the entire scan often point to a failed detector or detector module (depending on thickness of the line). Conversely, a solid but occasional vertical line may be indicative of a missing pulse from the accelerator or a synchronization problem between the accelerator and detector array.

5. Traditional Radiographic Testing and Inspection Applications

Radiographic testing is only one choice among a wide variety of NDT methods. However, it is often the method of choice when examining thick sections or searching for hidden defects within materials. As discussed earlier (Sec. 1.2), X-ray tube sources fill the need for examining smaller and less dense objects. With outputs up to 450 keV, tube sources are capable of penetrating about three inches of steel depending on the allowable exposure time and a number of other factors.

Because accelerators are becoming more economical and can produce significantly higher energy X-rays, they are playing an increasingly important role in radiographic NDT applications. The advantages of accelerators over tube sources of X-rays include:

- *Practicality of thick-section examination.* High energy radiography can examine thick sections with relative ease using traditional imaging techniques. In contrast, other inspection technologies often fail to identify artifacts hidden within thick sections or fail to completely penetrate the material under test.

- *Minimization of distortion.* High energy sources allow the source-to-object distance to be increased, creating a more uniform illumination over a broader area, which then reduces the distortion created by beam nonlinearities.
- *Faster throughput rates.* High energy sources also have much larger flux rates; hence, imaging can often be completed in much shorter times.
- *Higher resolution.* The robust fluxes produced by accelerators provide strong illumination for inspection, which facilitates high resolution imaging across a wide range of material densities.

5.1. *The radiographic inspection facility*

Radiography departments are usually identifiable by the facilities that are designed to carry out radiographic inspection work. This is particularly true for high energy (megavolt) radiography areas, which are noted for their heavily shielded bays or bunkers that enclose accelerators and manipulator equipment. These bays are traditionally quite large, not just because of the space required for equipment and the objects being inspected but also for reducing the X-ray scatter associated with accelerator outputs. Because accelerator leakage may be as high as 0.1% of the primary beam, distances of the scatter paths should be considered along with shielding and collimation options. This often results in examination bays with tall ceilings on the order of 30 ft and wall-to-wall distances of 40 ft or more. The facility shown in Fig. 12 has such a large bay. In addition, the wall behind the detector may be modified with an anti-scatter grid or a beam pipe to further reduce scatter. For a more detailed discussion of high energy X-ray bays, see Berry *et al.*[16]

The "reading room" is often overlooked as another critical part of an X-ray inspection facility. Reading rooms have traditionally been almost lightless spaces with film viewers and filing cabinets where workers painstakingly inspect images for subtle defects and discontinuities. Quality standards often require that each image be interpreted by two different people who must each find the same artifacts. The data on agreement rates are then documented as a quality control check on the process.

With the transition to digital imaging and filmless X-rays, reading rooms have become computerized, and filing cabinets no longer consume precious space. Now inspectors perform the same analyses on computer screens and benefit from new image enhancement software tools. This innovation has required new standards to ensure that electronic displays can adequately reveal the features present in the actual image. To this end, the medical

Fig. 12. Rocket motor inspection facility. (Picture courtesy of US Air Force, 510 ICBMSS, Hill Air Force Base, Utah.)

industry has led the transition from film viewers to computer screens and made recommendations for the type and size of displays (e.g., see Samei[17]).

Control rooms and equipment rooms are also important components of the radiographic facility. They typically contain the X-ray console that allows full operational control over the X-ray system. Most control rooms are also equipped with video monitors to observe the inspection bay via closed-circuit video and sometimes include controls to remotely manipulate the objects being tested or to position other equipment. While the inspection bay contains the X-ray head, it is often recommended that the modulator be housed outside of the bay in an environmentally controlled area that is free of radiation and maintained at room temperature. This equipment may be located in or directly adjacent to the control room. Smaller modulators are typically internally air cooled, but they produce at least a few kilowatts of heat, which may require additional facility cooling capacity. Conversely, some larger modulators are equipped with liquid cooling and deposit less heat load to the surrounding air. For both types of modulators, sufficient clearance is needed for service access and air flow requirements.

5.2. *Radiographic principles*

Nondestructive radiographic testing involves a number of practical steps that relate directly to the imaging process itself. This includes placement of the source (accelerator), object, and film (or detector) such that a sharp and well-defined exposure can be obtained over the area of interest. The shape or geometry of the object must be considered, as well as the projection of the object onto the recording medium. For most applications, the goal is to achieve a precise image on film that uniformly reflects the actual object being imaged. That is, measurements from one point to another on the X-ray image are linear mappings of the object's actual characteristics. In other cases, radiographers may choose to use a specific geometric placement to provide magnification of certain features. While conceptually simple, some of the variables involved are non-linear, creating a challenge for operators to design the test setup. Consequently, radiographic imaging is a highly skilled process that involves much more than simple operation of an accelerator and detector or film plate. Much like the work of photographers who set up their studios by positioning a camera, lighting, subject, and background, radiographers configure the inspection area to establish the optimal geometry and select the appropriate equipment to obtain the best images possible.

Once the geometry is established, setting the exposure parameters can still be quite complex. Radiographers must select an X-ray film that will match the geometric setup and have appropriate sensitivity for the intensity of the penetrating beam. X-ray films are similar to optical films with variations in parameters like grain size, grain density, and emulsion thickness. Therefore, once a film is selected, other equipment such as filters and masking screens are utilized by the radiographer in creating the optimal image. Some of the typical considerations in radiographic imaging are summarized below:

- *Scatter*. Ideally, X-ray imaging would result entirely from the direct rays passing through the object under inspection and continuing in a straight line path to the recording medium. However, the object itself produces scatter known as "object scatter" that results in noise (with digital detectors) or fogging (with film).[18] Other sources of scatter are collectively termed "background" scatter and are the composite effects of the X-ray leakage from the accelerator, along with scatter from the X-ray cabinet, collimators, other equipment in the room, and the walls themselves. Some have classified these sources of scatter as leakage radiation, improper pre-object collimation, post-object collimation, and back-wall scatter (see Ref. 16). Because the different types of scatter can have significantly different effects on the resulting image, it is important to identify the different sources of scatter and their geometry with respect to the source and detector.

- *Intensifying screens and filters.* Lead foil can be placed directly in front of (touching) the X-ray film emulsion to intensify the primary beam. The intensification effect is obtained from the emission of photoelectrons and Compton effects directly at the film surface. Intensifying screens also have a filtering effect to absorb a greater proportion of lower energy photons, which are more likely to arrive through scatter. Filters are used to improve image quality, primarily when the object under inspection produces excessive scatter. They also may help compensate for large thickness or density differences within the object.
- *Object placement.* Object placement (orientation) is important when objects are irregularly shaped or have non-uniform internal structures. With complex shapes and internal variations, some projections will reveal more information than others. Consequently, optimal orientations are determined partly by the object itself and partly by the type of defect being sought. Cracks, voids, density variations, and other artifacts may have spatial variations that combine with the material profile to make them difficult to identify in certain orientations.
- *D/T ratio.* Radiographers often use the calculation of a "D/T" ratio as a guide for effective imaging. This ratio represents the source-to-object distance (D) divided by the object-to-film distance (T). When the film is placed directly against the object, T is normally defined as the thickness of the object. Ratios of at least 4 are desirable, but radiographers must consider a number of factors that affect the exposure.
- *Geometric enlargement.* Geometric enlargement is the increase in the size of a feature on a radiograph that results from increasing the object-to-film distance in the exposure setup. Enlargement or magnification allows the inspector to view features that would otherwise be too small to distinguish. This technique of creating geometric enlargement also has a quality impact in terms of reducing the image definition.
- *Radiographic sensitivity.* Radiographic sensitivity is the general term used to describe the smallest detail that can be seen using a radiograph. The term also refers to the ease with which images can be seen or details can be detected. In other words, radiographic sensitivity is the amount of information that can be found on a radiograph. When using digital detectors, this equates to the minimum detectable level along with the minimum incremental value reported at the digital output (see prior section on detectors.) For film, radiographic sensitivity is typically determined

from sensiometric charts, which comprise a group of curves that compare exposure times for different films with the amount of flux from the source.

- *Film contrast.* Film contrast is determined by both the film characteristics and the chemical process used for processing the film. More technically, film contrast is defined as the slope of the film's characteristic curve. Increasing contrast will bring out very faint anomalies but also has the effect of obscuring relative differences between slight gradations. Refer to standard BS EN 584-1:2006[19] for more information on film contrast.
- *Film grain.* Film grain is the size and density of silver bromide crystals per square centimeter of emulsion on the film surface. A larger grain size results in faster (more sensitive) film but reduces the observable detail. In general, finer grain film is desirable and can be used when the illumination is sufficiently high for good exposures.
- *Detector operation.* Digital detector systems are increasingly replacing film with the effect of simplifying the exposure process, since many variables associated with film use are no longer relevant. However, digital detectors still require some preparation. This primarily involves calibrating the imaging chain by creating a baseline exposure without an inspected part present. In this step, detector elements can be normalized to the intensity of the beam, and variations due to a number of variables can be neutralized. A second operation may be performed to further adjust the detector system. This involves exposing a phantom or standard test object and then normalizing the detector outputs to the standard.

5.3. *Radiographic procedures*

The radiographer is responsible for managing the entire testing process, which usually begins with planning and documentation and includes setup, exposure, and verification steps. Once the configuration requirements are established, the equipment is placed in the desired location. For new setups, test images are often taken with phantoms (simulated test objects), but actual test parts are also used. Image quality indicators (IQI) are placed in proximity to the test object and serve as verification checks for the setup. Because linacs typically require a 20–30 min warm up time before they can produce a beam, operators frequently use this time for test preparation.

Exposure records and verification results are a critical part of radiographic work and provide auditable process checks. These records are also used to troubleshoot ineffective radiographs and provide information about what parameters need to be changed. Once the setup is verified, actual radiographs can be made to locate defects or meet other intended objectives.

While these basic inspection principles apply to all radiographic processes, each of the following applications has unique requirements that must be addressed.

5.3.1. *Radiography of castings*

Large steel and aluminum castings are made from pouring molten material into molds. Some castings are complex combinations of multiple cores and parts, creating asymmetrical densities and shapes throughout the final product. In addition, some regions in castings will be especially critical because additional machining operations will be applied to create openings, fastener sites, sealing surfaces, and the like. Consequently, the X-ray examination of castings must include the search for a variety of potential defects, including but not limited to porosity, delaminations, cracks, voids, inclusions, tears, and core shifts. To ensure that all of these defects can be identified, radiographers may choose to include multiple views or perform computed tomography on the parts.

5.3.2. *Radiography of welds*

Radiography is often used to verify the quality of welds, especially those where deep penetration of the adjoining materials is required to meet safety and performance specifications. In these cases, the inspection process is usually specified in detail with precise accept and reject criteria. These usually include parameters such as weld penetration, porosity, fusion, voids, and cracks. Because the weld path is visible at the surface, radiographers will use this detail to orient the object for examination. Depending on the geometry and dimensions of the inspected object, this may require additional images from different orientations. Note that the projection of a linear crack along one axis may be seen as a single point on a perpendicular axis, providing little information for examination. Therefore, curvilinear welds cannot be optimally seen from a single view.

Researchers continue to seek improvements in the inspection process to ensure that defects are correctly identified. For example, one group of researchers has proposed a new technique that uses gray level profiles transverse to the weld bead.[20] This approach uses neural networks to classify discontinuities among the most common defects, such as lack of fusion, lack of penetration, undercut, crack, slag inclusion, and porosity. The researchers have concluded that this method has potential for accurate classification and application to automated radiographic testing systems.

5.3.3. *Radiography of assemblies*

It can be difficult to inspect assemblies that have internal features, yet high-reliability and safety-critical assemblies, such as certain aircraft flight-rated components, commonly require 100% inspection to ensure that they were properly fabricated. While radiography of a completed assembly can frequently supply the needed verification, this can be challenging when the assemblies have complex internal geometries and varying densities that make radiographic examination more difficult. Therefore, an analysis of the design is important to understand the effects of these differences, and some special considerations should inform the process.

First, as they would for any asymmetric part, radiographers should consider the projection of the assembly onto the imaging plane. Specific or multiple orientations may be required to see all of the areas of interest. When sections are curved or vary in dimension, the potential distortion of their projections must also be considered.

A second problem relates to the densities of the internal structures, which could be overexposed or underexposed. In general, the X-ray energy needed will be defined by the thickest or most dense section. Also, a contrast analysis is important to ensure that the area of interest can be differentiated from the adjacent areas.

Third, too much energy can overexpose all parts of the object. A rule of thumb is that the material being examined should be in the range of 4–7 HVLs. This allows enough photon absorption to differentiate between density variations while obtaining enough penetration to create a well exposed image.

Finally, when assemblies have irregular shapes or vary in density, the radiographic setup can also have significant influence on the quality of the resulting images. As described earlier, larger D/T ratios are desirable to avoid distortion, while the use of filters and screens can produce improved sharpness and contrast. The optimal selection of film types and related exposure settings can also contribute to imaging quality.

5.3.4. *Radiography of rocket motors*

Industrial NDT accelerators were first used to image solid rocket motors, and this application remains an important role for high energy radiography. Traditionally, the cross section of a solid rocket motor includes a steel case, an insulating layer, a propellant liner, and the propellant itself. In addition, the nozzle and other related components make up the entire motor and are a crucial part of the examination. From a radiographic perspective, contemporary motors are similar to early versions, except that tolerances are more critical and new composite materials sometimes replace traditional steel housings.

The most often cited requirement for rocket motor quality is that the propellant burning process must be predictable and result in well-defined pressures over the entire flight. From the radiographic perspective, this means that the supporting components must operate as planned, and the propellant must be free of irregularities such as voids, bubbles, or separations. Because propellant often has a molasses-like consistency when poured into the motor cavity, there is an opportunity for such irregularities to occur. In some motors, the propellant is adhesively bonded to the liner and serves to add structural rigidity to the assembly. Additionally, the center core of the propellant can be hollow and designed with a specially shaped profile to optimize the burn rate and pressure. Collectively, these quality concerns create a critical and continuing role for accelerator-based radiography.

6. Security Inspection Applications

Security applications, primarily cargo screening, quickly emerged and have dominated the demand for industrial NDT accelerators in the last two decades. This work began in the early 1990s when manufacturers produced and deployed the first full-sized cargo screening X-ray systems at a time when global security was only a peripheral concern. Initially, manifest verification was the driving force for adopting this technology. The cost of X-ray screening was justified from the recovery of VAT and other tariffs from highly valued imports that were surreptitiously mislabeled. At the same time, contraband materials could be reliably discovered even when carefully concealed deep within large containers, tankers, or trucks.

In response to the early deployments of fixed X-ray inspection systems, smugglers quickly learned to avoid the few ports and borders thus equipped, creating a very robust market for mobile X-ray scanning. Consequently, authorities devised and deployed a combination of mobile and fixed installations to remain a step ahead of the perpetrators.

6.1. *X-ray cargo inspection*

While manufacturers created a number of customized systems for specific locations, X-ray cargo inspection systems have become standard products that are typically available in the four basic types described in Table 3. All of these can provide good quality images. They differ in how their

Table 3. X-ray cargo inspection system types.

Type	Description
Mobile	A truck-mounted and self-contained X-ray scanning system. It includes a complete data-analysis facility within the scanning vehicle.
Gantry	An X-ray scanning system that moves past a stationary vehicle. These typically operate bi-directionally, improving throughput by scanning in both directions. Some gantry systems can be disassembled and relocated to another site.
Portal	An X-ray scanning system that remains fixed while the vehicle under inspection is driven or towed through the X-ray beam. Some systems may allow drivers to remain in their vehicles. Some systems can be disassembled and relocated to another site.
Fixed	An X-ray scanning system that is optimized to work in a dedicated structure that is not intended to be relocated. The X-ray imaging system inside the structure may be designed in either a gantry or portal style.

designers have addressed certain system objectives related to their intended use, installed location, life-cycle costs, and overall throughput performance. From an operational perspective, these can also be classified as stationary or mobile.

Mobile systems are fully contained in vehicles or trailers that can be moved to a location, set up, and ready to scan within about 30 minutes. Advantages of mobile systems include their ability to be operated at temporary locations and rapidly moved from place to place, creating a dynamic and strategically unpredictable inspection capability across a large area. They are also smaller and sometimes less expensive than stationary systems, which typically require an enclosed operating area with more substantial shielding due to their proximity to the public.

Stationary systems are available in various configurations. Some have relocatable components that can be disassembled and moved to other locations, while others are designed into a permanent structure. Another difference among stationary systems is their shielding configuration. Some are heavily shielded near the X-ray source, which minimizes the requirement for structure shielding. Others have less source shielding but require more substantial structure shielding.

Following the terrorist attacks of 9/11, the newly created US Department of Homeland Security (DHS) recognized that high-energy X-ray scanning was particularly well-suited to address the inspection needs for millions of containers that enter the country each year. The agency therefore established X-ray capability as part of a "layered" security strategy frequently supplemented by other technologies and methods, providing authorities some discretion in how to inspect incoming vehicles.[21] For example, some X-ray screening systems allow scanning at different speeds so that enhanced images can be obtained from a particularly suspicious vehicle or container. Consequently, system throughput performance must be judged in terms of meeting specified objectives rather than simply comparing average inspection times per vehicle.

Notwithstanding the political and economic arguments about the need for full deployment of cargo screening systems, the second decade of X-ray-based cargo screening has been primarily characterized by three factors: (1) a consensus for the proven effectiveness of existing technology, (2) the need for this technology to address homeland security concerns, and (3) some notable innovations in high-energy X-ray systems technology specifically relating to cargo imaging.

By the end of 2009, an estimated 500 high-energy mobile and fixed site systems were operating throughout the world on a daily basis. Figure 13 shows a typical cargo scan without magnification. Actual examination of these images involves panning across a magnified image using software tools that enhance objects against their backgrounds. Customs and border protection services have validated the effectiveness of these systems and have incorporated them as a routine component in their inspection repertoire.

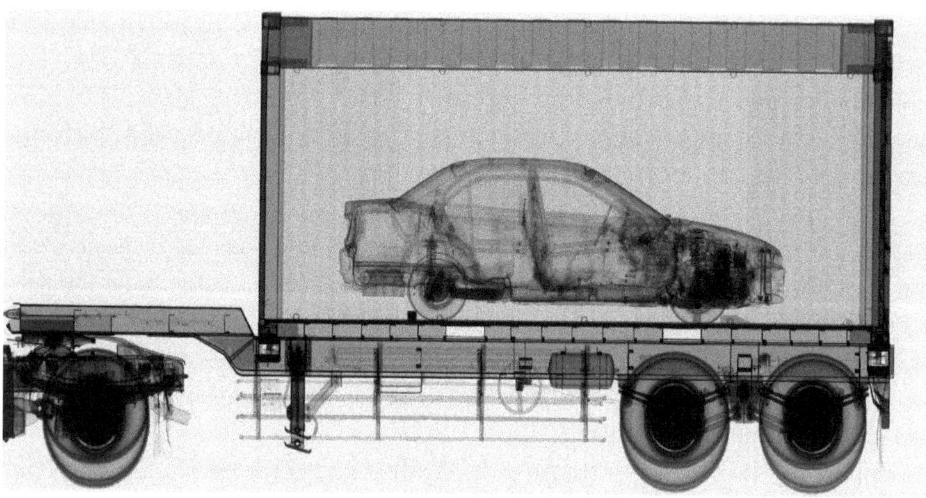

Fig. 13. Complete cargo scan using 5 mm pitch detectors.

This is evidenced by a number of well-documented contraband seizures that have been reported in the press and elsewhere. While other technologies play an important role to complement the use of high-energy X-ray systems, no other proven technology is available to completely image the contents of a loaded container in such a cost-effective and reliable manner.

The effective deployment of high-energy X-ray cargo inspection systems for homeland security concerns has been vital for at least two significant reasons. First, terrorists have the time and resources to carefully conceal special nuclear materials and other weapons of mass destruction (WMDs) using materials that obscure them from other modes of inspection. Thus, anything less than a full penetration scan will severely limit the inspection effectiveness. For example, using high-energy X-ray scanning, authorities have found contraband in engine compartments, truck chassis, and other areas adjacent to high density materials. In contrast, backscatter and isotope-based systems with limited penetration can be less effective, especially when perpetrators conceal contraband with intentional shielding.

Second, the potential devastation caused by WMDs calls for an inspection technology with an unparalleled probability of detection while simultaneously providing a very low false alarm rate. The most commonly deployed technology currently available for this purpose is passive detectors that sense radiation or other substances emanating from the suspect object. Yet many passive detection and surface detection technologies are subject to significant Type 1 and Type 2 errors. Type I errors (false negatives) occur when actual threats are not detected. This might result from improper instrument operation, lack of sensitivity, or too little time when the sensor is close to the threat material or because the threat materials are effectively shielded. Type II errors (false positives) are alarms that falsely identify a threat. These can occur due to instrumentation functionality or calibration problems, as well as sensitivity to low levels of naturally occurring radiation that is present in agricultural products, building materials, or other legitimate cargo. Therefore, the seriousness of WMD smuggling, along with the importance of high reliability detection, has become another reason for deploying high-energy X-ray screening systems.

6.2. *Recent innovations*

The first generation of accelerator-based cargo security products was characterized by physically large systems that were slow to set up and operate. In addition, they required a substantial amount of manual calibration, operation, and maintenance. Current cargo screening systems represent a second generation, incorporating more advanced accelerators, detector arrays, and software that are designed to work in concert. Consequently, today's systems are smaller, faster, more reliable, and capable of producing extremely sharp images. Three developments in this new generation of products are particularly noteworthy.

6.2.1. *Interlaced dual-energy accelerators*

A major new development has been the introduction of *interlaced dual-energy* linacs for the purpose of material discrimination. When operating an interlaced dual-energy accelerator, information about the type of material can be displayed rather than just the composite density as seen with traditional X-rays. The term "interlaced" means that two or more distinctly different energies can be alternated on a pulse-by-pulse basis with a repetition rate of several hundred Hz. While this functionality has been available in medical and baggage screening applications for some time, the technology to rapidly switch high-energy X-rays is a more recent innovation.

In cargo screening systems, interlaced operation occurs by alternating pulses at two distinct and well-separated energies, such as 4 and 6 MeV. The resulting images contain crucial differences that reflect the nonlinear absorption rates of different materials. This is shown in Fig. 14, which simulates

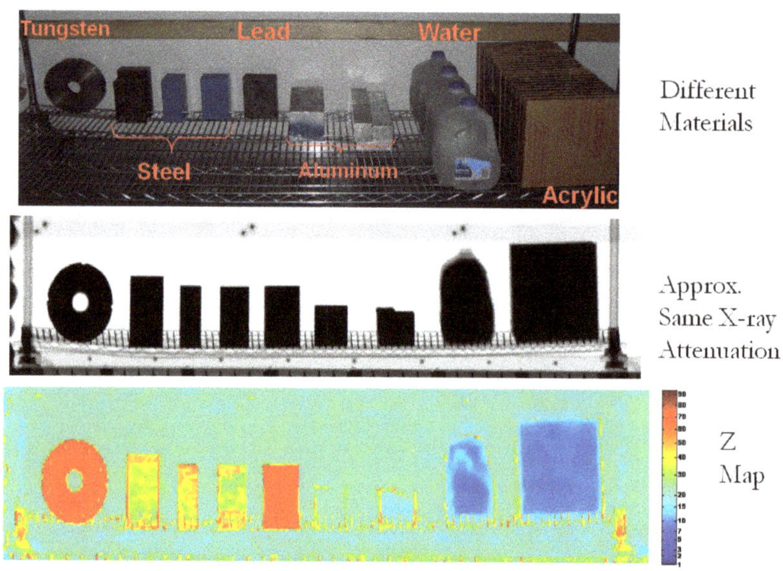

Fig. 14. Three visualizations of the same objects: A photograph (top), a normal radiograph (middle), and a dual-energy radiograph (bottom).

the differences between images produced by standard X-rays and images that are possible from computing the ratios of absorption rates at distinct energies. The top row is a photograph of different materials: tungsten, steel, lead, aluminum, water, and acrylic. Note that different thicknesses (front to back) of these test objects were used to purposely simulate radiographically similar densities. Because traditional radiographs record only total density, all of the objects look very similar in the middle row. However, the third row illustrates how dual-energy material discrimination can identify the differences between these objects.

The third row is a computer generated "Z-map" (density map) that has different colors assigned to each object based on its density. Blue colors represent low density materials such as water and acrylic. Higher density materials are represented by green and yellow (e.g., aluminum and steel), while the highest density materials are shown in red (lead and tungsten). To obtain these representations, specialized software algorithms can analyze the differences between images obtained separately (at each energy) in order to discriminate between materials of different densities. This material discrimination feature is possible because X-ray absorption is nonlinear relative to material type. With an interlaced dual-energy system, the imaging software derives a composite image that highlights the distinctiveness of objects based on their unique densities. In practice, this allows objects to be more clearly identified, even when they are in-line with objects of similar densities. For example, a small block of aluminum would be uniquely visible even when positioned behind several inches of steel. This feature is particularly important to detect the smuggling of special nuclear materials that perpetrators may intentionally shield with lead or steel to evade passive detection systems. With dual-energy discrimination, radioactive materials such as uranium or plutonium could be uniquely identified with a high degree of certainty, even within a heavily lead-shielded enclosure. For users, this means that dual-energy systems can offer improved object recognition and the necessary functionality for assisted or automatic recognition of threat objects.

Dual-energy cargo screening can address a number of the more challenging problems of finding contraband in cargo containers. For example, when compared to baggage screening, the objects that exist within containers vary much more widely in their composition, orientation, and size. While baggage can be easily opened or repositioned, containers are often fully loaded and have a fixed position with respect to the scanning device. Dual-energy systems address this by more clearly identifying specific contents and better discriminating those items of interest from the background.

While perpetrators and smugglers have unlimited time to disguise contraband, the screening process is severely time-limited, so a single X-ray scan is often the primary layer of defense for identifying such a threat. Because of this, inspection systems are increasingly designed for computerized image analysis and automatic identification of serious threats. Dual-energy X-ray systems play an important role in this process by creating richer data that can be automatically processed to identify unique materials.

6.2.2. *Advanced detector technology*

The second notable innovation is the vast improvement in detector technology combined with its adaptation to the cargo screening market. New detectors are noted for their improved resolution, greater sensitivity, and lower cost. By combining multiple detector modules (Fig. 15 shows a typical unit with 32 channels), engineers have optimized the manufacturability of complete detector arrays while simultaneously addressing other important functional parameters. For example, this modular technique has allowed the use of narrower detector elements to create finer resolution images. At the same time, improvements have been made in sensitivity and dynamic range.

Fig. 15. A typical detector module, ~$250 \times 148 \times 55$ mm, with 32 channels and 18 bit resolution.

The modular nature of these detector arrays has allowed system designers to optimize their performance in cargo screening systems by configuring them in L-shaped or curved panels. In comparison to straight-line assemblies, these receive broader coverage and more accurately capture image data about the contents being scanned. The individual module approach also provides easier replacement and much greater control over the calibration and operation of each detection region, allowing it to be optimized based on its individual location in the array.

6.2.3. *More powerful software*

The third major innovation for cargo screening X-ray systems relates to the functionality that results from the use of more powerful imaging software. Newer systems allow images to be "painted" across the screen as they are being acquired, thereby increasing the amount of viewing time available for each image. Once an image has been rendered, a number of software algorithms are used to position, enhance, and adjust the image to better reveal the contents of the container being inspected.

Early user interfaces limited the operator to fairly basic controls, such as pan and zoom, and displayed the X-ray image in traditional black and white. Screening officers relied on their vigilance and experience to identify threat objects that could be obscured by other materials within the container being examined. In contrast, today's newer software algorithms provide features that are more efficient and effective in helping the user identify potential threat objects. Some of these features are described below.

- *Image colorization* is the computer assignment of predefined colors to the gray scale images based on the density of objects. This feature is available on most single-energy systems. Users may find this easier to read than the subtle gradations in gray, which can still be seen when switching to the monochromatic mode. Systems typically offer a number of color bands to meet user preferences, although this feature may be locked-out for consistency across different work stations. While the colorizing process is helpful, there can remain some ambiguity when objects of similar densities are positioned in line with the X-ray beam. A number of other algorithms that can help in these situations are available.
- *Edge detection filters* determine the edges of objects in horizontal, vertical, and oblique orientations. Edges can be computer-constructed from gradations that are too small to be easily recognized by visual inspection.
- *Edge enhancement filters* increase the contrast across object boundaries to more clearly identify the edges of each object.

- *Smoothing filters* use specific algorithms to average adjacent pixels for the purpose of reducing noise that can be superimposed on the object of interest. Different smoothing options allow choices such as uniform smoothing, distance smoothing, or Gaussian smoothing. These differences reflect the weighting of the adjacent pixel averaging process to achieve the clearest image for the particular scanning environment. Smoothing filters generally provide a tradeoff between contrast resolution and spatial resolution with the effect of improving contrast at the cost of detecting differences in materials of similar density.
- *Erosion filters* reduce the boundary definition to make objects appear thinner.
- *Dilation filters* typically make an object appear thicker by enhancing its boundary definition.
- *Log filters* reveal images based on the logarithms of their densities with the effect of reducing attenuation effects of less prominent objects. This may help operators identify objects that are hidden behind denser objects.
- *Density highlighting* is an advance over traditional colorization of objects. It is typically a programmable option that allows the user to assign custom colors to objects of specific density. This can help the operators identify specific objects more quickly.
- *Region-of-interest* (ROI) tools allow users to assign a specific area on the display as a region of interest for further analysis.

Even more advanced software algorithms are being employed to achieve automated detection. These combine features inherent in dual-energy systems with software algorithms that are tuned to detect specific threat objects. Automated detection features may also be used to identify and alert an operator to areas of interest that conflict with other intelligence, such as cargo manifests, or that have unique profiles, such as certain nuclear materials. Manufacturers have also developed "assisted detection" algorithms to complement human detection efforts. These include an interactive component that allows users to specify the types of data that represent a threat and cause the computer to search for objects with those specific parameters. Future advances in both assisted and automated detection will provide increasing confidence in visual inspections while also reducing the amount of time that containers must spend in the scanning area.

6.3. *X-ray cargo screening trends*

The security inspection market for industrial accelerators is undergoing both maturation and expansion. The market is maturing from the perspective that port and border authorities have gained significant experience in screening equipment and strategies and now have the expertise to specify requirements and performance objectives for their systems. However, government policies in many countries are still in flux, awaiting a consensus on inspection standards and practices that ensure the safety of imported cargo. Nevertheless, the growth of X-ray-based cargo screening systems throughout the world over the last decade has been substantial, due in part to the US government mandate of 100% cargo screening (US Cargo Security Initiative[22]).

The further expansion of accelerator-based security screening will be driven by the increasing expectation of fulfilling multiple missions, including the ability for systems to detect WMDs, interdict traditional contraband materials, and identify contents for tariff purposes. While some related technologies such as portal monitors and container seals may have the potential to complement X-ray screening, they are individually insufficient to address the broader scope of homeland security requirements.

For accelerator-based systems to achieve all the security objectives outlined above, a growing percentage of newer systems are expected to be "dual-view", utilizing two accelerators per system (one on each axis normal to the scanning direction). With dual-view systems, operators can more quickly and reliably identify the contents of a container as well as the nature of any threats present. This, along with increased screening speeds, will be important to cope with the flow requirements and container backlogs that are likely at peak unloading times. Of course, the deployment of additional screening systems can also increase throughput. This helps to both alleviate bottlenecks during peak traffic periods and minimize the impact of any equipment maintenance or downtime.

The new innovations in dual-energy technology along with the improved imaging chain in X-ray container screening systems have been driven by the increased focus on security concerns following the events of 9/11. While the need for high-energy X-ray screening to perform manifest verification and contraband interdiction remains strong, the additional emphasis on WMD detection has rapidly advanced this new technology. International efforts are on track to utilize and extend this technology as homeland security concerns continue to play an increasing role in container inspection throughout the world. This is likely to result in more sophisticated versions of material discrimination through the availability of multiple-energy or intensity modulated systems.[23]

Future systems will continue to strive for faster scanning rates and better overall throughput since port congestion will continue to be an operational concern for authorities. Although scanning speeds have improved from the

2–3 min required for first generation gantry systems, the current scan rates of 30–40 s are still considered too long when the additional overhead times are added.

Advances should also be made in the electronics used in scanning systems. The electronics system provides the control of the mechanical gantry or vehicle, the conversion of detected signals to digital data, the computer processing and storage of data, the user display, and the transmission of data to alternate locations. These have already been improved to keep up with the faster scanning rates and decreased total processing and analysis times, resulting in a shorter overall screening cycle. However, future systems will continue to provide more advanced graphics processor chips, faster analog to digital converters, and more parallel processing. For larger installations, fiber optic interfaces will be more common, improving the connection to computer networks, high speed servers, and storage devices. User displays will likely become larger in order to provide more simultaneously displayed information, and some will have integrated touch screens for ease of use. Data encryption and access control will also receive more attention as a greater emphasis is placed on data and physical security in these installations. Overall, the trend of data integration from multiple sources will continue as officials seek to utilize cargo screening data as a key source of security intelligence.

7. High Energy Industrial CT Applications

Computed tomography (CT) of large objects such as car engines (or even entire cars) has become possible due to the confluence of three technologies: (1) reliable linacs that generate the necessary energy to penetrate dense objects, (2) digital X-ray detectors that provide the capability to capture immense amounts of image data and rapidly transfer it to computer files, and (3) the enhanced computational power and associated software that allows for three-dimensional image construction, section highlighting, and defect recognition. The specially designed computers and software packages involved often use multiple hardware graphics chips and networked processors to perform the parallel processing required for real-time or near-real-time imaging. Other software applications include the availability of X-ray metrology to accurately measure internal artifacts, clearances, and defects that could not be measured in any other way. In addition, such measurements are automated and computer controlled, facilitating their use on automated production lines.

Computed tomography systems are increasingly designed to provide more accurate digital reconstructions of the parts being scanned. To accomplish

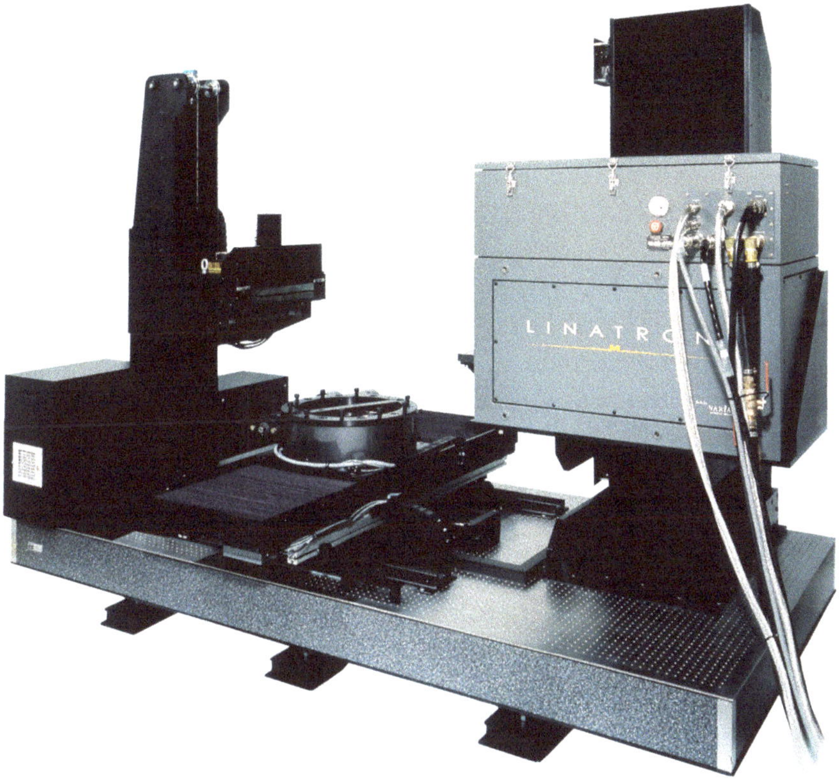

Fig. 16. A 2 MeV industrial CT system with the capacity to scan 600 mm diameter parts.

this, designers have to ensure that the mechanisms are inherently accurate and robustly built. In addition, some systems employ a routine calibration process to fine-tune the data as it is acquired or allow for corrections in post-processing. Such calibration schemes can be fairly simple or involve complex algorithms that correct a number of systemic errors.

Figure 16 shows a contemporary 2 MeV CT system where both the source and detectors move vertically while the object under inspection rotates. This system can image dense engine parts and other intricate objects, allowing for prototype verification, reverse engineering, and the recognition of hidden defects.

The availability of this accelerator-based CT imaging technology has expanded a number of radiography markets, some not directly related to the traditional concepts of testing and inspection. Among these are a number of engineering design applications, production process control, and advanced internal metrology.

7.1. *Engineering and design applications*

Reverse-engineering, especially the identification of internal artifacts, can be accomplished with CT scanning without altering or destroying the part under examination. This is especially valuable when used to evaluate prototypes, verify and validate designs, or diagnose failures. More broadly, CT scanning can be used to compare internal variations resulting from different materials or manufacturing processes or to compare similar parts from a variety of manufacturers. CT systems can also be selected to ensure the accuracy of CAD models by analyzing data on the parts made from those models. By incorporating the outputs of CT machines to create a three-dimensional representation of the actual part and comparing it directly with the CAD model, engineers can determine what design changes are necessary to obtain the desired part.

7.2. *Production environment*

While robust quality assurance programs can create very high confidence levels in part quality, certain products require 100% inspection. Generally, these are components affecting safety or those that are too costly to replace should they fail short of their needed service life. Cost and feature advances in CT have made it possible to use this technology in the production environment. For example, engine blocks and heads are most economically made from cast aluminum where the total casting mass must be minimized to reduce cost and weight. However, once these are assembled, hidden defects are very costly to discover and repair. Another example is the manufacture of turbine blades for aircraft. These must be 100% tested and certified to meet aviation safety requirements. CT not only fulfills the inspection requirement but also provides the data collection mechanism for maintaining a complete inspection record for each part.

8. Quality Standards and Measurements

Industrial radiography has traditionally depended on film to capture the images used for testing and inspection. Because the film-based radiography process is complex and the nature of testing and inspection has critical consequences, a number of standards have emerged over time to ensure image quality (e.g., see Ref. 19). While digital radiography is replacing many film applications, the same imaging quality concerns apply.

Industrial radiographic quality is typically measured in terms of the ability for a radiograph to reveal a small artifact or inclusion embedded within a larger object. This is referred to as "sensitivity" and checked with calibrated image quality indicators (IQI) known as penetrameters. There are different types of penetrameters, but they all generally serve the same purpose. The accepted standard in the US is a hole type penetrameter as specified by ASTM E-1025.[24] This standard is flexible, allowing the dimensions, width, and length to be altered as long as the modifications are documented and reported properly. This flexibility allows users to define and utilize penetrameters that are particularly useful for their specific needs.

Hole type penetrameters consist of holes of different diameters drilled perpendicularly through a material of fixed thickness. The penetrameter is ideally made from the same material as the specimen under test, and its thickness is a percentage of the specimen thickness (usually 1–2%). The holes in the penetrameter are identified as an integer number that relates the diameter to the thickness of the penetrameter itself. Three holes are normally present, representing the actual thickness (1T), twice the thickness (2T), and four times the thickness (4T) of the penetrameter.

During the actual sensitivity test, a radiograph is performed with the penetrameter placed on the object. The sensitivity of the radiograph is then determined by the smallest hole that is visible from the penetrameter. The outline of the penetrameter must also be observable. Radiographers typically report the results of their sensitivity tests in a format such as "2–2T" (read as "2 dash 2T"). This would indicate that a penetrameter whose thickness is 2% of the object's thickness can be seen along with a hole whose diameter is twice the thickness of the penetrameter. Many variations of this method are used by industry and are fully accepted by quality departments when the process is adequately documented and validated for the application.

Wire type penetrameters operate in the same way but utilize German D.I.N. wire diameter standards. These test devices have wires of specified material and of varying diameter suspended in an opening in the penetrameter body. The smallest wire size that can be identified corresponds to the sensitivity.

8.1. *Half-value layer energy measurements*

For industrial applications, users are often concerned with obtaining an accurate measurement of accelerator output energy. With this knowledge, radiographers can establish operating distances (shot geometry) for the equipment under test, determine exposure times, calibrate detectors, and the like. For X-ray energies in the 1–9 MeV range, output energy is typically measured in equivalent HVLs of steel. Since one HVL is the thickness of material required to absorb 50% of the incident beam, when a number of 1-HVL plates are stacked together, the intensity emanating from each plate is 50% of the intensity from the prior plate.

HVL methodology is used as a standard because the measure follows directly from the penetration process itself, is effective for a wide range of energies, and can be duplicated at different facilities with relative ease. The HVL can then be converted to MeV by using a chart such as the one in Fig. 17. Sources for this measurement technique can be found in NCRP Report 49,[25] the *Linatron High-Energy X-Ray Applications Manual*,[26] and the *Radiological Health Handbook*.[27] The HVL levels from these references are shown in Fig. 17. The curve shown for HVL in steel is derived from the Radiological Handbook for energies of 1 MeV and lower, and the Linatron Manual for values higher than 1 MeV. Note that the two references agree well at 1 MeV.

HVL measurements in this range are accurate and repeatable when measurement conditions are clearly specified and closely duplicated. The HVL

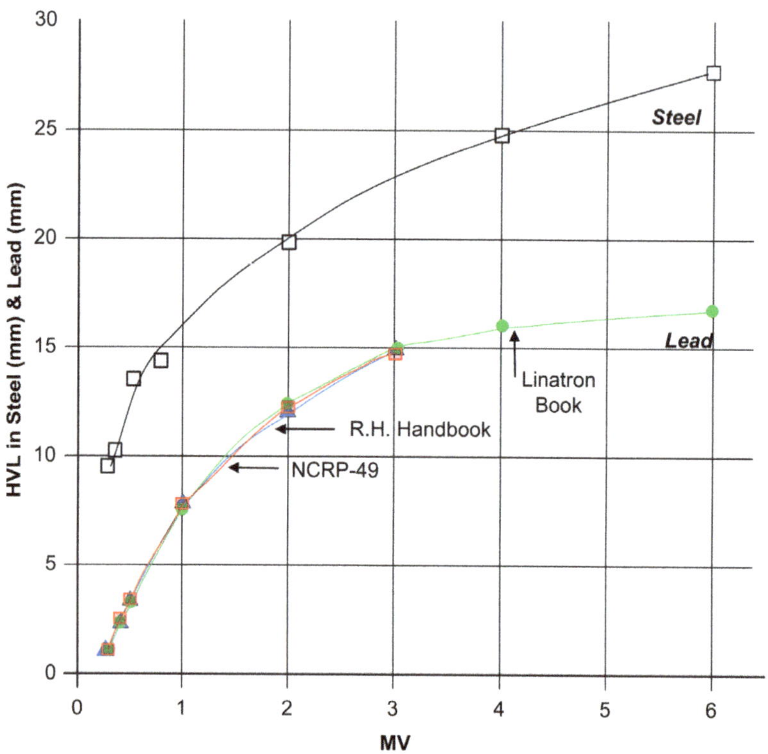

Fig. 17. Half-value layer as a function of X-ray energy under broad-beam conditions.

metric is generally considered to be a broad-beam measurement: therefore, it can be dependent on the geometry of the measurement process. Data from the NCRP report and the Radiological Health handbook are provided within the context of radiation shielding effectiveness and safety considerations. In other words, the standard intends that measurements will be made when detectors are placed directly behind shielding materials such as the outside wall of an exposure bay and away from materials that could be a significant source of scatter. Consequently, accurate HVL values require that (1) the source is not tightly collimated, (2) the attenuating material is close to the detector, (3) potential sources of scatter are eliminated, and (4) the X-rays are hardened.

In practice, HVL is usually determined by examining a sequence of exposures, with each exposure representing X-ray penetration through a thicker section of material. Ion chambers or probe-style sensors are used to determine when penetration is achieved with the number of HVL-equivalent sheets present. By convention, a 30° cone collimator is placed directly in front of the target to ensure broad-beam exposures while limiting out-of-plane scatter that could affect the measurement. During a typical HVL measurement process, an appropriately specified detector is placed inside a lead-lined container to further reduce out-of-plane scatter and to ensure that direct-path radiation is recorded. Then a series of steel plates are added in a sequential, cumulative fashion. Traditionally, one-inch thick steel is used for the plate thickness since it is close to the HVL thickness for energies in the 3–6 MeV range and is not too heavy to handle. Also, by convention, a 1/2-inch thick steel plate is placed at the beginning of the stack to harden the beam. The measurement process proceeds by adding one plate at a time and taking successive flux measurements that can be compared to the known HVL thickness of steel. The data are then reported in equivalent HVLs of the material used.

8.2. *Cargo and vehicle screening standards*

Customs and border authorities carefully examine X-ray images of vehicles and cargo containers, and within seconds, they make important decisions about which ones to interdict based on artifacts observed or not observed in the displayed image. Because these decisions can have important safety consequences for the public and legal consequences for the transporter, they must be based on credible performance standards. In addition, the available images and the software that creates them may differ substantially from one supplier to another and across inspection facilities. This product diversity, along with the critical consequences from operating these products, calls for an imaging standard tailored specifically for X-ray cargo screening.

The Institute of Electrical and Electronics Engineers (IEEE), under accreditation from the American National Standards Institute (ANSI), began this work when it published the first generally accepted standard for determining imaging performance of X-ray cargo screening systems. Known today as ANSI N42.46-2008,[28] this standard was created through the sponsorship of the National Committee on Radiation Instrumentation and was intended to be used to determine imaging performance of both X-ray and gamma-ray inspection systems. The intent of this standard was to focus on fundamental imaging capabilities directly related to the imaging chain. It established relatively simple tests with objective data in ranges to directly compare the results of competing systems. This standard is applicable to all types of vehicles, whether empty or loaded, including marine containers, air cargo containers, railroad cars, and palletized and unpalletized cargo larger than 1 m^2 in cross section. It also applies to systems with X-ray sources that operate at single and multiple energies as well as those systems intended for backscatter operation.

This standard defines both the process and the equipment required to accurately measure the performance of the imaging systems contained within its scope. Meeting the standard requires users to complete a set of tests that characterize the imaging performance of a screening system in terms of its ability to detect small-sized objects or phantoms (objects designed specifically for testing X-ray images). In addition to the specific tests, the standard specifies that the radiation field around the equipment be measured as part of the characterization of system performance. This provides a contour map of radiation related to the 100 mrem per year value that is considered safe for workers who may be present for 2000 h per year. It also specifies the measurement of other related radiation exposure levels for the purpose of comparing systems. Importantly, the standard does not define safe levels of radiation for those working around scanners; rather it simply provides a consistent method for comparing the differences of systems in normal operation with respect to these generally accepted values.

The primary purpose of the standard is to allow manufacturers, potential users, and other parties to have a consistent indicator of screening system performance when using the system for the inspection of actual cargo and vehicles. To achieve this, the standard directs that a test fixture be utilized with the capacity to detect a specified test object located behind a varying thickness of steel plates (see Fig. 18). Steel is the appropriate material for

Fig. 18. A test fixture with movable steel plates.

this standard because it is inexpensive, readily obtainable, and provides the required attenuation of megavolt range X-rays in a reasonable size.

With such a fixture, standard measurements are obtainable for a number of imaging test requirements, generally consisting of penetration, spatial resolution, wire detection, and contrast sensitivity as summarized below:

- *Penetration.* During the testing process, specified test objects are placed in various locations in the container being tested. The test objects are defined in the standard and consist of carbon steel in the shape of a kite-like arrow. A test fixture holds the test object at randomly assigned ordinal directions behind a specified thickness of steel. Systems successfully pass the test when operators can properly identify the arrow direction (see Fig. 19). To establish maximum penetration, the test is repeated with an increasing number of carbon steel plates until the arrow direction can no longer be identified.
- *Spatial resolution.* The spatial resolution test is intended to measure the minimum separation between the features of a test object that can be detected in an X-ray scanning system. It is conducted in air without blocking plates and must be performed in both horizontal and vertical orientations at specified points within the container. This is significant because cargo scanning systems use fundamentally different mechanisms

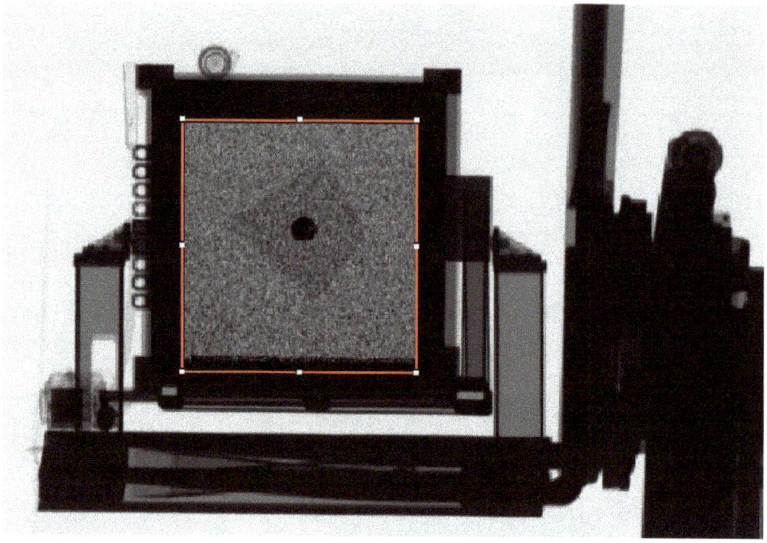

Fig. 19. A test arrow located behind steel plates.

for establishing resolution in the horizontal and vertical dimensions. In the horizontal direction, maximum resolution is determined by a combination of the transit speed and the X-ray pulse rate. Conversely, the resolution in the vertical dimension is fixed by the geometric spacing and number of the detector elements. The test object for this test is a steel line-pair gauge with three slots in a steel plate or with three square steel rods in a base. In both cases, the width of the material must be equal to the distance d between the plates or rods, which also have a thickness d. To complete the test, specified values of d are used, with "1" being a value that results in an indiscernible difference between the material and adjacent spaces.

- *Wire detection.* The wire detection test is used to determine the smallest diameter of wire that can be observed in an X-ray scan. The test can be performed in air or with steel blocking plates, as determined by the manufacturer (see Fig. 20). More specifically, the standard specifies the use of bare copper wires of varying diameters (typically 10–24 AWG) formed in a sinusoidal pattern. Again, the test object is to be placed at specified locations in the container and oriented in vertical and horizontal directions. Wires are considered to be detected when the majority of their lengths are observable on the X-ray scan.
- *Contrast sensitivity.* This test is designed to measure the minimum incremental thickness of steel that can be seen on an X-ray image. It utilizes a thin arrow-shaped steel test object located behind steel blocking plates of

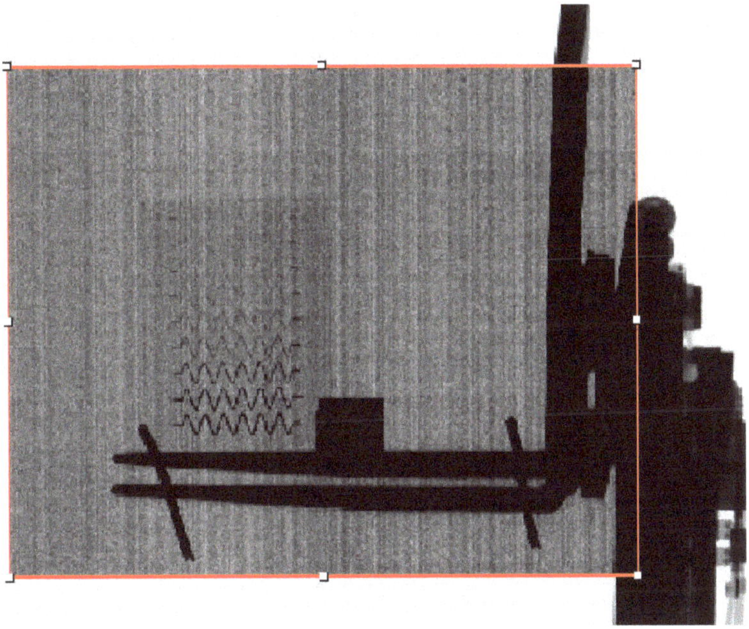

Fig. 20. Wire test behind steel plates.

a specified thickness. The ability to determine the orientation of the arrow on the X-ray is evidence of contrast sensitivity for the thickness used. The complete test requires three thicknesses of steel blocking plates, which correspond to 10%, 50%, and 80% of the maximum penetration thickness determined in the penetration test. The contrast sensitivity metric is expressed as a percentage and equates to the ratio of the minimum thickness of the test object to the total thickness of the combined test object and blocking plates.

A similar cargo screening standard was issued by the International Electrochemical Commission (IEC) in June 2010.[29] This standard (IEC 62523) closely parallels the IEEE standard for image quality of cargo inspection systems. For example, both standards specify similar test methods for steel penetration, wire detection, spatial resolution, and contrast sensitivity. However, the IEC standard goes beyond imaging standards to include a reference to other IEC standards for electrical safety and electromagnetic compatibility. The IEC standard also differs from the earlier one in the use of certain terminology and reference values, resulting in possible confusion when comparing both standards. For example, while both establish an isodose contour for measuring emitted radiation, the IEEE standard requires a radiation plot based on the reference value of 0.5 μ Sv as compared to 2.5 μ Sv in the IEC standard. Another similarity is that both standards reference material discrimination, but neither establishes a test method or a minimum level of performance for this feature. This is a case where the standards have not yet caught up with the rapidly emerging improvements in material discrimination functionality. It is also partly due to the complexity of material discrimination imaging and measurement in terms of the tradeoffs between material characteristics and energy levels. Consequently, there can be significant differences among manufacturers in how they design, calibrate, and implement this feature.

8.3. *Digital data standardization*

Industrial X-ray inspection and testing personnel have long utilized film for imaging, analysis, and archival storage. The film process is well defined and stable, and radiographers have extensive training and tools available for handling film-based X-rays (see Refs. 5 and 20). However, with advancing technology, industrial radiography has increasingly moved from film to digital images, creating the need for new standards in data formatting, storage, retrieval, analysis, and archiving. Users are also interested in transferring digital images across platforms and using software tools to aid in the comparison of like images over time.

Perhaps the most significant demand for data standardization has been driven by the security screening market, which utilizes high resolution X-ray images for law enforcement in the fight against terrorism. The resulting images are becoming forensic and evidentiary tools, requiring new standards for compatibility, transmission, security, and storage. To address this, the US DHS in conjunction with the National Electrical Manufacturers Association (NEMA) published the Digital Imaging and COmmunications in Security (DICOS) standard in August 2010.[30] This NEMA IIC 1 v01 standard, informally known as DICOS v01, provides a data interchange protocol and an interoperable, extensible file format for data interchange. It closely follows elements of the latest version of the original NEMA DICOM imaging standard published in 1985,[31] which is used in the medical field for X-ray images of patients. However, the new standard is specifically for industrial images, especially those relating to security applications such as baggage scanning and cargo screening. The data structures in this standard include special categories needed for security applications, such as identification documents, demographic information, X-ray and CT images, and related threat assessment data. Like the medical version, its intent is to achieve a manufacturer-independent digital file or group of files that can be accessed with third party computer software. The actual information embedded within these files also contains a significant body of metadata related to the equipment and the conditions under which the image was obtained, as well as security-specific information such as material identification, trace detection signatures, and threat assessments.

The challenge for future versions of this standard will be to adequately capture the very large data files associated with emerging X-ray imaging technologies. These include even higher resolution cargo images and multiple-view scanners. In addition, cargo scans are often performed at two or more energies, which multiplies the corresponding image file sizes. In addition, CT imaging techniques are increasingly being applied to air cargo screening, creating very large data files. These data will need to be encrypted, stored, and possibly transmitted across sites in real time for security or public safety purposes. For all of these reasons, the emerging data standards will continue to evolve and provide opportunities for third party developers to address these needs.

9. Summary

Electron accelerators for industrial NDT became standard products in the 1950s as the only practical inspection technology to ensure the safety and proper operation of large munitions. By providing a reliable source of highly energetic X-rays for imaging those materials, accelerators became a proven and practical technology for large scale industrial applications. This accelerator technology continued to gain momentum from its parallel use in medical applications, especially in oncology treatment systems. Because these medical and industrial applications shared many technical requirements, future industrial systems benefited from medical advancements and quickly became more reliable and cost-effective. Ultimately, this synergy between the industrial and medical markets resulted in better and more effective industrial accelerators that support the diverse range of applications seen today.

Throughout the 1970s and 1980s, industrial electron accelerators with energies as high as 15 MeV became common for the inspection of full-penetration welds, defect detection, material density, porosity measurements, and a host of other NDT applications. Some of these early systems have been in service for over 30 years and are still operating today. While business consolidation has eliminated the need for some of these machines, new applications such as CT scanning and a greater overall focus on product quality have maintained a steady demand for these products.

Beginning in the 1990s and accelerated by the 9/11 terrorist attacks of 2001, security applications for accelerators have skyrocketed and now account for hundreds of systems worldwide. The US government has led this effort with its Cargo Security Initiative (CSI), Secure Ports program, and others. With the goal of 100% screening of all containers entering US ports and without a practical technological alternative, it has been estimated that this market will continue to grow significantly in the years to come. By the end of 2007, the US Customs and Border Protection reported agreements in principle with 34 countries to expand the Secure Port Program and declared that it was already operational at 50 ports throughout the world. Since that time, the US Department of Homeland Security has placed more emphasis on a layered security approach with the intent of identifying and prioritizing high risk shipments rather than completing the plan toward achieving 100% inspection using non-intrusive techniques such as X-ray scanning. Nevertheless, X-ray-based cargo screening has become an essential technology, proven by its continued growth and acceptance in numerous countries around the world.

While the growth of accelerator-based cargo screening in the US has not yet met the levels anticipated by the US government's 2006–2011 Strategic Plan for US Customs & Border Protection, adoption in other countries has been far more extant. For example, countries as diverse as Russia, Japan, and Saudi Arabia have implemented very robust X-ray screening for incoming inspection of containers in their own countries. Moreover, officials in these regions have frequently implemented more advanced equipment utilizing dual-view and dual-energy accelerator configurations.

Similar growth is expected for the air cargo market as governments worldwide continue to improve air safety. While this market is still in its infancy, hundreds to thousands of accelerators could be needed as it matures. Two recent developments will likely portend a sea change for the use of accelerators in air cargo inspection. First, a smaller, more portable accelerator technology has just emerged, making 1 MeV accelerators available to manufacturers and systems integrators. Because this technology holds the potential for systems to be smaller, less expensive, and more modular, it is a good match for the needs of air cargo screening. In addition, the energy level of 1 MeV is low enough to operate safely without excessively large shielding but is also strong enough to image through pallets of materials that are commonly carried by air.

Second, the world has looked to the US to establish standards and certify products for aviation security. In mid-2010, the first accelerator-based air cargo scanning system was certified by the US government. These advances are important signs that this market will continue to develop and create a robust demand for low energy accelerators.

Looking forward, the decade beginning in 2011 should see the procurement and utilization of several thousand linear accelerators for nondestructive testing and inspection, especially for those applications related to homeland security, military defense, energy, and the environment. The number of applications will increase along with the technological advancements that deliver lower cost and smaller units. This number may also include a robust replacement market for several hundred first and second generation security scanning systems that have been deployed during the previous 15 years.

For the future, research scientists are looking closely at more compact field-portable accelerator systems, which will "take advantage of high-gradient acceleration schemes, energy recovery technology, and innovative radio-frequency systems" (quote from Ref. 6, p. 69). These initiatives will address many of the existing issues that limit field deployment of accelerators across a wide range of testing and inspection applications. Overall, the increasing importance of industrial accelerators, along with technology to make them more application friendly, will guarantee their prominent role as a significant tool for nondestructive testing and inspection for years to come.

Acknowledgements

Figure 12 is copyrighted by Hill Air Force Base and is used with permission. All other images in this chapter are copyrighted by Varian Medical Systems and are used with permission.

References

1. R. H. Bossi, F. A. Iddings, G. C. Wheeler and P. O. Moore, Eds., *Nondestructive Testing Handbook*, Volume 4, 3rd Edition (American Society for Nondestructive Testing, Columbus, 2002).
2. H. Berger, in *Nondestructive Testing Handbook*, Volume 4, 3rd Edition, Eds. R. H. Bossi, F. A. Iddings, G. C. Wheeler and P. O. Moore (American Society for Nondestructive Testing, Columbus, 2002), p. 24.
3. A. St. John and H. R. Isenburger, *Industrial Radiography* (John Wiley & Sons, New York, 1934).
4. H. H. Lester, "X-Ray Examination of Steel Castings", *Chemical and Metallurgical Engineering* **28**(6), 161 (1923).
5. *ASTM E94-04 Standard Guide for Radiographic Examination* (American National Standards Institute, Washington DC, 2004).

6. *Accelerators for America's Future* (US Department of Energy, Department of Science, Office of High Energy Physics, Washington DC, June 2010), http://www.acceleratorsamerica.org/index.html, accessed August 15, 2010.
7. J. E. Clayton, in *Proceedings of IPAC'10, Kyoto, Japan*, WEIRA06, 2461–2465 (JACoW, 2010).
8. C. Tang, in *Proceedings of IPAC'10, Kyoto, Japan*, WEIRA02, 2447–2451 (JACoW, 2010).
9. *Radiation Source Use and Replacement: Abbreviated Version* (Committee on Radiation Source Use and Replacement, National Research Council, 2008), ISBN: 0-309-11015-7.
10. *Energy Policy Act of 2005* (US Public Law 109–58, 2005).
11. J. Moss *et al.*, *Report on the Workshop on the Role of the Nuclear Physics Research Community in Combating Terrorism,* (US Department of Energy, Washington DC, 1992).
12. S. L. Caldwell, *The SAFE Port Act: Status and Implementation One Year Later*, GAO-08-126T (United States Government Accountability Office, October, 30, 2007), available at http://www.gao.gov/new.items/d08126t.pdf.
13. C. Koch, *Industry Partner Insights on Challenges and Opportunities*, Keynote Speech, American Assoc. Port Authorities Annual Spring Conf. (2005).
14. B. Mongelluzzo, *J. Commerce On-line — News Story*, Jan 11, 2010, 6:14PM GMT (2010), http://www.joc.com/maritime/container-imports-increase-after-30-month-decline, accessed September 15, 2010.
15. T. Wangler, *Principles of RF Linear Accelerators* (John Wiley & Sons, New York, 1998).
16. P. Berry, K. Vansyoc, and D. Summa, *Materials Evaluation* **11**(65), 1099 (2007).
17. E. Samei, *PACS Displays at 2006 RSNA Conference* (Radiological Society of North America, Chicago, 2006).
18. D. J. Schneberk and H. E. Martz, in *Nondestructive Testing Handbook*, Volume 4, 3rd Edition, Eds. P. O. Moore, R. H. Bossi, F. A. Iddings and G. C. Wheeler (American Society for Nondestructive Testing, Columbus, OH, 2002), p. 348.
19. *Non-Destructive Testing — Industrial Radiographic Film, Part 1: Classification of Film System for Industrial Radiography,* BS EN 584-1:2006, (British Standards Institute, London, 2006), ISBN: 0-580-48344-4.
20. G. X. De Padua *et al.*, *Materials Evaluation* **11**(65), 1139–1145 (2007).
21. G. D. Kutz and J. W. Cooney, *Security Vulnerabilities at Unmanned and Unmonitored U.S. Border Locations* (GAO report, released September 27, US Government Accountability Office, Washington DC, 2007).
22. *U.S. Customs Container Security Initiative*, Fact Sheet 8/08/02.
23. G. J. Langeveld, W. A. Johnson, R. D. Owen and R. G. Schonberg, *IEEE Trans. Nucl. Sci.* **56**(3), 1288–1291 (2009).
24. *Annual Book of ASTM Standards*, Vol. 03.03, Nondestructive Testing (ASTM International, West Conshohocken PA, 2006), ISBN-10: 0803140967.

25. *National Council on Radiation Protection and Measurements Report NCRPM 49* (1976).
26. *Linatron High-Energy X-Ray Applications Manual* (Varian Medical Systems, 2006), http://www.varian.com/us/security_and_inspection/resources/technical_information.html.
27. *Radiological Health Handbook*, US Department of Health Education and Welfare, 1970.
28. *American National Standard for Determination of the Imaging Performance of X-Ray and Gamma-Ray Systems for Cargo and Vehicle Security Screening*, ANSI N42.46-2008 (American Society for Nondestructive Testing, Columbus, 2008).
29. *Radiation protection instrumentation — Cargo/vehicle radiographic inspection system*, IEC 62523:2010 (International Electrochemical Commission, 2010).
30. *DICOS IODs Standard*, NEMA Standards Publication IIC 1 v01 (National Electrical Manufacturers Association, Rosslyn, 2010), http://www.nema.org/stds/iic1.cfm.
31. *DICOM Standard*, ACR-NEMA Standards Publication No. 300-1985 (National Electrical Manufacturers Association, Rosslyn, 2009), http://dicom.nema.org/.

Chapter 8

INDUSTRIAL USE OF SYNCHROTRON RADIATION: LOVE AT SECOND SIGHT

Josef Hormes* and Jeffrey Warner†
Canadian Light Source Inc., University of Saskatchewan
101 Perimeter Road, Saskatoon, SK S7N 0X4, Canada
**josef.hormes@lightsource.ca*
†jeffrey.warner@lightsource.ca

Synchrotron radiation (SR) has become one of the most valuable tools for many areas of basic and applied research. In some cases, techniques have been developed that rely completely on the specific properties of synchrotron radiation; in many other cases, using synchrotron radiation has opened completely new and exciting opportunities for conventional techniques. In this chapter, the challenges, problems, and advantages of the industrial use of synchrotron radiation will be highlighted, in an admittedly subjective way, based on the experience of the authors at various synchrotron radiation facilities. "Typical" examples of industrial use of SR will be discussed for all areas of industrial activities, i.e., production, quality control and control of regulatory requirements, and research and development. Emphasis will be put on examples from R&D as this is the most intensively used area. Because this field is much too broad for a complete review here, examples will focus on applications from just three major sectors: biotechnology, pharmaceuticals and cosmetics, and automotive and mining. Environmental research is a fourth area that will be partly covered in the section on regulatory requirements.

1. Introduction

Synchrotron radiation (SR), often referred to as "synchrotron light", is the electromagnetic radiation that is emitted from electrons. At this time, there are about 50 operational, dedicated synchrotron radiation facilities with several more in the planning and/or construction phase (see *www.lightsources.org*). Keeping in mind that the costs for the construction of one of these facilities range between ~US\$30M and US\$1B (with operation costs being ~10–15% of the construction costs), synchrotron radiation is obviously an extremely valuable tool for a broad range of applications, from pure basic research (e.g., spectroscopy of atoms in the gas phase) to use-oriented investigations (e.g., *in situ* investigations of catalytic reactions), and even to fabrication. Many of these applied aspects will be covered in this chapter in the Industrial Applications section. Before we proceed, we believe that the title of this contribution needs some explanation. In a very rigid sense, we would prefer to summarize under "industrial applications/industrial use" those utilizations that are connected with a direct flow of cash from industry to the SR lab where the research is carried out or the product is fabricated. However, giving an overview on just this type of industrial application would be difficult because these examples are very often exactly the applications that are of real interest for industry and from which results are never published in the open literature for competitive reasons. Thus, we will mainly cover examples of industrial research that are based on joint projects between academia/research laboratories and industry that are funded by various funding agencies — meaning that the direct financial contribution from industry might be limited. For a similar reason, it is also difficult to quantify the monetary flow from industry for this SR work, although it is well known that almost every SR facility also has a number of industrial users.

At least in principle, SR can be used in all areas of industrial activities (excluding administration): Production, Quality Control (including control of regulatory requirements), and Research and Development (R&D). There are in fact examples for the use of SR in all of these areas, and we will present some illustrative examples for each of them. As one would expect, most of the published applications are in the R&D category, and thus examples from this area will be the main focus of this chapter. However, with the increasing awareness of environmental problems, there is a rapid development of an additional area of SR applications of industrial relevance — namely the control of products and processes to comply with regulatory requirements. In the Quality Control section, we will also discuss some examples from this fast developing area where SR-based technologies have tremendous advantages when it comes, for example, to the detection of very low concentrations of elements and/or the speciation of elements in an X-ray amorphous environment where standard diffraction techniques cannot be applied.

From the experience of the authors, there are several requirements that must be fulfilled for a successful collaboration with industry:

- Competent support in all phases of a project.
- Clear and, if possible, quantitative results with a direct link to the actual problem.

- Secrecy of work and results, and clear (industry-friendly) arrangements regarding intellectual property.
- Short turn-around time, especially for problems connected to production.
- Competitive prices.

In many cases, satisfying all of these requirements at once is a challenging task for SR labs, and as a consequence several different access routes to SR use have been designed by various labs for industrial customers that fulfill at least some of these requirements. In general, there are three distinct routes:

(1) *Academic mode.* In this mode, industry follows the approach that is used for academic users, i.e., beam time is allocated through a review committee based solely on scientific merit. This access is free of charge, but the results are "open" and are expected to be published. In addition, an even more important drawback can be the significant waiting time for some beamlines.
(2) *Proprietary mode.* This is the mode in which industry pays for all costs in exchange for confidentiality and in general the required rapid access.
(3) *Construction and use of a dedicated beamline.* In this mode, the company "rents" a port and builds and operates its own beamline, either on its own or in collaboration with academic and/or other industrial partners.

There are several hybrid modes, but for real-world industrial problems the proprietary mode seems to be the most suitable. Within this mode, there are various options of how the SR facility interacts with the customer, ranging from simply providing access to a suitable beamline to providing what we call a 100% analytical service. From our experience, this latter concept has the best potential to fulfill the industrial requirements discussed above, and it consists (for R&D projects) of the following six steps:

(1) Discussing the problem and developing an experimental "strategy" to solve that problem together with the potential customer.
(2) Preparing the samples for the actual measurements.
(3) Measuring the samples (and suitable reference compounds).
(4) Analyzing the data and interpreting the results with regards to the actual problem.
(5) Preparing a written report (with further recommendations, if necessary).
(6) Sending the report (together with the invoice).

This concept seems to be accepted by industry. For example, it has been quite successfully used at the Canadian Light Source Inc. (CLSI), and it could be the basis to finally intensify the industrial use of SR.

The examples that we will discuss in this chapter are chosen based on the subjective preferences of the authors, and for the sake of the scope of this review not all possible areas of applications will be able to be covered in the R&D section. Those readers who are interested in additional information on what can be done together with industry using synchrotron radiation should consult the articles published by various SR labs, e.g., the National Synchrotron Light Source (NSLS, Brookhaven, USA),[1] SPring-8,[2] or the European Synchrotron Radiation Facility (ESRF, Grenoble, France).[3,4] Broad ranges of applications are also discussed on the corresponding web pages of various facilities; an excellent starting point is the web site *www.lightsources.org* which is supported by most major synchrotron radiation facilities.[5] We would like to mention just two: the web site of the Diamond Light Source,[6] and the NSLS web site describing science highlights of its industrial user program.[7] A very good summary of the problems that exist when industry makes use of user facilities such as SR laboratories (and some interesting suggestions on how to overcome these problems) is provided in a recent report by the US National User Facility Organization.[8]

2. Synchrotron Radiation: History and Properties

With some broad-mindedness, one can claim that the history of synchrotron radiation goes back more than 100 years to the year 1897 when J. J. Lamor calculated the radiation emitted by a non-relativistic accelerated electron and/or to 1898 when Liénard calculated the radiation emitted by such an electron moving in a circular orbit. Schott came significantly closer to today's description of synchrotron radiation when in 1907 to 1912 he calculated the emission of a relativistic electron moving in a circular orbit in an attempt to explain the observed discrete atomic spectra.[9] The theory of synchrotron radiation as we are using it today was developed in 1946 by Julian Schwinger[10,11] and independently by D. Ivanenkow and coworkers in Russia based on their earlier work on "the maximal energy attainable in a betatron".[12]

For the experimental observation of the radiation from accelerated relativistic electrons, the corresponding high energy accelerators were necessary, and it was only in the 1930s that the opportunity arose when the first betatrons were built by Kerst and Veksler. In 1946, J. P. Blewett tried unsuccessfully to observe synchrotron radiation at General Electric's (GE) 100 MeV betatron in Schenectady, NY (USA): he looked at the wrong wavelength range with his detectors and could only observe the shrinking of the electron orbit due to the radiation losses.[13] About a year later, synchrotron radiation was observed for the first time by pure chance at the 70 MeV synchrotron at GE's research laboratory in Schenectady by two technicians (Floyd Haber and Gerald Knowlton) when they were looking for sparks that occurred quite often in these machines because of the applied high voltage and bad vacuum conditions.[14,15] More details of this story are described, for example, in an interesting article in *Physics Today*.[16] Worth noting in the context of this article is surely the fact that the history of synchrotron radiation started in an industrial laboratory. However, in these early days, the value of synchrotron radiation (especially the value for industrial applications) was not recognized, and it took a few more years before some groups started the systematic investigations of the properties of SR at the FIAN synchrotron in Moscow (1953/56) and the 300 MeV Cornell synchrotron (1952/53), both accelerators having been built for nuclear research. The first spectroscopic investigations using SR were carried out in 1956 by D. H. Tomboulian and P. L. Hartmann when they measured absorption spectra at the Be K- and the Al LII/LIII-edges.[17] In 1962, K. Codling and R. P. Madden had been the first to use SR for systematic experiments at the 180 MeV synchrotron at the National Bureau of Standards (now NIST) in Washington, DC.[18]

More or less at the same time, activities started in Europe and Asia. At the Frascati laboratory near Rome (Italy), researchers began measuring absorption in thin metal films using a 1.15 GeV synchrotron. In 1962, scientists in Tokyo (Japan) formed the INS–SOR (Institute for Nuclear Studies–Synchrotron Orbital Radiation) group and by 1965 were making measurements of soft X-ray absorption spectra from a 750 MeV synchrotron. In 1964, the 6 GeV Deutches Elektronen-Synchrotron (DESY) in Hamburg (Germany) began operating for both high-energy physics and synchrotron radiation. In 1968, the first SR spectrum to be measured at a storage ring, specifically at the Tantalus I storage ring at the University of Wisconsin, clearly showed the superiority of storage rings as SR sources (a synchrotron is a "flickering" light source with intensity variations from pulse to pulse whereas a storage ring is a "dc" source with a very slow decrease of the intensity due to the loss of electrons). Soon after, other storage rings were used for SR experiments, e.g., the ACO ring (Orsay, France) in 1971, the SURF II ring at NBS, and the SPEAR ring at Stanford (CA, USA) in 1974, with several more to follow. The first dedicated storage ring for SR experiments was put into operation in 1981 in Daresbury (UK) as a first machine of the "second generation facilities", i.e., facilities built just for SR applications. Third generation storage rings that are now the leading sources for SR experiments are characterized by the quality of the electron beam (spatial stability of the beam and the "size" of the beam that determines the emittance) and by the fact that most experiments use arrays of alternating magnets, known as insertion devices, as sources for radiation that provide significantly higher intensities as compared to the radiation from the bending magnets. A summary of the early history of synchrotron radiation with a lot of references can be found in an article by A. L. Robinson in the *X-Ray Data Booklet.*[19]

At the beginning of the 1980s, there were at least three storage rings built mainly for industrial applications, namely for the fabrication of microelectronic devices. In Germany, the Berliner Elektronenspeicherring-Gesellschaft für Synchrotronstrahlung mbH (BESSY) was founded in 1979 with the objective of carrying out basic research in conjunction with German and European industry. A central motive for forming BESSY was to strengthen European technology in the field of microelectronics by developing SR lithography as a technique for the production of microelectronic components. During 1991, Oxford Instruments' compact synchrotron X-ray source Helios I was installed at IBM's Advanced Lithography Facility at the East Fishkill (USA) site following factory tests in Oxford during 1990.[20] Since January 1992, this machine has been run routinely for lithography development work. Finally, the 1.5 GeV storage ring at the Louisiana State University Center for Advanced Microstructures and Devices (CAMD) was built in the early 1990s for research in VUV-X-ray lithography for the fabrication of microelectronic devices. However, by the end of the 1990s, it had become clear that industry could fabricate high density microelectronic devices with improved conventional techniques, and the research activities at SR facilities were abandoned. In lieu thereof, several facilities started activities in the area of microfabrication of microelectromechanical (MEMS) devices. Both areas will be discussed later in this chapter.

As compared to other light sources, synchrotron radiation has some extraordinary properties that are also the basis for its value for industrial applications. The most important property is that SR provides a continuous spectrum in the range between ~1 eV and 100 keV, i.e., from the far infrared range to the very hard X-ray range, whereas the useful high energy limit depends on the energy of the corresponding electron accelerator and/or the magnetic field strength of the insertion devices that are used as the source for the radiation. In the vacuum ultraviolet and in the X-ray range, SR is by orders of magnitude the most intense light source that is available; in part of the infrared range, SR is also superior to conventional light sources. The second important property is the high collimation of the radiation that is emitted with an opening angle of ~1 mrad; i.e., at 10 m away from the light source (the electron orbit), the height of the radiation is ~1 cm. In this respect, SR is comparable to a laser pointer. These two properties are the reason for the extremely high intensity that is obtained on the corresponding samples, even when the radiation is monochromatized, and for the opportunity to obtain very small spot sizes at the sample. The most important properties of synchrotron radiation can be summarized as follows:

- Continuous spectrum — 1 eV to 100 keV; SR provides the most intense spectrum in the VUV and X-ray region.
- Collimation — 1 mrad.
- Linear polarization — >95% in the plane of the electron orbit.
- Circular polarization — up to 90% above and below the plane of the electron orbit.
- Partly coherent.
- Absolute calculability.
- "Clean" source — emission of radiation takes place under ultra-high vacuum conditions.
- Time structure — allows time resolved experiments (mainly given by the length of the "electron bunches" in the accelerator and its circumference).

3. SR-based Techniques for Industrial Applications

At least in principle, all of the spectroscopic and imaging techniques that are used for basic research using SR can also be used for industrial applications. In many cases, the methods have been optimized to such an extent that industrial applications are possible without further improvements. However, one should not underestimate that some of the questions coming from industry will push the technical limits of the corresponding techniques: industry in general tries to solve problems with in-house capabilities whenever possible, and only the most challenging problems will go to a SR lab. Because synchrotron radiation provides a continuous spectrum, spectroscopic techniques from all ranges of the electromagnetic spectrum can be and are used with SR as a radiation source. Table 1 presents an overview of the most common techniques used in SR facilities. These techniques are ordered according to the typical photon energy used. However, it would also be possible and reasonable to group the techniques according to their "information depths":

- Long range order — diffraction (powder, single crystal).
- Medium range order (10–1000 Å) — small angle scattering, tomography.

Table 1. Summary of the most common techniques used with synchrotron radiation.

Measurement Techniques
Fourier Transform Infrared Spectroscopy
Infrared Microspectroscopy
Circular Dichroism
UV–VUV Photoelectron Spectroscopy (ESCA)
VUV-Microspectroscopy
Powder Diffraction
Surface Diffraction
Small Angle X-ray Scattering + Wide Angle X-ray Scattering (SAXS–WAXS)
Protein Crystallography (PX)
Microtomography
X-ray Fluorescence (XRF) and X-ray Microscopy
X-ray Absorption Fine Structure Spectroscopy (EXAFS, XANES)
Fabrication Techniques
UV–VUV Lithography for Microelectronics
X-ray Lithography (LiGA) for MEMS (sensors, gears, etc.)

- Short range order (atomic scale and up to ~10 Å) — IR and photoelectron spectroscopy, X-ray absorption spectroscopy.

Most of the techniques summarized in Table 1 are also standard laboratory techniques used with conventional radiation sources, such as globars in the IR range, and X-ray generators, such as rotating anodes in the X-ray range. However, by using SR as the radiation source, the capabilities of most techniques can be dramatically increased, offering in many cases completely new opportunities. To illustrate this, here are some of the experiments that are possible only with SR:

- Microanalysis on a micrometer (sub-micrometer) scale (diffraction-fluorescence-absorption spectroscopy).
- Real-time studies down to millisecond and even microsecond resolution.
- *In situ* analysis of processes as a function of pressure, temperature, shear stress, chemical reactions, etc.
- Experiments with samples under extreme conditions of pressure and temperature (300 GPa, 2000°C).
- Studies of ultra-dilute systems (down to the ppb range).
- New imaging techniques such as element specific microtomography, phase contrast imaging, and X-ray microscopy (resolution below 50 nm).

- Analysis of microcrystals with dimensions as small as 10 μm^3 using diffraction techniques.

As was mentioned earlier, there are just a few techniques in Table 1 that rely completely on the properties of synchrotron radiation, of which the most important one is X-ray Absorption Fine Structure (XAFS) spectroscopy. XAFS is also one of the most-used workhorses for industrial applications in most SR facilities, being second only to protein crystallography. Because of the importance of this technique as well as the fact that it is a typical SR-based technique, it will be explained here in some more detail, and we will show several examples of where it is used.

XAFS spectroscopy — the more general term X-ray Absorption Spectroscopy (XAS) is sometimes used interchangeably with XAFS — measures the energy dependence of the X-ray absorption coefficient μ(E) at and above an inner shell absorption edge of a selected element, i.e., it probes the energy dependence of the photoabsorption coefficient. Distinct energies for the core electron excitation in the absorption process provide element selectivity for different atom species. Figure 1 shows a "typical" XAFS spectrum, specifically the spectrum of a Pt foil measured at the LIII-edge. In general, the XAFS spectrum is divided into two parts as indicated in the figure: (1) the X-ray Absorption Near Edge Structure (XANES) that starts ~50 eV before the "edge" (i.e., the jump of the absorption coefficient when the energy of the

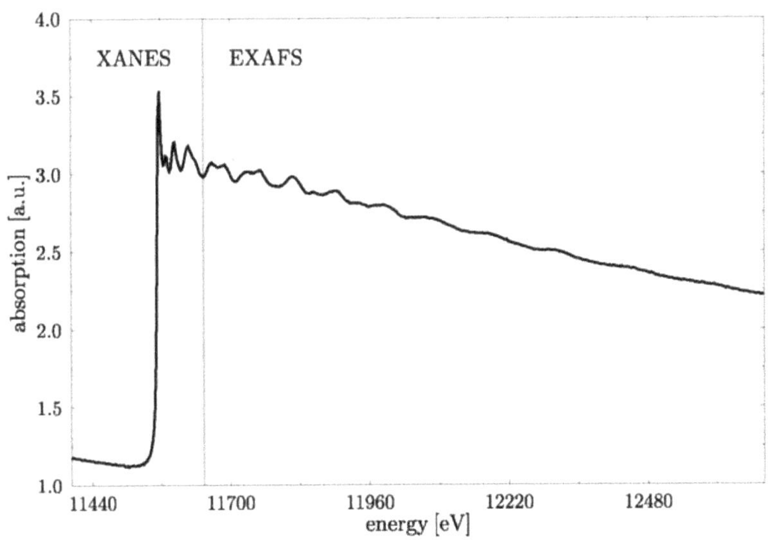

Fig. 1. The LIII-edge-XAFS spectrum of a Pt foil showing the XANES and EXAFS regions.

incoming photon is high enough to excite the corresponding inner shell electron into an empty orbital or into the continuum) and reaches up to ~50–100 eV above the edge, and (2) the Extended X-ray Absorption Fine Structure (EXAFS) that is visible up to ~1 keV above the edge. XANES contains (mainly) information on the electronic structure of the local vicinity of the absorbing atom species. Hence, the energy and shape of the absorption edge depend sensitively on the chemical environment of the absorbing atom. By comparison to appropriate reference systems, for example, the chemical valence state can be determined and conclusions can be drawn on the type of chemical neighborhood, the local symmetry, and possible charge transfer between binding partners.

Information can be extracted from EXAFS about the local geometric structure, i.e., about bond lengths and coordination numbers, and the "type" of neighbor (chemical species) in the first coordination shells. As the energy dependence of the cross section is measured, the tunability of synchrotron radiation is a crucial precondition for XAFS spectroscopy. When compared to traditional diffraction techniques, XAFS spectroscopy has a few decisive advantages that are also of importance for many industrial applications:

- XAFS is non-destructive and element specific.
- No long range order is required.
- XAFS is extremely sensitive (ppm-range or even better).
- There are no special requirements for the samples; samples can be solid, liquid, or in the gas phase.
- Quantitative analysis of spectra is possible.
- No vacuum is required.
- Time resolved *in situ* experiments are possible.

The theory of XAFS spectroscopy, the various modes for measuring XAFS spectra, the data reduction processes, and numerous applications from a broad range of research areas are described in several text books and review articles, and in the proceedings of the corresponding conferences.[21–25]

4. Synchrotron Radiation for Quality Control and the Control of Regulatory Requirements

4.1. *Quality control*

Quality control is a process employed to ensure a certain level of quality in a product or service. For the manufacturing industry, the basic goal of quality control is to ensure that the products meet specific requirements and are of constant quality when they go the final customer and consumer.

To achieve this goal, an important step is the inspection and characterization of incoming and outgoing materials and products.

In principle, there are several SR-based spectroscopic techniques that have potential for quality control, such as X-ray microdiffraction to determine strain in samples, grazing incidence X-ray reflection measurements to determine the roughness of surfaces, and small angle scattering or X-ray tomography to check the homogeneity of a sample, just to name a few. However, for most industrial applications and especially for mass-production, quality control has to be fast and cheap — requirements that are very difficult to achieve when synchrotron radiation is used.

In spite of this, there are a few isolated examples of the usefulness of several SR-based techniques for quality control of industrial samples or processes. For example, X-ray microdiffraction was used on several occasions by Lucent Technologies (Agere Systems) at the Stanford Synchrotron Radiation Lightsource (SSRL, Stanford, CA, USA) facility to map crystallographic strain and/or multilayer thickness in micro-optoelectronic devices,[26,27] and in at least one other case by CINVESTAV (Mexico) at NSLS.[28] An interesting overview of the potential of this technique is given in a review by the corresponding SSRL group.[29] High resolution X-ray imaging with a resolution in the 50 nm range could also be a powerful technique for quality control of micro- and nano-electronic products as was demonstrated by Zschech *et al.*[30] However, there are no further publications that show "real" industrial applications of these techniques. There seems to be just one technique that fulfills the "fast and cheap" requirement for industrial use of SR for quality control, and it has in fact been used by industry at several SR facilities on a more or less continuous basis for several years — namely, Total X-ray Reflection Fluorescence (TXRF).

A more general overview of the applications of SR-induced TXRF was recently published by Streli *et al.*[31] In the semiconductor industry, TXRF is a very common analytical technique for monitoring surface contaminations on non-patterned Si-wafers. The detection limit for laboratory-based instruments using conventional X-ray generators as an excitation source is $\sim 10^9$ atoms/cm^2. This is still roughly an order of magnitude higher than the limit expressed in the SEMATECH roadmap.[32] As for most fluorescence techniques, the sensitivity of TXRF can be increased by increasing the incoming photon flux and/or reducing the spectral background. Both options have been exploited by applying synchrotron radiation, which provides a high incident flux combined with low divergence and a linearly polarized beam. This leads to an increased fluorescence signal while simultaneously reducing the elastically scattered background when sample and detector are properly arranged relative to each other. Early test experiments were carried out in 1994, more or less simultaneously, at HASYLAB (Desy, Hamburg, Germany) and SSRL, and a bit later at the ESRF.[33] All of these experiments indicated that it should be possible to increase the sensitivity by at least an order of magnitude. In the following years, systematic investigations of the potential of TXRF for the semiconductor industry were carried at SSRL by P. Pianetta and coworkers.[34,35] They investigated the detection limits for both low Z elements ($Z < 13$)[35] and heavier elements.[34] For Na, Mg, and Al, the detection limit (at a bending magnet beamline) was $\sim 2 \times 10^{10}$ atoms/cm^2 and thus still below the requirements of the SEMATECH roadmap (5×10^8 atoms/cm^2); for heavier elements, a detection limit at $\sim 5 \times 10^7$ atoms/cm^2 was observed with an integration time of ~5000 s. Figure 2 shows a typical SR-TXRF spectrum of a wafer that was cleaned using state-of-the-art technology.

The quantitative determination of trace impurities on the surface of a wafer is the first crucial step in quality control; however, the knowledge of the chemical state of the impurity allows a better judgment of the effect of the impurity and helps to determine the true source of the contamination problem. This information cannot be obtained by conventional laboratory-based TXRF instruments, but it can be obtained by combining

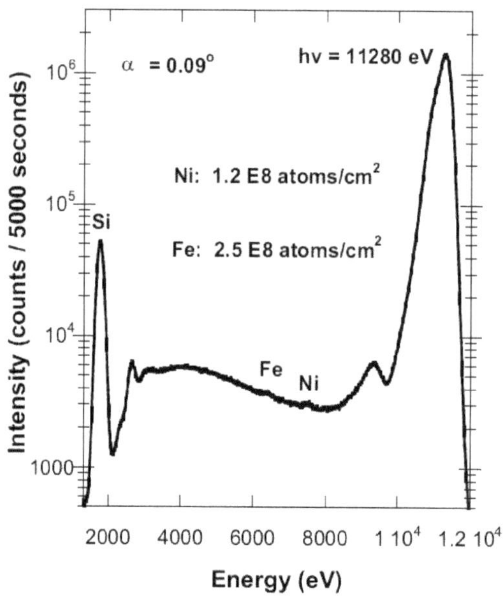

Fig. 2. TXRF spectrum of a cleaned Si-wafer. Cl, Fe, and Ni are contaminations on the wafer surface. The strong peak at 11280 eV is from elastically scattered photons (K. Baur, P. Pianetta, private communication).

TXRF with XANES measurements. Corresponding test experiments have been carried out again by the SSRL group for Cu impurities.[36] A more detailed study to determine if this technique is suitable for "real" contaminations as they occur in a microelectronic VLSI production lab was recently carried out by Meier *et al.*[37] To finish this section, it is worth mentioning that angle-resolved TXRF is also used by some companies for the analysis of depth profiles of impurities in silicon wafer, which gives additional valuable information about the effect of contaminations.[38,39]

4.2. *Control of regulatory requirements*

There is a long history of using synchrotrons to obtain information for industrial clients on the chemical speciation of compounds in the environment. This mirrors the general rise in academic and government use of synchrotrons in the environmental sector, which is stimulated mainly by two factors: (1) the increasing awareness of general public environmental problems (pollution of water and soil, radioactive waste, dangers in the workplace, etc.) and the parallel efforts of the corresponding bodies to regulate environmental issues to the last detail, and (2) requirements for chemical analyses that are at the corresponding detection limit. XAS spectroscopy, in the form of XANES and EXAFS, is an excellent technique for environmental investigation because of its sensitivity to element oxidation state and molecular structure. This sensitivity to local chemical form is critical for environmental studies because of its direct relevance to toxicity. It is the molecular identity of a contaminant that may have deleterious biological effects, not just the presence or absence of a particular metal. The local coordination environment is important for contaminant species because of the ubiquity of amorphous chemical species in nature. XAS is also one of the few techniques that can distinguish between precipitated species, adsorbed species, and chemically bonded species. The advent of detectors able to detect environmental concentrations of many elements has enabled measurement of contaminants in samples without extraction or pre-concentration procedures.

To illustrate the types of regulatory control services that SR facilities can provide in this area, we will present some examples of work done for industrial clients at the Canadian Light Source (CLS, Saskatoon, Saskatchewan, Canada). While similar services are offered at other synchrotron facilities (e.g., ESRF, Diamond Light Source, and ANKA), we are of course most familiar with those in Canada. The CLS is well-placed to perform industrial measurements related to the mining sector because of its proximity to uranium deposits in northern Saskatchewan and more generally the large mining sector in Canada. In 2008, over three quarters of the industrial work performed at the CLS was of an environmental nature.[40] Since 2008, the percentage of beam time allocated to regulatory control issues has decreased significantly as product development type projects have increased.

In the following we discuss two "typical" examples of environmental/regulatory research carried out at the CLS for industrial customers. These case studies involve (1) the speciation of workplace aerosols in a nickel refinery and smelter, and (2) the disposal of arsenic-containing tailings in a uranium mine tailings management facility.

4.2.1. *Workplace aerosols*

Nickel is used in many types of stainless steel and non-ferrous alloys, as well as in electroplating, with applications in almost all industrial sectors. As a result of its importance in industrialized nations, the toxicity of nickel was recently re-evaluated in the European Nickel Risk Assessment. This assessment focused on evaluating and controlling exposure to select nickel compounds: nickel metal, nickel sulphate, nickel chloride, nickel nitrate, nickel carbonate, nickel sulphides (Ni_3S_2 and NiS), and nickel oxides (NiO, NiO_2, Ni_2O_2, and Ni_2O_3). This focus on specific nickel compounds requires a method of accurately determining nickel speciation in a variety of environmental samples. A standard method of speciating nickel for nickel producers is the sequential extraction-based Zatka method, which separates the nickel into four operationally defined categories.[41] This technique suffers from many of the same drawbacks as any extraction technique: (1) dependence on surface characteristics (surface area, surface reactions, and surface coatings), (2) inherent non-selectivity, (3) interferences from other chemical species, and (4) the tendency to modify the extraction procedure to non-validated sample sizes or matrices. XAFS spectroscopy was recently used to identify the nickel species occurring in several fly ash filter samples.[42]

In the first example, samples were collected on IOM (Institute of Occupational Medicine) cassettes with Versapore filters from various locations within the nickel refinery and smelter at Copper Cliff, Ontario. Measurements at the CLS indicated that significant differences existed between the nickel speciation of aerosol samples reported from the Zatka method compared to results from SR-XANES. The XANES results were more consistent with the likely speciation based on the nickel processing unit and results from scanning electron microscope analysis. Figure 3 illustrates the linear combination fit of a typical nickel-containing workplace aerosol sample collected.

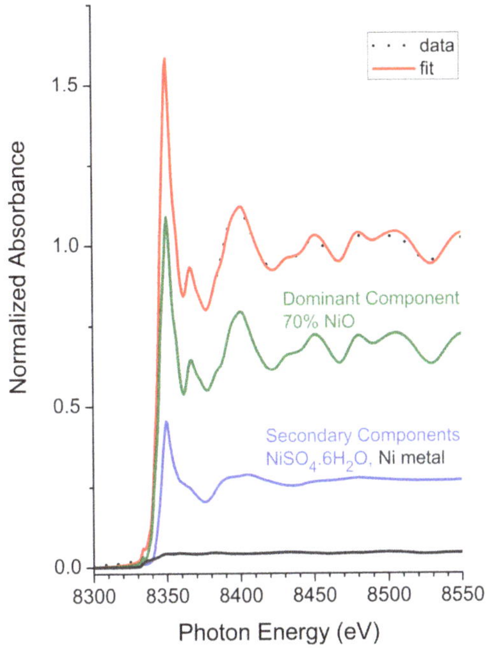

Fig. 3. SR-XANES spectrum of a personal aerosol filter collected from a nickel refinery. The bottom curve indicates a small amount of nickel metal as the third component.

4.2.2. *Tailings management*

Areva Resources Canada Inc. (Areva) operates the JEB Tailings Management Facility (TMF) as an integral part of its uranium mining operations in northern Saskatchewan. The uranium ore processed at the McClean Lake operation may contain arsenic at concentrations up to 2% by weight, posing a potential risk to surrounding waters. Areva has designed the JEB TMF to minimize the migration of soluble constituents from the TMF through physical containment and geochemical controls (see Fig. 4).[43] The effectiveness of the geochemical controls is monitored through periodic drilling campaigns into sediments obtained from the TMF.

The tailings contain arsenic removed during processing of the ore as well as residual arsenic-containing primary minerals and raffinate (an acidic sulfate waste solution). Prior to disposal, arsenic is neutralized using lime and precipitated with ferric iron. The tailings disposal facility is monitored for soluble arsenic through an examination of pore waters. XAS is used to characterize solids in the tailings management facility as a function of depth. Figure 5 was constructed based on the ability of XANES to quantitatively determine the relative amounts of arsenic species in the solid state.

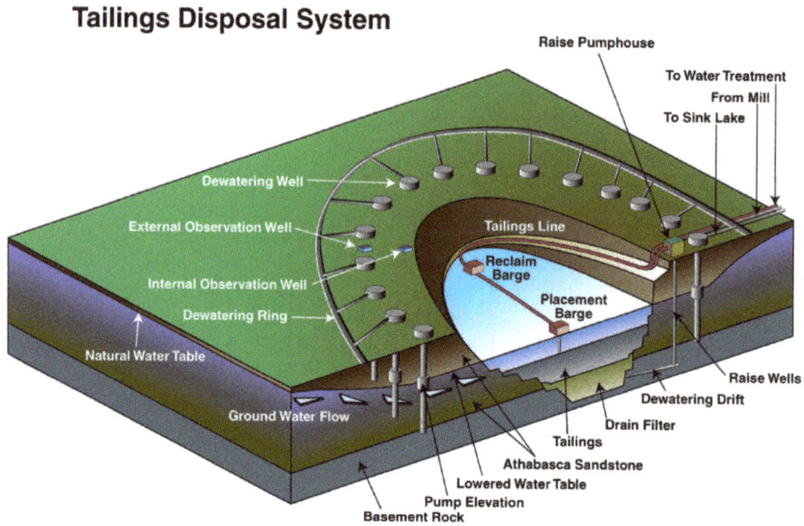

Fig. 4. Design and operational features of the JEB Tailings Management Facility (TMF). (The authors and editors wish to thank the following for permission to reproduce copyrighted material: Taylor and Francis from *Synch. Rad. News* **20**(3) (2007)).

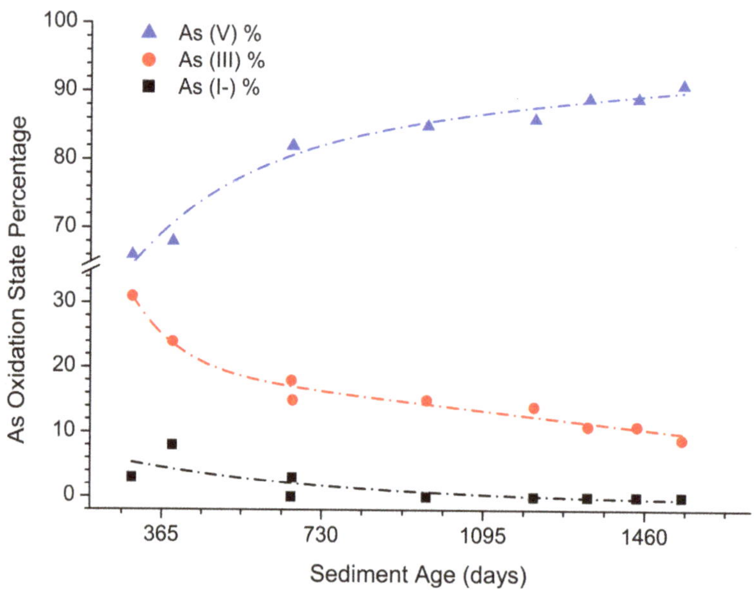

Fig. 5. Relative amounts of each arsenic oxidation state plotted against the tailings sediment age. (The authors and editors wish to thank the following for permission to reproduce this copyrighted material: Taylor and Francis from *Synch. Rad. News* **20**(3) (2007)).

Interpretation of the data taken over four years of measurements suggests that mixed tailings materials — composed of (1) leach residue [primarily rammelsbergite ($NiAs_2$), gersdorfite (NiAsS), and niccolite (nickeline, NiAs)], (2) clays (primarily illites), and (3) neutralized raffinate solids (essentially gypsum and Fe, Al, and Si-rich precipitates) — participate in oxidation reactions. The oxidation products were identified using XANES as thermodynamically stable, As(V)-containing minerals. Further analysis using EXAFS showed the As(V)-containing mineral to be consistent with a poorly crystalline scorodite.[44]

Supporting data were obtained from pore water sampling, which show As(III) concentrations exhibiting a time-dependant rise and fall in concentration, stabilizing after four years at under 1 mg/L. Total arsenic concentrations are below the action limit of 2 mg/L after sediment aging. Based on these data, it appears that arsenic is forming a stable As(V) mineral species, which is keeping soluble arsenic concentrations low in the TMF.

5. Synchrotron Radiation for Production

There are two areas in which synchrotron radiation is/was used by industry for production processes: (1) VUV-X-ray lithography for the fabrication of microelectronic devices and (2) deep-etch X-ray lithography as the first step of the LiGA process. LiGA is a German acronym that stands for Lithographie (Lithography), Galvanoformung (electroplating) and Abformung (molding). This technique has been developed for the production of MEMS devices.

5.1. *X-ray lithography for the fabrication of microelectronic devices*

X-ray lithography was first reported by IBM in 1972,[45] and X-ray proximity lithography using synchrotron radiation was also reported by the IBM group in 1976 when Spiller and coworkers used the SR facility at DESY for their feasibility studies.[46] At this time, industry was convinced that the "next generation" of microelectronic devices at 130 nm ground rules (and later on even smaller ones) would need the soft-X-ray radiation from SR facilities, and thus the first experiments initiated a broad range of activities to develop X-ray resists, X-ray masks, X-ray steppers, and finally also dedicated soft X-ray SR facilities. It might be worth noting that the problems connected with "private storage rings" built and operated by industry were actually discussed in a *Science* article.[47] There was actually a little bit of hype when in 1989 J. B. Murphy counted in his review at least seven compact machines based on superconducting magnets and five conventional machines that were in the planning or construction phase for X-ray lithography.[48] In a more technical overview, S. Hector summarized the industrial activities and achievements in this field and pointed out that there were hardly any leading semiconductor companies without activities in this area.[49] However, of the machines discussed in Murphy's review, just a few had been built, and as was mentioned in an earlier section, by the end of the 1990s major industrial activities were given up and the conventional machines such as BESSY and CAMD became "multi-purpose" SR facilities. The superconducting machines were either used much longer (HELIOS at IBM) or are still in use (AURORA at Sumitomo Heavy Industries) for research related to microelectronics. A summary of the results obtained at AURORA, for example for a 70 nm device manufacturing system, are reviewed, e.g., by Miyatake *et al.*[50]

As there seems to be a limit at around 180 nm for optical lithography, it is still not clear if synchrotron radiation could play a role when it comes to nanolithography, i.e., fabricating devices with ground rules smaller than 100 nm. This is at least one of the reasons why there have been, and still are, always some "rest activities" in SR labs. In the US, most of these activities were focused at the University of Wisconsin's Center for NanoTechnology facility at the Aladdin storage ring, and many of those activities had been supported by industry in an indirect way via SEMATECH. Some of the research focused on development and testing of equipment necessary for nanolithography, e.g., stepper systems,[51] but new techniques had also been developed, e.g., interferometric lithography.[52,53] There are also some ongoing activities at the Pohang Light Source (PLS) in Korea and at some of the SR facilities in Japan.

5.2. *Deep etch X-ray lithography and microfabrication*

The LiGA process has been developed from the mid-1980s starting with the activities in the former Nuclear Research Center in Karlsruhe (Germany) for the inexpensive mass production of micro-structured devices.[54] As compared to other microfabrication technologies, LiGA has some significant advantages:

- Very high structures up to several millimeters can be fabricated in one step.
- The smallest lateral dimensions are a few micrometers (or even less) with structural details in the submicrometer range.

- A broad range of materials can be used (polymers, metals, alloys, ceramics, nanomaterials, etc.) at various steps within the fabrication processes.
- Real three-dimensional structures can be fabricated.
- Extreme vertical side walls with very low surface roughness (rms roughness better than 20 nm) can be achieved.
- Possibilities exist for low cost mass fabrication by parallel processing of parts and/or molding as a typical mass fabrication technique.

All of these advantages plus the possibility of fabricating fascinating structures make LiGA very attractive for SR facilities; as a consequence there are hardly any SR facilities without LiGA activities. Figure 6 shows two examples of "typical" LiGA structures. Figure 6 (left) shows a "simple" honeycomb structure fabricated by direct exposure in PMMA, Poly(methyl methacrylate). The height of the whole structure is $\sim 400\,\mu$m, the diameter of each individual structure is $\sim 80\,\mu$m (one should keep in mind that the diameter of a hair is $\sim 100\,\mu$m), and the thickness of the walls is $\sim 4\,\mu$m. Figure 6 (right) shows a more complicated structure — a metallic planetary micro-gear fabricated from an SU8 (SU8 is a commonly used epoxy based negative photoresist) master by electroplating.

However, most of these LiGA-related activities are of a R&D nature carried out by academic research groups, in some cases in collaboration with industrial partners who define, for example, the specifications of a specific device. Although LiGA now has a 25-year history, and in spite of the fact that potential industrial applications were demonstrated very early in its development, commercialization of LiGA products has been slow to materialize.

Fig. 6. Examples of the application of deep etch X-ray lithography: (left) a honeycomb structure made from PMMA; (right) a metallic planetary gear made using LiGA technology in a conventional ball-bearing.

There is still no real visible economic success based on what is called in industry a "killer product".

There are a few products fabricated using LiGA technology that have been, or still are, commercially available, e.g., mechanical gear systems,[55] optical fiber connectors, and microalignment flexures for photonics applications.[56,57] Perhaps the most interesting example of such a commercial LiGA product is a microspectrometer for the visible and IR range developed at the Research Center in Karslruhe and now distributed by Boehringer–Ingelheim. This spectrometer was used in its IR version for monitoring flammable and toxic gases in industrial applications by the Italian company Felcom. The visible spectrum version has been used for several applications as well, including a non-invasive bilirubin test for new born babies by SpectRx, the precise determination of tooth colors, and the quality control of colors.[58] There have also been a few spin-off companies with a focus on LiGA-based products, such as Mezzo Technologies with a focus on heat-exchangers,[59] Micromotion with a focus on microgears,[55] and AXSUN Technologies[57] and Boehringer–Ingelheim with a broader spectrum of products.[58]

A comprehensive overview of the possible industrial applications of LiGA technology is given in reviews by Khan Malek and Saile,[60] and by Saile *et al.*[61] These reviews also cover emerging applications such as microfluidics (Lab-on-a-chip) and the various foundry activities that still exist at SR facilities such as ANKA and CAMD. For example, in order to make access to the LiGA technology easier for industry, a study has been made on the effects of standardizing the exposure process at a number of synchrotrons throughout the world and the special masks that would be required.[62] There have also been studies of the potential use of X-ray lithography as a nanolithographic tool, for example, by Chen *et al.* at the Singapore Synchrotron Light Source (SSLS).[63] Nanostructures with feature sizes down to 200 nm and aspect ratios up to 10 have been successfully achieved by this approach. Keeping all of these ongoing activities in mind, the future of LiGA as an industrial tool seems still to be open.

6. Synchrotron Radiation for Research and Development

As one would expect, research and development are the most important areas for industrial use of synchrotron radiation. The ten most striking reasons for industry using SR-based analytical techniques are summarized on the web site of the HASYLAB (DESY, Hamburg) SR facility.[64] Besides the advantages of SR-based techniques already discussed in this chapter, perhaps the most convincing one is that SR offers a wide range of analytical tools. Thus, in most cases, the most appropriate method for investigating a special problem is available at an SR lab, and this can provide new insights into the sample and/or the process of interest.

In the remainder of this section, we will discuss some typical examples of SR-based R&D for industrial applications. The examples chosen here are somewhat subjective and will be presented for specific applications/industries in the following areas:

- Biotechnology, pharmaceuticals, and cosmetics
- Automotive
- Mining, oil, and gas

Examples of applications in the semiconductor industry and for environmental research have already been discussed in the section on Quality Control and Control of Regulatory Requirements. As one would again expect, most industrial applications in the R&D area concentrate on "materials" in the widest sense. However, as these aspects are covered in several excellent text books and reviews, we have decided not to include them in this overview, and instead refer the reader to the corresponding references.[65–69]

6.1. *Biotechnology, pharmaceuticals, and cosmetics*

The "Life Sciences" industry is a long time user of SR facilities because the available techniques are particularly well adapted to a number of this industry's needs. For example, cosmetic companies study:

- The molecular structure of skin and hair and their changes after the application of cosmetic products.
- The structure, stability, and aging of "typical" products such as emulsions, mousses, and gels.
- Pigments, powders, pastes, and additives that are used in their products.

At the present time, most if not all large pharmaceutical companies utilize SR for their R&D work, focusing mainly on protein crystallography (PX) as a tool for developing new drugs. However, other techniques are also used, e.g., for characterizing drugs in powder form on an atomic level (X-ray absorption spectroscopy), and in some cases for characterizing tissue and bones and their interaction with medications. In the following sub-sections, we will discuss three applications of SR in industrial biomedical research.

6.1.1. *Cosmetics and the human hair*

Synchrotron radiation is a bright source of infrared (IR) photons. Thus, it is possible to carry out IR microscopy experiments with spot sizes close to the diffraction limit. This high spatial resolution in combination with the excellent spectral quality allows biological samples to be investigated in more detail. Some life science applications are discussed in a review article by Dumas and Tobin.[70]

A biological sample of interest to industry is the human hair as there is a tremendous market for cosmetic products for coloring and bleaching hair, for permanent waving, and for repair of the damages caused by these "aggressive" processes. A first detailed IR-microscopy analysis of a hair was done by J.-L. Bantignies *et al.*[71] Hair consists mainly of keratin, a protein crosslinked by the amino acid cystine. Both of these chemical/biological compounds show several characteristic vibration structures between 4000 and 900 cm^{-1}, the range in which spectra were taken with a resolution of $\sim 4\,\text{cm}^{-1}$.

Spectra from the medulla (the center of the hair), cuticle (the outside of the hair), and cortex (the space in between) showed clear differences in the IR spectra as is shown for cuticle and cortex in Fig. 7. Thus, it is possible to record two-dimensional chemical images (i.e., the cross section) of a hair by plotting the intensity of a specific vibration band as a function of position on the hair. A typical "image" for the amide A band (3290 cm^{-1}) is shown in Fig. 8. Bantignies *et al.* could also show that bleaching causes changes in the chemical/biological structure that are different at different positions

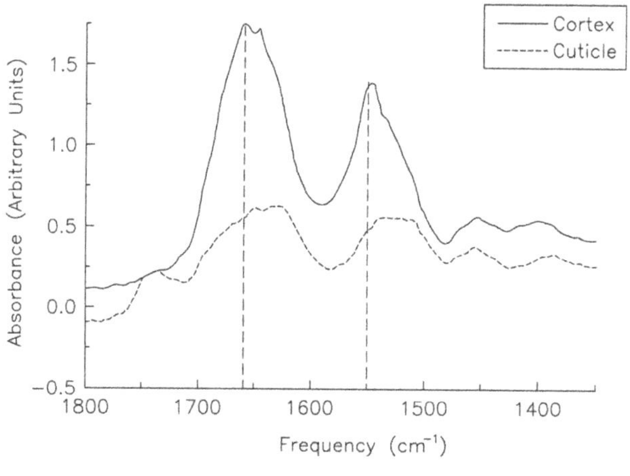

Fig. 7. IR microscopy spectra for cortex and cuticle of a human hair.

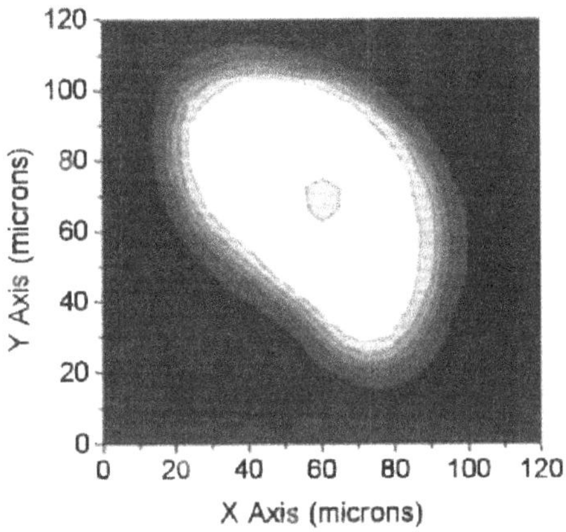

Fig. 8. Typical "image" for the amide A band (3290 cm^{-1}) in a human hair.[71]

within the hair — a result of great importance for the cosmetic industry. In a follow-up study, the same group (with L'Oreal as an industrial partner) could show that there are also significant differences in the structure of hair from Caucasian and Afro-American origin.[72] Similar IR studies have been carried out at the SPring-8 SR facility in collaboration with a Japanese cosmetic company (Kracie Home Products).[73]

Another industrial group at SPring-8 (Kanebo Cosmetics) used X-ray microtomography to image damaged and undamaged hairs.[74] This technique provides the three-dimensional microstructure of a hair without preprocessing. The three-dimensional image of a hair without chemical treatment obtained by this technique is shown in Fig. 9.

Similar X-ray experiments have been carried out at the ESRF facility. Here, Cloetens *et al.*[75] recorded a holotomography image of a hair, i.e., a combination between tomography reconstruction and phase imaging resulting in a three-dimensional distribution of electron density, and hence of mass density in the sample. This technique also allows distinguishing very clearly between the various parts of a hair.

6.1.2. *Metals containing drugs*

X-ray absorption spectroscopy is a valuable tool for the speciation of metals. Thus, XAS also has a broad range of applications in bioinorganic systems such as inorganic metal-based drugs, nutritional supplements, and media for diagnostic applications. An overview of these applications is given in a short review by Nicolis *et al.*[76] In the following, some special EXAFS studies of

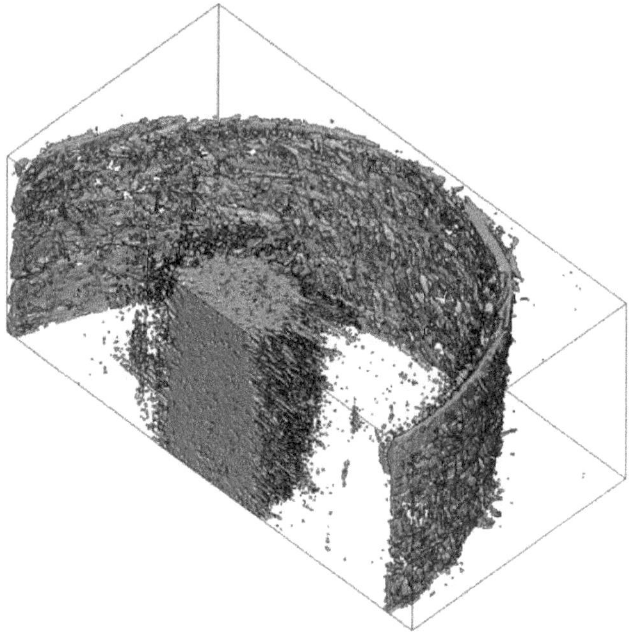

Fig. 9. Three-dimensional image of a hair without chemical treatment.[74]

platinum-based anticancer drugs will be discussed as a typical example of using XAS for the investigation of drugs.

Platinum complexes such as cisplatin, carboplatin, and oxaliplatin, constitute a special group of DNA-damaging anticancer drugs that are widely used in clinical applications. These drugs react with sulfur nucleophiles, especially with sulfur-donor proteins or peptides (e.g., methionine and/or glutathione), leading to unwanted side effects. To improve the understanding of the interaction of platinum drugs with S-donor proteins, Obata *et al.* have carried out a time resolved EXAFS study at the Pt LIII-edge of the corresponding reactions.[77] They concluded that all original ligands are substituted by S-atoms from reduced glutathione, but they observed significant differences in the reaction rate for the various Pt compounds. A similar study was carried out by Provost *et al.* for carboplatin and oxaliplatin in a solution with a large excess of methionine.[78] Figure 10 shows the Pt LIII-EXAFS spectra of the starting materials and their degradation products in the solution after one month of reaction time. The analysis of these spectra indicates that the reaction leads to different Pt–methionine complexes for different drugs.

In many cases, it is of tremendous interest for the pharmaceutical industry to determine in detail the uptake mechanisms of a drug on a molecular level and its interaction with individual cells. This is, of course,

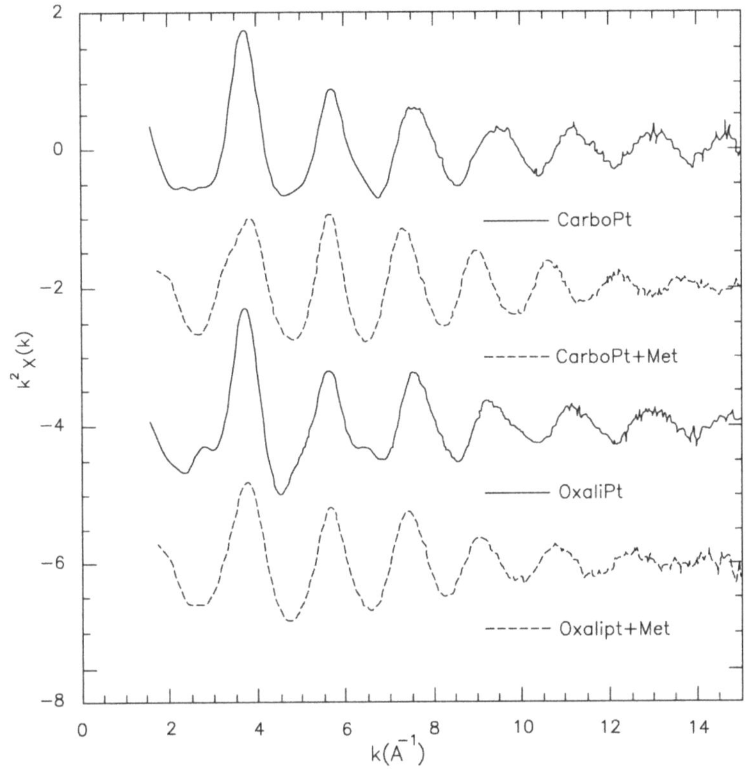

Fig. 10. Pt LIII-EXAFS spectra of starting materials and their degradation products in the solution after one month reaction time.

also possible using X-ray fluorescence and X-ray absorption spectroscopy, as was demonstrated, e.g., in a publication by Kiyozuka *et al.*[79] They investigated the intracellular distribution of cisplatin in human ovarian cancer cells using a transmission X-ray microscope at BESSY. From their results, they concluded that cisplatin inhibits not only the DNA replication but also disturbs the mitochondrial function and RNA synthesis in the nucleolus. A more general overview of the various techniques that can be used for imaging and speciation of trace elements in biological samples and a comparison of their sensitivities is given in a publication by Lobinski *et al.*[80] As compared to other techniques, XAFS spectroscopy provides information not just about the oxidation state of the corresponding element but also detailed information about the metal site and its geometric structure.

6.1.3. *Protein crystallography*

The determination of the structure of protein crystals is of high importance in the design of new drugs for medical therapy. Not only does the sequence of amino acids determine the functioning of a protein, but its complex three-dimensional structure is of crucial importance. X-ray crystallography is a very suitable method of determining the arrangement of atoms within a crystal and thus its three-dimensional structure. Three-dimensional pictures of the density of electrons within the crystal are derived from the angles and intensities of the diffracted beams. The mean positions of the atoms in the crystal can be determined from this electron density map, as well as their chemical bonds, their disorder, and other information. It also allows studying how proteins interact with other molecules, how they undergo conformal changes, and how they perform catalytic reactions in the case of enzymes. This is a basis for designing novel drugs that target a specific protein or for engineering an optimized enzyme for a specific industrial process. The impact of what is now called "structure-guided drug design" on drugs that are either already on the market or in the clinical testing phase is described in a paper by Hardy and Malikayil.[81] They list more than 40 examples ranging from HIV to influenza and glaucoma, for which structure-guided drug design has had a significant impact on the development of the corresponding drugs.

Traditionally, protein crystallography (PX) has been regarded as an expensive and time consuming technique. Thus, it had hardly any impact on drug development when conventional X-ray sources were used or in the early days of experiments with synchrotron radiation. However, that has changed dramatically over the last years with the implementation of major changes in the methodologies and the emergence of the concept of "high-throughput crystallography".[82] Advances have been made in all stages of the process, starting from molecular biology and protein expression to improved methods for crystallization, and finally to more effective structure determination using SR facilities. In contrast to laboratory X-ray generators, the use of SR enables investigations of larger molecules or complexes of molecules, and at the same time the structure information obtained is more precise. In addition, measurement with SR is much faster than with lab equipment, thus avoiding the problem of sample degradation. A review of the various technologies involved in high-throughput protein crystallography is given in a tutorial review by Tickle *et al.*[83]

Another critical advantage of using SR is the fact that very small crystals can be measured. This is crucial because in many cases it is very difficult or impossible to grow large crystals of good enough quality. Protein crystallography with monochromatic microbeams has become a routine tool at several third-generation SR facilities,[84] and several protein structures have been solved using a 10–30 μm beam. However just recently, high resolution spectra of xylanase II have been obtained with a focused SR beam of 1 μm using a crystal volume of $\sim$20 μm^2, which corresponds to $\sim 2 \times 10^8$ unit cells.[85] These experiments open completely new opportunities in scanning protein microcrystallography and make random data collection from microcrystals a realistic possibility. In PX, as in other applications of diffraction and scattering, the tunability of SR leads to anomalous diffraction from substituted heavy atoms. By applying anomalous diffraction, one can derive the phase information needed for the full structure determination using only one native or derivative structure.

Protein crystallography is an actual (and perhaps still the only real) success story of the industrial use of SR. In 2002, for example, $\sim$80% of beam time purchased by industry at the ESRF was purchased by the pharmaceutical industry, and of this the majority was used for protein crystallography.[3] At this time, it is not unrealistic to claim that all major pharmaceutical companies worldwide are using SR-based PX and that all major SR labs are offering a PX service for industrial customers. Here are just a few examples of how industry uses SR facilities for PX measurements:

- At the Advanced Photon Source (APS) at Argonne National Lab, a consortium of six pharmaceutical companies has for more than 15 years operated the IMCA (Industrial Macromolecular Crystallography Association) CAT beamline. The goal is to provide an outstanding support environment for macromolecular crystallography by developing and operating an efficient, reliable, high-throughput facility at which both confidential and proprietary experiments can be readily and easily accommodated.[86]
- At SPring-8, a group of 21 Japanese pharmaceutical companies (the Pharmaceutical Consortium for Protein Structure Analysis) in affiliation with the Japan Pharmaceutical Manufacturers Association (JPMA) is also operating a dedicated beamline.[87]
- At MAX lab in Sweden, NovoNordisk (in collaboration with Astra Zeneca) is using a (at least partly dedicated) PX station at the Cassiopeia beamline.[88]

The importance of PX for medical applications, and thus also for the pharmaceutical industry, is underscored by two additional service enhancements:

(1) *Mail-in Services.* Most of the leading SR facilities have established a mail-in service for PX samples. To make use of this service, the customer sends his crystals (normally cooled to liquid N_2 temperature) via a suitable carrier to the corresponding facility where the measurements

are carried out by qualified beamline scientists. Thus, the customer saves time and money on travel. In many cases the data are then transmitted "online" to the customer to avoid delays.[89,90] In the near future, several facilities will also offer remote control of PX experiments giving users better control of the data collection. The CLS and a few other facilities offer a similar mail-in service for other techniques, and the remote control feature will be added in the near future.

(2) *Automation.* For many PX experiments, especially those designed to screen large numbers of samples, mounting and dismounting of the crystals and alignment of the crystals in the beamline is a time consuming bottleneck. To overcome these problems, SR facilities have put a lot of effort into developing the "fully automated sample mounting and alignment system".[91] In a first step, automated mounting and dismounting systems have been developed that are in most cases based on small industrial robots and special compact cassettes that can hold large numbers of samples.[92] In a second step, the tools that allow the automatic aligning of samples in the beam have also been developed.[93]

An interesting overview of the use of SR in pharmaceutical research with a main focus on PX was presented in a short 2006 review by L. N. Johnson.[94] In this review, she highlighted the fact that there are at least seven drugs on the market (actually from seven different companies) that have been developed on the basis of structure-based drug design. One can assume that today this number is significantly higher. The first of these drugs targeting hypertension was already developed in the early 1980s, the other six by the end of the 1990s. Three of those drugs targeted HIV, two were for influenza, and one was for glaucoma.

Although PX is an extremely valuable technique for bio-medical and pharmaceutical research, it does not provide direct information for instance about the metal site in a protein and its function. Here a combination of techniques, e.g., XAS and PX, can provide the structural details necessary to understand the chemical mechanisms involved in the active states of metalloproteins. Some examples for such combined studies are described by Strange and Hasnain in a chapter of a book that describes a broad range of techniques (not all SR-based) for the investigation of protein-ligand interaction.[95]

A synchrotron radiation-based technique for pharmaceutical research that is not as well known as protein crystallography is circular dichroism (CD), a technique in which the difference in the absorption of right and left circularly polarized light is measured.[96] CD spectroscopy enables the analysis of protein structure, dynamics, and folding in solution. Because of its sensitivity to conformational changes, CD spectroscopy has a significant potential for drug discovery by screening, for example, for ligand and drug binding. In their review on the application of CD spectroscopy as a tool for drug discovery, Wallace and Janes discuss several very impressive examples,[97] such as the binding of the anti-tumor drug Taxol to the anti-apoptosis protein Bcl-2.[98]

6.2. *Automotive*

The automotive industry is a major industrial and economic force worldwide. It makes ~60 million cars and trucks each year, and it employs ~4 million people directly and many more indirectly. However, automotive traffic is also responsible for almost half of the world's consumption of oil and is a major contributor to air pollution in most industrialized nations. Due to the increasing demand for environmental protection, more stringent regulations that can only be met by improved catalysts have been imposed on the automotive industry in recent years. Thus, investigations of catalysts are the most important application of SR-based techniques in the automotive/transportation sector. The second area of activity in this sector involves the investigation of the properties of materials used in the corresponding industry, e.g., new light and strong composite materials, hydrogen storage materials, alloys, and rubber. Here we have chosen to discuss the investigation of rubber as a typical example. A third area (not covered in this chapter) is the investigation of stress in metallic parts and the analysis of malfunctions of various parts.

According to the International Rubber Study Group (IRSG), rubber consumption in 2006 was pegged at 21.4 million tons. This represents an increase of 2.2% compared to the previous year. Of the total output, 9.2 million tons were accounted for by natural rubber and 12.2 million tons by synthetic rubber. This means demand for rubber is still strong and actually increasing. Although the vulcanization of rubber was invented by Charles Goodyear more than 150 years ago, and rubber technology is very advanced when it comes to tailoring the properties to specific applications, there are still a lot of open questions when it comes to understanding the various processes (e.g., vulcanization and aging) and the interpretation of mechanical and other properties on an atomic/molecular level. One reason for this is that rubber is what is called an X-ray amorphous material without long range order for which the application of standard diffraction techniques is not possible.

6.2.1. *Automotive catalysts for emission control*

About 90% of chemical manufacturing processes and more than 20% of all industrial products in the US employ underlying catalytic steps. For petroleum refining, over 80% of the processes involve catalysis. This explains the preeminent importance of catalysts and research of catalytic processes for a broad range of industries. While the role of industrial SR research in this area could have been discussed in other parts of this chapter, we have chosen to discuss catalytic research here because the noble-metal automotive (three-way) catalysts can be regarded as "typical" for a much larger group of catalysts and because automotive companies are very actively involved in SR-based research of these catalysts.

Automobiles make our daily lives more convenient, but on the other hand they have a significant impact on our natural resources and on the global environment. Due to an increasing awareness of the importance of environmental protection, regulations for automotive exhaust emissions are getting more stringent, and as a consequence, automotive companies are trying to improve the catalysts that are used to purify harmful exhaust. Catalytic converters transform the three main pollutant emissions (nitrogen oxides, carbon monoxide, and unburned hydrocarbons) into nitrogen, carbon dioxide, and water. These catalysts consist of precious metals such as Pt and Rh as active components and a ceramic support from Al_2O_3/ceria-based oxide as an oxygen storage component. As with most catalysts, the two major directions for research are higher catalytic efficiency (mainly determined by the active catalytic component) and increased longevity (strongly determined by the "oxygen storage material"). Both problems can be addressed for example by time resolved *in situ* XAS, as was done by the TOYOTA research group.[99,100] They investigated a CeO_2–ZrO_2 supported Pt auto exhaust catalyst as a function of various parameters, such as particle sizes and temperature, by time resolved *in situ* EXAFS and XANES measurements at the Pt K- and LIII-edges, and at the Ce K-edge.

As a typical example of that type of research, Fig. 11 shows the compilation of a set of Ce K-XANES spectra recorded at temperatures from 50–500°C. These spectra clearly demonstrate a change of the valency of Ce during the oxygen release/storage process. The shift of the absorption edge to lower energies as a function of increasing temperature and the isobestic points in the serial XANES spectra indicate that the valency changes directly from Ce^{4+} to Ce^{3+}. Based on these experiments, the group developed a catalytic system with "redispersion" properties; i.e., the loss of active surface area by "sintering" that is observed for many catalysts at very high

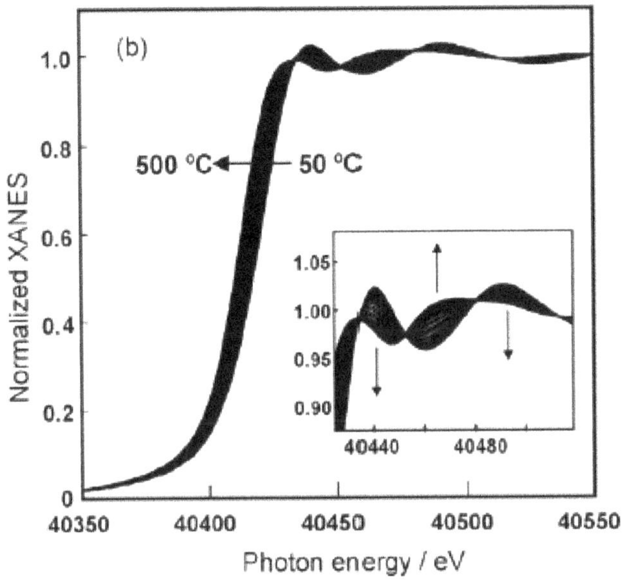

Fig. 11. Normalized Ce K-XANES spectra taken as a function of temperature during the oxygen release of Pt/CZ55 in flowing 5% H_2/He.

temperatures (above ~800°C) can be avoided and "reversed". A competing group from Daihatsu Motor also used EXAFS and XANES measurements in the development and characterization of an alternative system where the self-regeneration of the Pt (and Rh) catalysts is ensured by a "tailored" support structure — here a special perovskite.[101]

X-ray absorption spectroscopy is an extremely valuable tool for the detailed investigation of all types of catalysts (homogeneous and heterogeneous) and catalytic reactions because it allows element specific, time-resolved *in situ* investigations. A broad range of examples are discussed in a book on this topic edited by Y. Iwasawa[102] and in a recently published special issue of *Catalysis Today*, which summarizes the contributions from a 2008 ESRF workshop on "Time-resolved and *in situ* study of heterogeneous catalysts and catalytic processes using X-rays: current possibilities and future prospects".[103] In this volume, just to name two examples of industrial relevance, J.-D. Grunwaldt and coworkers highlight with several examples the tremendous potential of XAS when going from integral to spatially resolved measurements,[104] and P. J. Chupas and coworkers describe new opportunities for catalysis research using high energy X-ray scattering and the analysis of Pair Distribution Function.[105]

An important new type of high resolution fluorescence spectroscopy (hard X-ray photon-in, photon-out) is described by P. Glatzel *et al.*[106] By applying this technique, it is (at least in principle) not only possible to take valency-specific XAS spectra but also make site-specific measurements, meaning that it could be possible to measure, e.g., XANES spectra of "just" the surface atoms of a catalyst that are the catalytic active ones — a dream for catalysis research.

The importance of the research described here for the automotive industry is emphasized by the fact that Toyota, for example, operates its own beamline[107] at SPring-8 and that the opportunities offered by SR-based techniques are discussed in its own research journal.[108]

6.2.2. *Rubber research*

As was already mentioned, rubber is an X-ray amorphous material without long range order for which the application of standard diffraction techniques is not possible. Thus, rubber is a model case for the application of XAFS spectroscopy for which no such long range order is required. The mechanical properties of sulfur-vulcanized rubber depend strongly on its crosslink distribution and crosslink density. Information about these sulfur crosslinks is very difficult to obtain, and the usual wet-chemical approach (swelling) destroys the sample and offers only very limited precision. Here, S K-edge XANES spectroscopy represents a very promising alternative.

Already in 1994, Chauvistre *et al.* followed the vulcanization process of natural rubber by time-resolved *in situ* XANES measurements at the S K-edge. In these experiments, one can not only see the "opening" of the S8-ring but can follow the shortening of the polymer chain, and when the reaction is not stopped "on time", one can also see the beginning of the aging process.[109] In two later publications, these aging processes were investigated again using S K-edge XANES spectra in more detail as a function of various parameters, e.g., sulfur links of different types and densities, filler material, and the use of antidegradants.[110–112] These investigations led to several sometimes quite surprising results, e.g., that antidegradants "just" delay the thermo-oxidative aging without changing the aging resistance of the sulfur bonds, whereas carbon black (an active filler material) enhances aging resistance considerably. Carbon black also aids in the formation of long sulfur chain crosslinks. The studies quoted here are all based on what is called the fingerprint analysis of the XANES spectra; i.e., the spectra of interest are compared with those of well known and characterized model compounds. This is a very common approach in XAFS spectroscopy; however, in many cases the researchers do not properly consider the sensitivity of XANES spectra to the detailed chemical environment so that reference compounds are used that lead inevitably to wrong results. The XANES spectra of sulfur, as perhaps one of the most "sensitive" elements, show sensitivity up to the third coordination shell. Thus, reference compounds are required for an analysis of S K-XANES spectra in which sulfur has the same chemical environment that is "expected" in the sample of interest up to the third coordination shell — a sometimes extremely difficult challenge.[113]

Although qualitative results derived from spectroscopic studies can already be of significant interest for industry, quantitative results are in most cases the final goal because these can be more easily associated with macroscopic properties. For the investigation of rubber, the authors can show that it is possible to also analyze the spectra quantitatively. Figure 12 shows a typical set of spectra in the early phase of the aging process in which one can monitor the shorting of the crosslinks going from S3/S4 and longer sulfur chains to monosulfidic crosslinks (this process is reflected in the shift of the "white line" at ~2472 eV to higher energies), and then later to the first oxidized species, for which the corresponding "white lines" are at significantly higher energies. Figure 13 shows the quantitative analysis of the content of monosulfidic chains as a function of the aging time for two different samples that exhibit a very different behavior during the aging process. This is a very illustrative example of how it is possible to monitor and analyze processes of industrial importance, namely the life of a tire from vulcanization to the end of the aging process in just 24 hours!

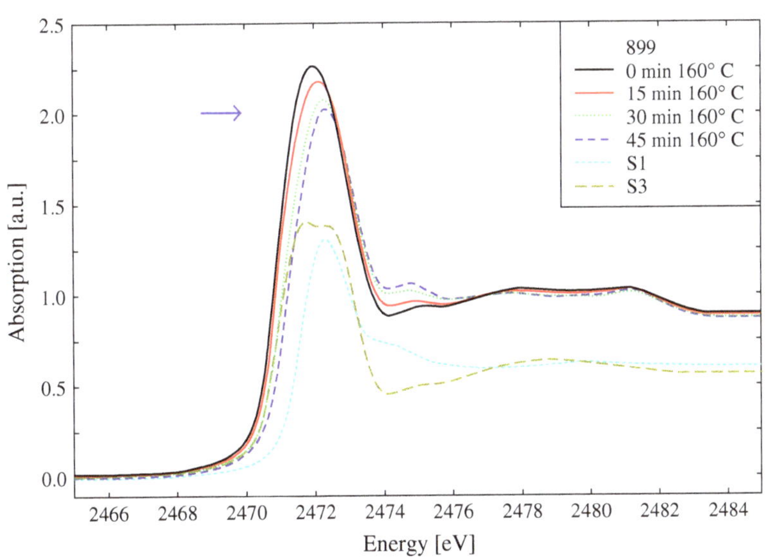

Fig. 12. Typical S K-XANES spectra of rubber in the early phase of the aging process.

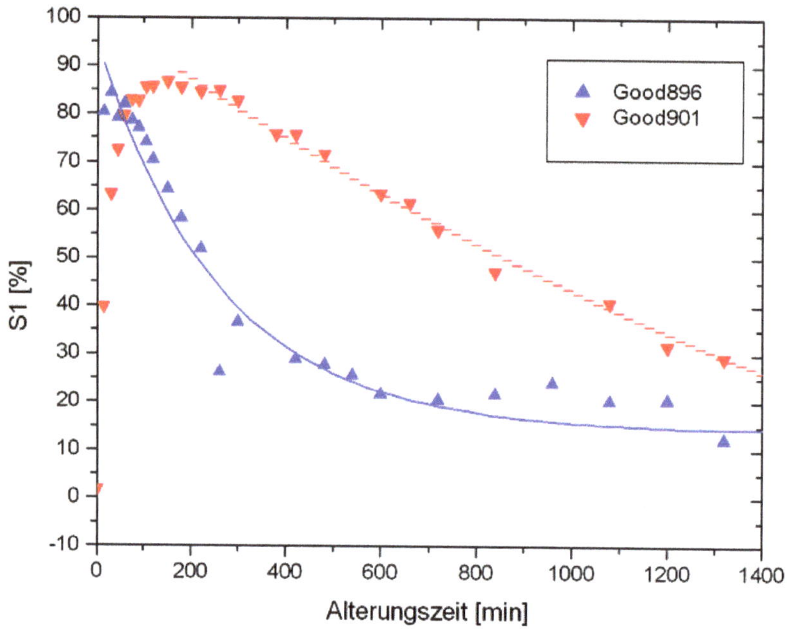

Fig. 13. Quantitative analysis of the content of monosulfidic chains as a function of the aging time ("Alterungszeit") for two different samples.

6.3. *Mining*

Because of the enormous importance of mining in many countries, particularly Canada, and because the mining industry is rarely covered when it comes to industrial applications, we have decided to highlight in the last part of this section some of the applications of SR-based techniques for the mining industry. At least in principle, SR-based techniques could be used in all three major areas connected to mining: (1) exploration, (2) mining and extraction, and (3) waste management and environmental problems.

The basic questions of importance for exploration and mining/extraction are in general similar, and they include, for example, information about[114]:

- The structural environment and the type of chemical bond of trace elements in minerals.
- The environment of minor and trace components in disordered mineralogical systems.
- The speciation of heavy metals in sediments, soils, and aqueous solutions.
- The sorption of aqueous metal ions on high surface minerals.
- Nucleation and crystallization processes.
- Kinetic processes, e.g., structural modifications during phase transformations.

All of these questions are of course problems in Earth Sciences, i.e., geology and mineralogy, and these are areas which in the past have already made extensive use of the analytical capabilities of SR-based techniques.[114,115] Thus, there is an excellent basis for applying these techniques for the mining industry.

At the time of this writing, the gold mining industry is experiencing boom times with gold prices reaching record heights. Thus, interest in new gold mining projects is picking up. The mineralogy of the gold-bearing ores, i.e., the speciation of Au in the various ores, is the key factor for predicting the best way of processing. Because of the high gold prices, low-grade ore bodies are also mined, and thus the detailed characterization of submicroscopic or "invisible" gold is required. In a very detailed overview article, Goodall and Scales compare the various techniques that can be used for the determination of gold mineralogy,[116] such as electron probe microanalysis, laser ablation inductively-coupled plasma mass spectroscopy, and synchrotron X-ray fluorescence. They emphasize all the advantages of using SR-XRF in the characterization of minerals, for example, the low detection limit for elements ($\sim$0.1 ppm) and the high spatial resolution ($\sim 2\,\mu$m) for the mapping and determination of "invisible" gold; however, they are afraid that the possible demand from industry could outstrip the supply. Keeping in mind that most SR facilities are competing for industrial customers, this seems to be an unfounded fear.

An example of the value of not just applying X-ray fluorescence for element detection but also using XANES spectroscopy for the speciation of the elements in a mineral is discussed in a publication by Cabri *et al.*[117] These authors use (micro-) XANES for the speciation of Au in gold-bearing arsenopyrites on grains well characterized by standard techniques. They find the "invisible" gold in two different chemical forms: chemically bound and elemental. These findings have of course direct implications for the extractive metallurgy. Typical Au LIII-XANES spectra of samples from three different mining sites, together with those of three reference compounds representing the different valencies of Au (Au^0, Au^{1+}, and Au^{3+}), are shown in Fig. 14. The XANES spectrum of the Olympiada samples has no "white line" and strongly resembles the spectrum of elemental gold; the spectra of the other two samples both show a "white line" as a clear indication of chemically bound gold. The differences between the spectrum of Au_2S and the spectra of the two samples where Au is chemically bound show that the Au-bearing phase in these samples is surprisingly not Au_2S.

In a similar study, Simon *et al.* investigated the oxidation state of gold and arsenic in gold-bearing arsenian pyrite minerals from a gold-mining site

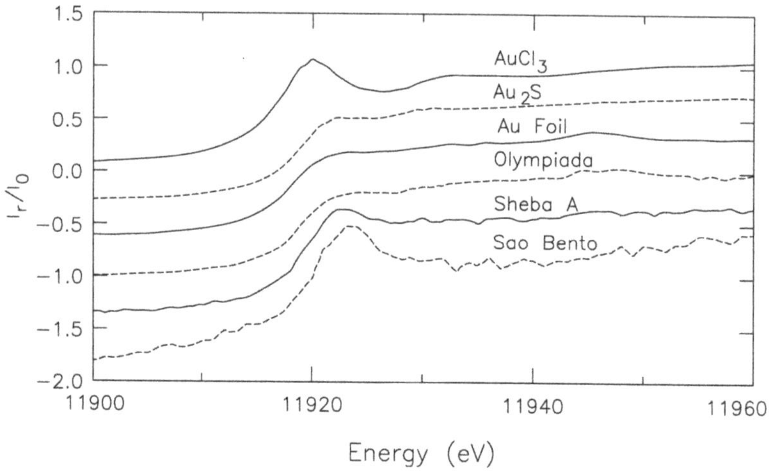

Fig. 14. Au LIII-XANES spectra of three arsenopyrite samples and three reference compounds representing the three different valencies of Au.

in Nevada.[118] These authors also observe Au as both Au^0 and Au^{+1} and arsenic as As^{-1}, and they conclude that Au^{+1} is twofold (as in Au_2S) and fourfold coordinated. For the fourfold coordinated structure, they propose a Au–As–S compound in which Au is bound to sulfur and arsenic atoms.

Tracking the oxidation state of ancient hydro-thermal fluids is the basis for a better understanding of the physical–chemical conditions of ore transport and deposition for many metals such as Cu, Au, Pb, and U. The oxidation state of trace elements, especially rare earth elements, incorporated in hydrothermal minerals is often regarded as an additional indicator for estimating the redox state of ancient fluids. For investigating this hypothesis, Brugger *et al.* investigated the valency of europium in hydrothermal scheelite, which is a common accessory mineral in various types of gold mineralization, using X-ray micro-fluorescence and micro-XANES spectroscopy.[119] The authors can show that it is possible to obtain information about the spatial distribution and the valency of Eu using SR-based XRF and XANES, although the concentration was in the 100 ppm range. This study could be the basis for more systematic studies of the oxidation states of rare earth elements in hydrothermal minerals and their correlation with the corresponding metals to be mined.

A systematic evaluation of the value of various SR-based microanalytical methods for the study of minor and trace elements in apatite and other minerals was recently carried out by Rakovan *et al.*[120] to obtain detection limits for the various techniques. They used XRF for the analysis of trace element concentration and distribution, XAFS for structural and electronic information (e.g., valency), XRD for phase identification, and fluorescence microtomography for mapping the internal structure of porous or composite materials.

There are not (yet) many studies focusing on the extraction process itself. However, there is a recent study by Scarlett *et al.* that very clearly shows the potential of SR-based experiments.[121] The authors of this study have used time resolved diffraction studies to investigate the mechanism and kinetics of pressure acid leaching of nickel laterite ores. By comparing their results with laboratory-based experiments, they can show that SR-based experiments allow a much more detailed analysis, e.g., by including poorly diffracting laterite components into their analysis.

In spite of the tremendous potential of SR-based techniques such as powder diffraction, XRF, and XAFS for improving exploration and mining techniques, most of the studies where industry is using these techniques focus on environmental problems. Thus, there is a broad range of publications dealing with the various pollutants connected to the various mining activities. There are publications dealing with arsenic,[122] mercury,[123] and zinc/lead.[124] However, as we have already discussed those environmental applications in Sec. 4.2.2, further examples will not be presented here, and the interested reader is directed to consult the references for more information.

7. Concluding Remarks

Synchrotron radiation is an exceptional tool not just for basic research but also for all types of applied research. It is not easy for synchrotron radiation labs to attract industrial customers because industry's requirements are sometimes difficult to satisfy in a "user research facility". This is true not just for the short turnaround times very often required when production problems are involved that require flexibility and improvisation, but also when it comes to "Intellectual Property" issues, and last but not least for the proper utilization and interpretation of results when the customers are small- and medium-sized businesses without their own R&D departments — a fact that is very often ignored by politicians. However, over the last years there have been apparent successes in attracting industrial users, and protein crystallography is for many labs a real success story. About 150 companies have been using the DoE SR facilities in the US — not a bad number.[125] In many cases there is also significantly more industrial use of an SR facility than the official lab statistics tell, because industry does not always contact the SR facility directly but instead will often utilize academic SR users who are "experts" in a specific research area of interest for the corresponding company. It is also difficult to find the right criteria for measuring the success of industrial use. A straightforward measure is very often the cash flow to the SR facility; however, the real economic impact that political supervisors would like to see goes far beyond this because industrial users are using their results from SR research to develop new products, improve their competitiveness in the market, and secure and create new jobs. The industrial use of synchrotron radiation remains challenging for all involved parties; however, there are finally some real success stories after years of struggle, and it could be that the industrial use of synchrotron radiation is indeed love at second sight.

References

1. K. Nasta and C.-C. Kao, *Synchrotron Radiation News* **20/4**, 7 (2007).
2. Y. Watanabe, *Synchrotron Radiation News* **20/5**, 40 (2007).
3. J. Doucet, *Nucl. Instr. Meth. Phys. Res.* **B199**, 10 (2003).
4. J. Doucet and K. Fletcher, Eds., *A Light for Industry*, ESRF (October 2002), http://ww.esrf.eu/files/Industry/Industrialbrochure.pdf.
5. http://www.lightsources.org/cms/.
6. http://www.diamond.ac.uk/Home/industry/casestudies.html.
7. http://www.nsls.bnl.gov/industry/highlights/.
8. http://nufo.org/files/NUFO_Industrial_Workshop_Report.pdf.
9. G. A. Schott, *Electromagnetic Radiation* (Cambridge University Press, Cambridge, 1912).
10. J. Schwinger, *Phys. Rev.* **70**, 798 (1946).
11. J. Schwinger, *Phys. Rev.* **75**, 1912 (1949).
12. D. Ivanenko and I. Pomeranschuk, *Phys. Rev.* **65**, 343 (1944).
13. J. P. Blewett, *Phys. Rev.* **69**, 87 (1946).
14. F. R. Elder, A. M. Gurewitsch, R. V. Langmuir and H. C. Pollock, *Phys. Rev.* **71**, 829 (1947).
15. F. R. Elder, A. M. Gurewitsch, R. V. Langmuir and H. C. Pollock, *J. Appl. Phys.* **18**, 810 (1947).
16. G. C. Baldwin, *Physics Today* **28**, 9 (1975).
17. D. H. Tomboulian and P. L. Hartmann, *Phys. Rev.* **102**, 1423 (1956).
18. K. Codling and R. P. Madden, *Phys. Rev. Lett.* **10**, 516 (1963).
19. L. Robinson, http://xdb.lbl.gov/Section2/Sec_2-2.html.
20. J. P. Silverman, C. N. Archie, D. E. Andrews and A. J. Weger, *Nucl. Instr. Meth.* **A347**, 31 (1994).
21. D. C. Koningsberger and R. Prins, Eds., *X-Ray Absorption: Principles, Applications, Techniques of EXAFS, SEXAFS and XANES* (Wiley-Interscience, Hoboken, 1988).
22. J. Stöhr, *NEXAFS Spectroscopy*, Corrected Edition (Springer, Berlin, 2003).

23. B.-K. Teo, *EXAFS: Basic Principles and Data Analysis* (Springer, Berlin, 1986).
24. *14th International Conference on X-ray Absorption Fine Structure XAFS14*, J. Phys. Conf. Series 190 (2009).
25. *X-ray Absorption Fine Structure — XAFS13: 13th Int. Conf.*, AIP Conf. Proc. Vol. 882 (American Institute of Physics, Melville, 2007).
26. Z. Cai *et al.*, *Appl. Phys. Lett.* **75**, 100 (1999).
27. N. Tamura *et al.*, *Appl. Phys. Lett.* **80**, 3724 (2002).
28. F. S. Aguirre-Tostado *et al.*, *Phys. Rev.* **B70**, 201403(R) (2004).
29. N. Tamura *et al.*, *J. Synchrotron Rad.* **10**, 137 (2003)
30. E. Zschech, W. Yun and G. Schneider, *Appl. Phys.* **92**, 423 (2008).
31. C. Streli, P. Wobrauschek, F. Meier and G. Pepponi, *J. Anal. At. Spectrom.* **23,** 792 (2008).
32. SEMATECH roadmap: http//www.itrs.net.
33. K. Baur *et al.*, *Nucl. Instr. Meth. Phys. Res. A* **467–468**, 1198 (2001).
34. K. Baur, S. Brennan, P. Pianetta and R. Opila, *Anal. Chem.* **74**(23), 610 A (2002).
35. C. Streli *et al.*, *Spectrochim. Acta B* **58**, 2105 (2003).
36. A. Singh *et al.*, *MRS Proceedings* **716**, 23 (2002).
37. F. Meier *et al.*, *Surf. Interface Anal.* **40**, 1571 (2008).
38. B.-K. Huh *et al.*, *Spectrochim. Acta* **B58**, 1445 (2003).
39. G. Pepponi *et al.*, *Spectrochim. Acta* **B59**, 1243 (2004).
40. Canadian Light Source Activity Report 2008, http://www.lightsource.ca/science/activity_reports.php.
41. V. J. Zatka, S. Warner and D. Maskery, *Environ. Sci. Technol.* **26**, 138 (1991).
42. F. Huggins *et al.*, *Environ. Sci. & Technol.* **45**(14), 6188 (2011).
43. J. Warner and J. Rowson, *Synch. Rad. News* **20**(3), 14 (2007).
44. N. Chen *et al.*, *Geochim. Cosmochim. Acta*, **73**, 3260 (2009).
45. D. L. Spears, *Electron. Letters* **8**, 102 (1972).
46. E. Spiller *et al.*, *J. Appl. Phys.* **47**, 5450 (1976).
47. L. Robinson, *Science* **199**, 413 (1978).
48. J. B. Murphy, in *Proc. IEEE 1989 Part. Acc. Conf.* Vol. **2** (IEEE, New York, 1989), p. 757.
49. S. Hector, *Microelectronic Engineering* **41/42**, 25 (1998).
50. T. Miyatake *et al.*, *J. Vac. Sci. Tech.* **B19**, 2444 (2001).
51. Q. Leonard *et al.*, *J. Vacuum Sci. Tech.* **23**, 2896 (2005).
52. F. Jiang, Y. C. Cheng, A. Isoyan and F. Cerrina, *J. Micro-Nanolithography MEMS and MOEMS* **8**, 021203 (2009).
53. A. Isoyan *et al.*, *Optics Express* **16**, 9106 (2008).
54. E. W. Becker *et al.*, *Microelectron. Eng.* **4**, 35 (1985).
55. *Micromotion:* http://www.mikrogetriebe.de.
56. See for example: U. Wallrabe *et al.*, *Microsystems Technology* **8**, 83 (2002).
57. *AXSUN Technologies*: http://www.axsun.com.

58. http://www.boehringer-ingelheim.de/produkte/mikrosystemtechnik/microtechnology/microoptics.htm, and references given on that site.
59. http://www.mezzotech.com.
60. Ch. Khan Malek and V. Saile, *Microelectron. Jour.* **35**, 131 (2004).
61. V. Saile, U. Wallrabe and O. Tabata, Eds., *LIGA and its Applications* (Wiley-VCH, Weinheim, 2008).
62. R. A. Lawes, *Int. J. Nanomanufacturing* **2**, 572 (2008).
63. A. Chen, G. Liu, L. K. Jian and H. O. Moser, *COSMOS* **3**, 79 (2007).
64. http://hasylab.desy.de/user_info/industrial_user/reasons_for_analytical_research/index_eng.html.
65. H. Saisho and Y. Gohshi, Eds., *Applications of Synchrotron Radiation to Materials Analysis* (Elsevier, Amsterdam, 1996).
66. T. K. Sham, Ed., *Chemical Applications of Synchrotron Radiation, Part I and Part II*, in Advanced Series in Physical Chemistry, Vol. 12 (World Scientific Publishing, Singapore, 2002).
67. T. K. Sham, *Int. J. Nanotechnology* **5**, 1194 (2008).
68. J. Hormes and H. Modrow, in *Analytical Advances for Hydrocarbon Research*, Ed. C. S. Hsu (Kluwer Academic, New York, 2003), pp. 421–454.
69. W. Reimers, A. R. Pyzalla, A. K. Schreyer and H. Clemens, Eds., *Neutrons and Synchrotron Radiation in Engineering Materials Science* (Wiley-VCH, Weinheim, 2008).
70. P. Dumas and M. J. Tobin, *Spectroscopy Europe* **15/6**, 17 (2003).
71. J.-L. Bantignies *et al.*, *J. Cosmet. Sci.* **51**, 73 (2000).
72. I. Kreplak *et al.*, *Intern. J. Cosmet. Sci.* **23**, 369 (2001).
73. S. Inamasu, T. Moriwaki and Y. Ikemoto, *SPring-8, Research Frontiers 2007*, 162 (2007).
74. K. Takehara, T. Inoue and K. Uesugi, *SPring-8, Research Frontiers 2008*, 156 (2008).
75. P. Cloetens *et al.*, *Europhysics News*, 26 (March/April 2001).
76. I. Nicolis, E. Curis, P. Deschamps and S. Bénazeth, *J. Synchrotron Rad.* **10**, 96 (2003).
77. M. Obata *et al.*, *Chem. Pharm. Bull.* **57**, 1107 (2009).
78. K. Provost *et al.*, *J. Phys. Conf. Series* **190**, 012206 (2009).
79. Y. Kiyozuka *et al.*, in *X-ray Microscopy: Proc. of VI International Conference*, AIP Conf. Proc. Vol. 507 (American Institute of Physics, Melville, 2000), p. 153.
80. R. Lobinski, C. Moulin and R. Ortega, *Biochimie* **88**, 1591 (2006).
81. L. W. Hardy and A. Malikayil, *Curr. Drug Discov.* **3**, 15 (2003).
82. A. Sharff and H. Jhoti, *Current Opinion in Chemical Biology* **7**, 340 (2003).
83. I. Tickle *et al.*, *Chem. Soc. Rev.* **33**, 558 (2004).
84. C. Riekel, M. Burghammer and G. Schertler, *Current Opinion in Structural Biology* **15**, 556 (2005).

85. R. Moukhametzianov *et al.*, *Acta Crystallographica Section D: Biological Crystallography* **D64**, 158 (2008).
86. http://www.imca.aps.anl.gov.
87. http://www.spring8.or.jp; beamline BL32B2.
88. Th. Ursby *et al.*, in *Proc. 8th Int. Conf. on Synchrotron Rad. Instr.*, AIP Conf. Proc. Vol. 705 (American Institute of Physics, Melville, 2004), p. 1241.
89. See for example: N. Okazaki *et al.*, *J. Synchrotron Rad.* **15**, 288 (2008).
90. See for example: http://www.px.nsls.bnl.gov.
91. E. Abola, P. Kuhn, Th. Earnest and R. C. Stevens, *Nature Structural Biology, Structural Genomics Supplement*, 973 (2000).
92. E. Cohen *et al.*, *J. Appl. Cryst.* **35**, 720 (2002).
93. G. Snell *et al.*, *Structure* **12**, 537 (2004).
94. L. N. Johnson, *Innovation in Pharmaceutical Technology* **21**, 16 (2006).
95. R. W. Strange and S. S. Hasnain, in *Protein-Ligand Interactions: Methods and Applications*, Ed. G. U. Nienhaus, (Humana Press, Totowa, 2005), p. 167.
96. N. Berova, L. Di Bari and G. Pescitelli, *Chem. Soc. Rev.* **36**, 914 (2007).
97. B. A. Wallace and R. W. Janes, *Biochemical Soc. Trans.* **31**, 631 (2003).
98. D. J. Rodi *et al.*, *J. Mol. Biol.* **285**, 197 (1999).
99. Y. Nagai *et al.*, *Angew. Chem. Int. Ed.* **47**, 9303 (2008).
100. Y. Nagai *et al.*, *Catalysis Today* **145**, 279 (2009).
101. H. Tanaka *et al.*, *Angew. Chem. Int. Ed.* **45**, 5998 (2006).
102. Y. Iwasawa, Ed., *X-ray Absorption Fine Structure for Catalysis and Surfaces* (World Scientific, 1996).
103. M. A. Newton, Ed., *Catalysis Today* **145**(3–4), 187–306 (2009).
104. J. D. Grunwaldt *et al.*, in Ref. 101, p. 267.
105. P. J. Chupas, K. W. Chapman, H. Chen and C. P. Grey, in Ref. 101, p. 231.
106. P. Glatzel, M. Sikora, G. Smolentsev and M. Fernandez-Garcia, in Ref. 101, p. 294.
107. http://www.spring8.or.jp; see: beamline BL33XU.
108. Y. Hirose, *R&D Review of Toyota CDRL* **38**, 1 (2003).
109. R. Chauvistre, J. Hormes and K. Sommer, *Kautschuk Gummi Kunstoffe* **47**, 481 (1994).
110. H. Modrow, R. Zimmer, F. Visel and J. Hormes, *Kautschuk Gummi Kunstoffe* **53**, 328 (2000).
111. H. Modrow, J. Hormes, F. Visel and R. Zimmer, *Rubber Chemistry and Technology* **74**, 281 (2001).
112. B. Brendebach and H. Modrow, *Kautschuk Gummi Kunstoffe* **55**, 157 (2002).
113. R. Chauvistre *et al.*, *Chem. Physics* **223**, 293 (1997).
114. L. Galoisy, *EMU Notes in Mineralogy* **6**, 553 (2004).
115. W. A. Bassett and G. E. Brown Jr., *Ann. Rev. Earth Planet. Sci.* **18**, 387 (1990).
116. W. R. Goodall and P. J. Scales, *Minerals Engineering* **20**, 506 (2007).

117. L. J. Cabri *et al.*, *The Canadian Mineralogist* **38**, 1265 (2000).
118. G. Simon *et al.*, *American Mineralogist* **84**, 1071 (1999).
119. J. Brugger *et al.*, *The Canadian Mineralogist* **44**, 1079 (2006).
120. J. Rakovan, Y. Luo and O. Borkiewicz, *Mineralogia* **39**, 31 (2008).
121. N. V. Y. Scarlett, I. C. Madsen and B. I. Whittington, *J. Appl. Cryst.* **41**, 572 (2008).
122. See for example: S. R. Walker *et al.*, *The Canadian Mineralogist* **43**, 1205 (2005), and references therein.
123. See for example: Ch. S. Kim, J. J. Rytuba and G. E. Brown Jr., *Appl. Geochemistry* **19**, 379 (2004), and references therein.
124. See for example: N. Schuwirth, A. Voegelin, R. Kretschmar and Th. Hofmann, *J. Environ, Qual.* **36**, 61 (2007), and references therein.
125. http://www.nslsuec.org/corporations.aspx.

INDEX

^{7}Li(d,n)^{8}Be reaction, 255
^{7}Li(p,n)^{7}Be reaction, 254
^{9}Be(d,n)^{10}B reaction, 257
^{9}Be(p,n)^{9}B reaction, 255
^{11}C, 141, 148
^{11}N, 141, 148
^{14}C, 141
^{15}N(^{1}H,$\alpha\gamma$)^{12}C, 231
^{15}O, 148, 171
^{18}F, 148
^{18}F-FDG, 145, 152, 153
^{18}F-tagged glucose analog, 142, 147
^{22}Na, 151
^{67}Ga, 146, 152
^{68}Ga, 144, 179
^{68}Ge, 151, 179
^{90}Sr, 149
^{90}Y, 149, 150
^{99}Mo, 146
^{99}Mo generators, 153
^{99m}Tc, 146, 152, 153
^{103}Pd seeds, 149
^{106}Ru, 149
^{111}In, 146, 150, 152
^{123}I, 145, 146, 152, 153
^{125}I, 149
^{144}Ce, 149
^{153}Sm, 150
^{188}Re, 150
^{192}Ir, 149
^{201}Tl, 142, 146, 152
^{211}At, 151, 172
^{213}Bi, 151
^{225}Ac, 151
^{239}PuBe, 276
^{241}AmBe, 245, 276
^{252}Cf, 244, 245, 276

absorbed dose, 90, 95, 97
accel-decel beamline, 18
accelerator designs, 16
accelerator mass spectrometry, 185, 186, 193, 201–208, 233
accelerator systems for ion implantation, 16
accelerator-based neutron source, 244, 246
active neutron interrogation, 290
advanced PV cells, 46
air cargo, 314, 360, 365, 366
  security, 318
air pollution, 399
American Society of Nondestructive Testing, 308
AP method, 287
AP-ENG, 270, 281, 282, 293
applications of TiO_2, 218
arc discharge ion source, 31
arc plasma ion source, 30
arsenic-containing tailings, 384
artificial radioactivity, 141
assay of spent nuclear fuel, 262
Auger electron, 194
automotive exhaust emissions, 400
aviation security, 287

barn, 168
beam-broadening, 64
beam deflection and wafer motion, 34
beam glow, 66
beam transport system, 178
beamline system types, 17
Bernas ion source, 29
betatron, 288, 309
biomaterials, 49, 51
biomedical, 282
biotechnology, 391
bomb disposal, 315
brachytherapy, 148
Bragg's rule, 213, 220
bremsstrahlung, 93, 320

bulk material analysis (BMA), 282
buried landmine, 287

^{14}C dating, 201
Calutron ion source, 27
carbon implants for tensile strained nMOS, 14
cardiology, 146
cargo inspection, 312, 344
cargo screening, 315, 317, 344, 347, 350
    initiative, 317
    standard, 363
    trends, 352
cargo and vehicle screening standards, 359
cascade-accelerator, 189–192
cascade-generator, 188, 189
catalysts, 400
catalytic converter, 400
catalytic processes, 400
CCD camera, 280
cesium-137, 96
charge-coupled device (CCD), 12
charged particle activation analysis (CPAA), 152, 185, 209, 231
chemical reaction, 176
chemical weapon analysis, 286
chicane decel electrode, 21
circular dichroism (CD) spectroscopy, 398
CMOS imagers, 12
cobalt-60, 88, 94, 122, 309, 315
Cockcroft–Walton, 18, 31, 106, 140, 261
cocktail implants, 13
C/O log, 278
commercial EB installations, 88
commercial radionuclide production, 142, 152
commercial sealed-tube neutron generators, 267
compensated neutron porosity logs, 277
complementary metal-oxide semiconductor (CMOS), 9
complex RBS spectra, 223
composition of a thin film, 222
compound nucleus, 166
computed tomography (CT), 313, 354
conductive heat transfer, 175
contraband, 281, 286, 347, 349, 353
    detection, 284, 287
control of regulatory requirements, 372, 383
controlatron neutron generator, 291
convection, 175
cosmetics, 391
Coulomb barrier, 165, 168, 170, 199, 201
counter-dope poly-Si gates, 12
CPAA, 185, 186, 193
cross-linkable polyolefins, 130
cross-linked polyethylene, 116
    PTFE, 130
    rubber sheet, 130
    SiC fiber, 129
    tire, 118
    wire systems, 116
cross-linking, 87, 106, 110, 115–117, 131
    applications, 88, 114
CT imaging, 355, 365
CT scan, 313, 356, 365
CT system, 313, 314, 355, 356
cured carbon-fiber, 129
curing and cross-linking of coatings, 117
curing of composite materials, 128
curing of inks, 88
curing of rubber, 106
curing thick composite products, 108
customs and border protection, 346
cyclotron, 140, 142, 146, 149, 151, 152, 155, 156–162, 232, 261
    components, 157
    equipment manufacturers, 153
    physics principles, 156
    versus linac, 163

ΔE detector, 225, 228
ΔE–E ERDA, 225, 229, 228
dc accelerator, 105
DD ENG, 270
DD fusion, 251, 270, 272, 273
deep proton implantation, 48
dees, D-shaped electrodes, 156
depth profile, 208
depth-dose curves, 90, 92
diagnosis of coronary artery disease, 146
diamond-like carbon (DLC), 53, 218
differential cross section, 210, 215
diffraction-fluorescence-absorption spectroscopy, 378
diffusion measurement, 232
digital data standardization, 364
digital detector, 314, 332, 333, 335, 339–341
    performance, 334

digital radiography, 322, 333, 356
digital X-ray detector, 354
diode gun, 60
direct current (dc) high-voltage generator, 99
direct food contact coating, 131
direct reaction, 165
direct voltage accelerator, 184, 185, 188
disinfection of food, 88
disinfestation, 88
dopant implants for CMOS, 12
doped Si ink, 46
dose indicators, 96
dose rate, 94, 97, 98, 105, 109, 121, 122
dose uniformity ratio, 123
dose–energy chart, 12
dosimeters, 96
double-bubble process, 116
DPF neutron source, 272
DRAM, 12, 25
drift tube, 162
drip melting, 79
drug development, 391, 396
drugs, 394, 395, 398
DT fusion, 253, 255, 264, 267, 269
DT neutron sources, 253
DTL (Drift Tube Linac), 155
dual-energy accelerator, 366
dual-energy cargo screening, 349
dual-energy technology, 353
dual-mode radiography, 287
dual-view X-ray system, 353
dynamic beam deflection, 64
Dynamitron, 106, 263

EB accelerator, 88, 89, 94, 118
  high-energy, 108–114
  low-energy, 98–104
  medium-energy, 104–108
EB additive manufacturing, 80
EB cutting, 74
EB dose rate, 94
EB drilling (EBD), 74
EB free form fabrication, 80
EB glazing type hardening, 76
EB heat treating, 76
EB melting and casting, 78
EB melting furnaces, 80
EB phase-transformation type hardening, 76
EB processing device, 59
EB stack gas treatment, 128
EB surface smoothing, 82
EB surface texturing, 80
EB tee-weld, 68
EB treatment of seeds and soil, 130
EB water treatment, 127
EBD production drilling, 76
EBraze, 68
EBW, 61–66
  capital equipment cost, 62
  process, high vacuum (HVEBW), 63
  process, non-vacuum (NVEBW), 63
  process, partial vacuum (PVEBW), 63
  system, high vacuum, high power, 66
elastic recoil detection analysis (ERDA), 185, 186, 208, 224, 228
electron accelerator, 88, 94, 314, 324
electron affinity, 205
electron beam (EB), 57
  equipment, 59
  processing, 57
  welding, 61; *see also* EBW
electron gun, 59
electron linac, 109, 307, 311, 321
  manufacturers, 312
  technology, 319
electron range, 89, 90, 94
electron spin resonance (ESR) spectrometer, 97
electron transformer-rectifier (ELV), 105
electronic neutron generator (ENG), 244
electrostatic accelerator, 188, 190–192, 204
endoergic reaction, 165, 167
endothermic reaction, 248
energy contamination, 20
energy straggling, 173
ENG, Active Cooling (AC-ENG), 269
ENG, Associated Particle (AP-ENG), 270
ENG, General Purpose (GP-ENG), 269
ENG, logging while drilling (LWD-ENG), 267
ENG, operational lifetimes, 245
ENG, Portable (P-ENG), 268
ENG, Wireline (W-ENG), 268
engine bearing surfaces, 233
environmental analysis, 291
environmental issues, 238

environmental problems, 372, 383, 407
environmental protection, 88, 399
environmental/regulatory research, 384
excitation function, 168
exoergic reaction, 165, 167
exothermic reaction, 248
exploration and mining/extraction, 404
explosive detection, 287
explosive material, 285
explosive screening, 287
extended X-ray absorption fine structure (EXAFS), 380, 383, 387, 393, 400
extraterrestrial exploration, 291

fast neutron radiography, 280, 281
fast-neutron resonance radiography, 283
FDG, 142, 145, 147, 152, 153, 179
filamentless ion source, 26
film contrast, 341
film grain, 341
film radiography, 322
film-based radiography, 356
fission signature, 288
fissionable material, 287
    detection, 288
flow analysis in pipes, 291
flux rate, 311
food irradiation, 95, 110, 111, 125, 131
    maximum absorbed dose, 126
    regulations, 126
    safety, 125
forensic applications, 316
Freeman ion source, 28
fuel cell production, 130

gas cluster ion beam (GCIB), 40
    doping, 41
gas target, 177
geophysical exploration, 273, 274, 292
glow discharge optical emission spectroscopy (GDOES), 233
gold mining, 405
goniometer, 236

half-value layer, 310; *see also* HVL
hard martensite surface, 78
hazardous materials, 238
    detection, 284
hazardous waste identification, 314
hearth melting, 79
heat loss by radiation, 175
heat-shrinkable film, 116
heat-shrinkable multi-layer plastic film, 104
heat-shrinkable PE, 116
heat-shrinkable polyethylene tubing, 116
heat-shrinkable tubing, 106
heat transfer, 175
helium-particle-only linac, 152
high beam transmission dc accelerator implanter, 49
high dose and low-energy implantation, 24
high energy industrial CT, 354
high explosives, 284, 287
high resolution fluorescence spectroscopy, 401
high resolution X-ray imaging, 381
high voltage dc accelerator, 188
high-current beamline, 18
high-energy proton accelerator, 49
homeland security, 262, 281, 315, 316, 324, 346, 353, 367
human hair, 392
HVEBW, 63, 64, 70
    example, 68
    multi-function unit, 69
HVL, 310, 311, 343, 357–359
hydrogen implants for SOI, 15
hydroxyapatite, 226, 228

IBA, 184, 185, 187, 188, 193, 207
IBA data furnace, 213
IEC neutron source, 273
IED, 315, 319
ILU accelerator, 104, 111
imaging standard, 360
implantation of nitrogen, 50
implants into metals, ceramics, and biomaterials, 49
improvised explosive device, 287, 315
indirectly heated cathode, 30
induced fission, 288
industrial electron beam accelerator, 98
industrial inspection and testing market, 312
industrial neutron production systems, 261
industrial radiographic quality, 356
industrial radiography, 308, 356, 364
insulating core transformer (ICT), 105

integrated circuits (IC), 9
interlaced dual-energy accelerator, 314, 322, 348
international unit of dose, 95
iodine-123, 144
ion beam analysis, 183, 184, 193
ion extraction, 159
ion implantation, 9, 53
    beamline, 17
    systems sold, 10
ion implanter, 9, 12, 33, 34
ion reactions, 247, 257
ion shower, 12
    implanter, 25
    tool, 32
ion source, 17, 158
    designs, 27
ionization of target material, 176
IR spectroscopy, 378
iridium-192, 309, 315
irradiation market segments, 99
isotope replacement, 315

keyhole welding process, 61
kinematic factor, 210, 219
klystron, 322

laboratory activation analysis, 281
large-area electron beam, 82
large area ion source, 32
lattice-strain measurements, 237
layered security strategy, 346
lead slowing down spectrometer (LSDS), 288
LiGA, 387, 388
linacs for radionuclide production, 163
linear accelerator, 22, 23, 53
linear cathode, 98
liquid target, 177

machine footprint, 235
magnetic-scan beamline, 35
magnetron, 321
manufacturers of electron beam equipment, 61
mass spectrometry, 202, 204, 205
material discrimination, 349
medical device sterilization, 88, 95, 110, 111, 119–124
medium-current beamline, 18
medium energy ion scattering (MEIS), 216
MEMS, 293, 376, 387
metal vapor vacuum arc (MEVVA) ion source, 50
metal-based drugs, 393
metrology, 314, 354
MeV beamline, 22
microanalysis, 378
microelectronic device, 387
microfabrication, 376, 388
microorganisms, 124, 127
minerals, 281
    exploration, 279
mining, 281
modified low dose SIMOX, 26
monochromatic microbeam, 396
monoclonal antibodies, 150
multi-dispersive ToF–E ERDA, 225
multi-MeV proton beam, 47
multicrystal detector, 197

nanolithography, 388
nanostructure, 390
NDT accelerator, 312, 320, 321, 324, 325, 328, 343, 344
NDT applications, 310, 331, 336, 365
NDT system, 321, 322, 325–327
negative ion, 158, 160, 188, 192, 203, 205, 206
    currents, 204
    extraction, 159–160
    source, 160, 192, 203, 205
neutron-based inspection, 317
neutron capture, 247
neutron die-away technique, 289
neutron discovery, 243
neutron energy, double-valued, 248
neutron gauging, 280
neutron inelastic scattering, 247
neutron interactions with matter, 274
neutron production, 247
neutron properties, 247
neutron radiography, 280
neutron resonance transmission analysis (NRTA), 288
neutron source, dense plasma focus (DPF), 272
neutron source, inertial-electrostatic confinement (IEC), 272

neutron time-of-flight, 260
neutron tube, 264
neutron yields, 249, 263, 268–270, 272, 273
nondestructive testing and inspection (NDT), 307, 312, 313, 319, 365
non-dopant ions, 12
nuclear absorption reaction, 166
nuclear medicine imaging, 145
nuclear microprobe, 185, 187, 197
nuclear physics technique, 184, 201
nuclear reaction, 140, 165, 170, 176, 187, 199, 201
nuclear reaction analysis (NRA), 185, 186, 209, 230
nuclear resonance broadening (NRB), 231
nuclear resonance profiling, 231
nuclear resonance radiography, 287
nuclear resonance techniques, 288
nuclear safeguards, 287, 293
NVEBW, 63, 64, 71, 73
  method, 70
  weld task, 71
  welding system, 73

oil and gas exploration, 279
oilfield applications, 275
open-vacuum accelerator, 288
open-vacuum ENG, 262, 273
open-vacuum system, 270, 291
operator skills, 234
organic light emitting diode (OLED), 223
oxygen implants for SOI, 15

PAI implant, 13
particle induced gamma-ray emission, 185, 187
particle induced X-ray emission, 185, 187, 194
pasteurization of fresh food, 88
Pelletron, 105, 122, 192, 206, 263
penetrameter, 357
P-ENG, 268, 293
penumbra, 329
PET, 142, 145, 147, 148, 151,
  /CT, 151
  clinical scanner resolution, 147
  equipment sales, 148
  future application, 179
  imaging, 144, 145–148, 179
  radionuclide, 148, 155
  radionuclide production, 152, 155, 156, 163
  radiopharmaceutical, 153, 179
  radiotracer, 146, 179
  single dose, 152
pharmaceuticals, 391
phase-synchronized linear accelerator, 23
photoelectron spectroscopy, 378
photofission reaction, 258
photoneutron source, 259, 271
photoneutron system, 288
photoneutron target, 258
photoneutron yields, 259
photonic master template, 16
photonuclear reaction, 258, 320
photovoltaic Si solar cell, 48
PIGE, 185, 186, 193, 197, 199–201, 230
PIII, 24, 33, 50
  chamber, 51
  demonstration unit, 50
PIXE, 185, 186, 193–201
planar CMOS transistor, 10
plasma immersion, 12
  ion implanter, 24
  systems, 49
PMMA, poly(methyl-methacrylate), 389
$POCL_3$ diffusion doping, 43
portable ENG, 291
portable inspection system, 315
portable neutron system, 285
positive ion, 188, 192, 203, 206
positron emission tomography, 142; *see also* PET
potential drop accelerator, 188
pre-amorphization, 12
pre-doping amorphization implant (PAI), 12
prompt γ-ray neutron activation analysis (PGNAA), 279, 282, 283, 285, 291
prostate cancer, 149
protein crystallography (PX), 391, 396
proton exchange membranes (PEM), 129
proton linacs for radionuclide production, 155
pulsed ENG, 282, 291
pulsed neutron generator, 274, 277
pulsed neutron source, 289
PV cell doping, 43
PVEBW, 63, 64, 70

example, 68
method, 68, 69
unit, 71
PX mail-in service, 397

Q value, 165, 167, 170, 230
quality assurance (QA), 234, 235, 356
quality control, 308, 337, 372, 380, 390
quality of welds, 342
quality standards, 356

R-theta scanning, 39
radiation effects testing, 284
radiation processing, 87, 89, 96, 106, 109, 111, 125
radiation protection, 238
radiation shielding, 161
radiation sterilization, 121, 123
radio frequency electron linac, 309
radio frequency quadrupole, 49, 155, 262; *see also* RFQ
radioactive carbon, 141
radioactive ion beam, 231
radioactive seeds, 148
radioactive tracers, 140
radiograph, 307, 308, 333, 340, 341, 356, 357
radiographic principles, 339
radiographic procedures, 341
radiographic sensitivity, 340
radiographic testing applications, 308
radiography of assemblies, 343
radiography of castings, 342
radiography of rocket motors, 343
radiography of welds, 342
radioimmunotherapy (RIT), 150
radioiodine, 144
radioisotope, 139
neutron sources, 244
radiometric method, 201, 202
radionuclide, 139
accelerator-produced, 144
decay constant, 171
general principles of production, 164
half-lives, 143
imaging, 145
production accelerator, 152
production accelerator choice, 163
production target, 161, 172
rate of production, 170
radiopharmaceutical, 142, 179
on demand, 179
radiotracer, 142, 143, 145, 146
rapid thermal process (RTP) anneal, 14
reaction cross section, 168, 169, 249
reaction energy, 167
reaction path, 165
reading room, 337
recoil spectrometry, 225
regulatory control services, 383
reliability, 234
requirements for producing radionuclides, 152
resonant neutron radiography, 281
resotron, 309
reverse-engineering, 356
RF linac, 17, 142, 152, 155, 162, 179, 261, 311
RF power, 322
RFQ, 49, 155, 262, 292
implanter, 49
linac, 155, 163, 263, 288
neutron generator, 262–264
Rhodotron accelerator, 111
ribbon ion beam, 32
rubber, 106, 115, 116, 118, 130, 399, 402, 403
RUMP, 213
Rutherford Backscattering Spectrometry (RBS), 185, 186, 199, 208, 215
Rutherford scattering, 215, 230

saturation, 170
factor, 171
scanning methods, 33
sealed-tube DT ENG, 275, 285, 289
sealed-tube ENG, 244, 246, 253, 258, 263, 264, 266, 273, 288, 291
seam tracking, 67
secondary ion mass spectroscopy (SIMS), 208, 215, 233
security inspection, 344
SEE, single event effects, 284
SEES (Secondary Electron Emission Sensing), 67
SEE testing, 284
segmented cathode, 100
self-shielded, 99, 103
cyclotron, 161
SEMATECH roadmap, 381
semiconductor industry, 381

SF_6, 238
shielding, 93, 100, 124
Si-based photovoltaic (PV) cell, 42
Si membrane cutting, 39
SIMNRA, 213, 224
SIMOX, 13, 15
  beams, 16
  implanter, 15, 25, 31
  ion source, 31
  process, 26
simulated multi-beam operation, 81
Single Photon Emission Computed Tomography, 152; *see also* SPECT
slab model, 214
small angle multiple scattering, 175
small angle scattering, tomography, 377
software algorithm, 335, 349, 351, 352
SOI, Si on insulator, 15
soil analysis, 291
solid target, 177
special ion sources, 31
special nuclear material, 347
speciation of aerosol samples, 384
speciation of Au, 405
speciation of compounds, 383
speciation of elements, 372, 405
speciation of heavy metals, 404
speciation of metals, 393
speciation of trace elements, 395
specific activity, 172
SPECT, 144, 145, 147
  imaging, 144–146
  procedures, 146
  radionuclides, 146
  resolution, 147
spectrum simulation codes, 213
spent fuel, 281
sputter-Auger electron spectroscopy (AES), 208
sputter-Auger/XPS, 215
sputter-X-ray potential spectroscopy (XPS), 208
SR facilities, 377, 379, 381, 383, 388, 390, 407
  collaboration with industry, 372
  for PX measurements, 397
SR for industrial activities, 372
SR in industrial biomedical research, 391
SR in pharmaceutical research, 398
SR, life science applications, 392
SR-based microanalytical methods, 406
SR-based R&D for industrial applications, 391
SR-based techniques, 377
  for mining, 404
  for quality control, 381
  in the automotive/transportation sector, 399
SRIM, 213, 215
SR-TXRF spectrum, 382
standing wave coupled-cavity, 320
standing wave electron accelerator, 321
stopping cross section, ϵ, 212, 220
stopping cross section factor, 221
stopping force, 211
  in a compound, 213
stopping power, 173
stopping-foil ERDA, 225, 228, 229
superconducting magnet cyclotron, 152
surface approximation, 219
synchrotron radiation (SR), 371, 377, 379–381, 392, 407
  facilities, 371, 374
  for production, 387
  history, 374
  properties, 377

tandem, 188, 190, 192, 193, 202, 204, 206
  accelerator, 22, 23, 53
  cascade accelerator, 152
target foil, 161, 164, 175, 176
targeted radionuclide therapy, 150
therapeutic radiopharmaceutical, 150
thermal neutron, 259, 263, 276, 279
  radiography, 280
thermal or epithermal neutron, 253, 257
thermal processing applications, 58
thick-target effect, 249
thick-target neutron yield, 251
thin layer activation (TLA), 151
thin Si PV membrane, 47
threshold energies for ionization, 30
threshold energy, 167
thyroid function, 146
tin oxide SnO_2, 223
TiO_2, 218
tissue sterilization, 130
ToF–E ERDA, 192, 225–231
total X-ray reflection fluorescence, 381, 382; *see also* TXRF

traceability, 235
transformer-driven dc accelerator, 106
triode gun, 60
tritium, 255, 257, 265, 267, 269, 292
 accumulation, 271
 build-up, 253
 contamination, 263, 271
 leakage, 254
tumor imaging, 146
tumor treatment, 148
turn-key installation, 235
TXRF, 381, 382
 angle-resolved, 383
 typical spectrum, 382
 with XANES, 383

vacuum system, 235
Van de Graaff, 105, 261, 263, 309
vehicle-borne explosive, 287
very large workpiece welding, 66

waste materials, 88, 127, 131
water-activation test, 279
water-cooled hearth, 79
water-cooled mold, 79
weapons of mass destruction (WMD), 347, 353
wear and corrosion resistance of metals, 50
well-logging tool, 275, 277
wheel and pendulum wafer scanning, 36
white radiation, 326
workplace aerosols, 384

X-ray absorption fine structure (XAFS), 379, 380, 406, 407
 spectroscopy, 379, 380, 384, 395, 402
 spectrum, 379
X-ray absorption near edge structure (XANES), 379, 380, 383, 385, 387, 400, 402, 403, 405
X-ray absorption spectroscopy (XAS), 378, 379, 383, 385, 395, 398, 400, 401
 for the investigation of drugs, 394
X-ray attenuation, 198
X-ray beam flattener, 327
X-ray beam properties, 325
X-ray beam symmetry, 328
X-ray collimation, 329
X-ray conversion, 94
X-ray crystallography, 396
X-ray discovery, 308
X-ray dose rate, 94
X-ray film, 339
X-ray fluorescence, 194, 378, 405; *see also* XRF
X-ray flux rate, 325
X-ray focal spot size, 328
X-ray imaging software, 351
X-ray inspection facility, 337
X-ray inspection system, 317
X-ray intensity, 94
X-ray linac, 311
X-ray lithography, 376, 387, 390
X-ray microscopy, 378
X-ray microtomography, 393
X-ray production, 320
X-ray scanning, 344, 345, 347, 350, 361, 362, 366
X-ray source, 309
X-ray spectrum, 326
X-ray tube, 307, 308, 310, 319
X-ray waste inspection system, 314
X-rays for NDT, 308
XRF, 194, 195, 378, 405, 406, 407

Z-map (density map), 349
zetatron, 264

www.ingramcontent.com/pod-product-compliance
Lightning Source LLC
LaVergne TN
LVHW070830220926
848486LV00030B/243